ADDITIVES IN POLYMERS

Analysis and Applications

ADDITIVES IN POLYMERS
Analysis and Applications

Edited by
**Alexandr A. Berlin, DSc, Svetlana Z. Rogovina, DSc, and
Gennady E. Zaikov, DSc**

AAP | APPLE
ACADEMIC
PRESS

Apple Academic Press Inc.	Apple Academic Press Inc.
3333 Mistwell Crescent	9 Spinnaker Way
Oakville, ON L6L 0A2	Waretown, NJ 08758
Canada	USA

©2016 by Apple Academic Press, Inc.

First issued in paperback 2021

Exclusive worldwide distribution by CRC Press, a member of Taylor & Francis Group
No claim to original U.S. Government works

ISBN 13: 978-1-77463-548-3 (pbk)
ISBN 13: 978-1-77188-128-9 (hbk)

Library and Archives Canada Cataloguing in Publication

Additives in polymers : analysis and applications / edited by Alexandr A. Berlin, DSc, Svetlana Z. Rogovina, DSc, and Gennady E. Zaikov, DSc.

Includes bibliographical references and index.
Issued in print and electronic formats.
ISBN 978-1-77188-128-9 (hardcover).--ISBN 978-1-4987-2859-1 (html)
1. Polymers--Additives. 2. Polymers--Analysis. I. Zaikov, G. E. (Gennadiĭ Efremovich), 1935-, author, editor II. Berlin, Alexandr A., author, editor III. Rogovina, Svetlana Z., author, editor

| TP1142.A34 2015 | 668.9 | C2015-906457-0 | C2015-906458-9 |

Library of Congress Cataloging-in-Publication Data

Additives in polymers : analysis and applications / [edited by] Alexandr A. Berlin, DSc, Svetlana Z. Rogovina, DSc, and Gennady E. Zaikov, DSc.

pages cm
Includes bibliographical references and index.
ISBN 978-1-77188-128-9 (alk. paper)
1. Polymers--Additives. I. Berlin, Al. Al., 1940- II. Rogovina, Svetlana Z. III. Zaikov, G. E. (Gennadii Efremovich), 1935-

| TP1142.A33 2015 | 668.9--dc23 | 2015035608 |

Apple Academic Press also publishes its books in a variety of electronic formats. Some content that appears in print may not be available in electronic format. For information about Apple Academic Press products, visit our website at **www.appleacademicpress.com** and the CRC Press website at **www.crcpress.com**

ABOUT THE EDITORS

Alexandr A. Berlin, DSc
Professor Alexandr A. Berlin, DSc, is Director of the N. N. Semenov Institute of Chemical Physics at the Russian Academy of Sciences, Moscow, Russia. He is a member of the Russian Academy of Sciences and many national and international associations. Dr. Berlin is world-renowned scientist in the field of chemical kinetics (combustion and flame), chemical physics (thermodynamics), chemistry and physics of oligomers, polymers, and composites and nanocomposites. He is the contributor to over 100 books and volumes and author of over 1000 original papers and reviews.

Svetlana Z. Rogovina, DSc
Professor Svetlana Z. Rogovina is a well-known specialist in the field of solid-phase modification of polysaccharides and production of blends of polysaccharides with synthetic polymers under conditions of the shear deformation. She is an author of about 100 papers and reviews dedicated to these problems. She has shown that in solid phase under conditions of shear deformations, some derivatives of cellulose (alkali cellulose and some others), as well as chitosan from natural polysaccharide chitin and its derivatives, may be produced. These results are very important because they allow scientists to realize that the polysaccharides modification in the absence of solvents by ecologically friendly method.

In recent years, Dr. Rogovina's interest lie with the creation of biodegradable compositions on the basis of polysaccharides and synthetic polymers. The blending of polymers under conditions of shear deformation allows to produce the materials with uniform distribution of components and as sequence with improving physico-mechanical properties and capacity to biodegradation.

Gennady E. Zaikov, DSc
Gennady E. Zaikov, DSc, is Head of the Polymer Division at the N. M. Emanuel Institute of Biochemical Physics, Russian Academy of Sciences, Moscow, Russia, and Professor at Moscow State Academy of Fine Chemi-

cal Technology, Russia, as well as Professor at Kazan National Research Technological University, Kazan, Russia. He is also a prolific author, researcher, and lecturer. He has received several awards for his work, including the Russian Federation Scholarship for Outstanding Scientists. He has been a member of many professional organizations and on the editorial boards of many international science journals. Dr. Zaikov has recently been honored with tributes in several journals and books on the occasion of his 80th birthday for his long and distinguished career and for his mentorship to many scientists over the years.

CONTENTS

LIST OF CONTRIBUTORS

S. H. O. Akhyari
Institute of Petrochemical Processes of National Academy of Sciences of Azerbaijan, Baku, Azerbaijan

N. A. Aksenova
Semenov Institute of Chemical Physics, Russian Academy of Sciences, Moscow, Russia.
E-mail: naksenova@mail.ru

K. V. Aleksanyan
Semenov Institute of Chemical Physics, Russian Academy of Sciences, Moscow, Russia

D. S. Andreev
Sebrykov Department, Volgograd State Architect-build University, Volgograd, Russia

V. A. Babkin
Sebrykov Department, Volgograd State Architect-build University, Volgograd, Russia

E. Bakuradze
Department of Biology, Faculty of Exact and Natural Sciences, Iv. Javakhishvili Tbilisi State University, Tbilisi, Georgia

A. A. Berlin
Semenov Institute of Chemical Physics, Russian Academy of Sciences, Moscow, Russia

A. E. Chalykh
Frumkin Institute of Physical Chemistry and Electrochemistry, Russian Academy of Sciences, Leninskii pr. 31, Moscow 119991, Russia

D. Dzidzigiri
Department of Biology, Faculty of Exact and Natural Sciences, Iv. Javakhishvili Tbilisi State University, Tbilisi, Georgia. E-mail: d_dzidziguri@yahoo.com

E. Erbay
Petkim Petrokimya Holding, Izmir, Turkey

V.K. Gerasimov
Frumkin Institute of Physical Chemistry and Electrochemistry, Russian Academy of Sciences, Leninskii pr. 31, Moscow 119991, Russia

N. Giorgobiani
Department of Biology, Faculty of Exact and Natural Sciences, Iv. Javakhishvili Tbilisi State University, Tbilisi, Georgia

A. Ya. Gorenberg
Semenov Institute of Chemical Physics, Russian Academy of Sciences, Moscow, Russia

A. V. Grachev
Semenov Institute of Chemical Physics, Russian Academy of Sciences, Moscow, Russia

V. A. Grigorovskaya
Semenov Institute of Chemical Physics, Russian Academy of Sciences, Moscow, Russia.
E-mail: valgrig3@ mail ru

V. G. Grinev
Semenov Institute of Chemical Physics, Russian Academy of Sciences, Moscow, Russia

N. A. O. Guliyev
Institute of Petrochemical Processes of National Academy of Sciences of Azerbaijan, Baku,
Azerbaijan

M. Iskakova
Department of Chemical Technology, Ak. Tsereteli Kutaisi State University, I. Chavchavadze Ave. 1,
0179 Tbilisi, Georgia. E-mail: marinaiskakova@gmail.com

N. F. O. Janibayov
Institute of Petrochemical Processes of National Academy of Sciences of Azerbaijan, Baku,
Azerbaijan. E-mail: j.nazil@yahoo.com

A. Jimsher
Institute of Macromolecular Chemistry and Polymeric Materials, I. Javakhishvili Tbilisi State
University, I. Chavchavadze Ave. 13, 0179 Tbilisi, Georgia

E. V. Kiseleva
Semenov Institute of Chemical Physics, Russian Academy of Sciences, Moscow, Russia

A. V. Kotova
Semenov Institute of Chemical Physics, Russian Academy of Sciences, Kosygina 4, Moscow 119991,
Russia

R. G. Kryshtal
Institute of Radio Engineering and Electronics, Russian Academy of Sciences (Fryazino Branch),
pl. Vvedenskogo 1, Fryazino, Moscow 141120, Russia

S. P. Kuznetsov
Lebedev Physical Institute, Russian Academy of Sciences, Moscow, Russia

O. P. Kuznetsova
Semenov Institute of Chemical Physics, Russian Academy of Sciences, Moscow, Russia

E. Markarashvili
Department of Chemistry, Iv. Javakhishvili Tbilisi State University, I. Chavchavadze Ave. 1,
0179 Tbilisi, Georgia

I. A. Matveeva
Semenov Institute of Chemical Physics, Russian Academy of Sciences, Kosygina 4, Moscow 119991,
Russia

A. V. Medved
Institute of Radio Engineering and Electronics, Russian Academy of Sciences (Fryazino Branch),
pl. Vvedenskogo 1, Fryazino, Moscow 141120, Russia

I. N. Meshkova
Semenov Institute of Chemical Physics, Russian Academy of Sciences, Moscow, Russia

I. Modebadze
Department of Biology, Faculty of Exact and Natural Sciences, Iv. Javakhishvili Tbilisi State University, Tbilisi, Georgia

G. Mosidze
Department of Biology, Faculty of Exact and Natural Sciences, Iv. Javakhishvili Tbilisi State University, Tbilisi, Georgia

O. Mukbaniani
Department of Chemistry, Iv. Javakhishvili Tbilisi State University, I. Chavchavadze Ave. 1, 0179 Tbilisi, Georgia; Institute of Macromolecular Chemistry and Polymeric Materials, I. Javakhishvili Tbilisi State University, I. Chavchavadze Ave. 13, 0179 Tbilisi, Georgia

F. A. O. Nasirov
Institute of Petrochemical Processes of National Academy of Sciences of Azerbaijan, Baku, Azerbaijan; Petkim Petrokimya Holding, Izmir, Turkey. E-mail: fnasirov@petkim.com.tr

L. A. Novokshonova
Semenov Institute of Chemical Physics, Russian Academy of Sciences, Moscow, Russia

L. A. Pevtsova
Semenov Institute of Chemical Physics, Russian Academy of Sciences, Kosygina 4, Moscow 119991, Russia

K. Yu. Prochukhan
Department of Chemistry, Bashkir State University, Ufa, Russia

Yu. A. Prochukhan
Department of Chemistry, Bashkir State University, Ufa, Russia

E. V. Prut
Semenov Institute of Chemical Physics, Russian Academy of Sciences, Moscow, Russia. E-mail: evprut@chph.ras.ru

L. N. Raspopov
Institute of Problems of Chemical Physics, Russian Academy of Sciences, Chernogolovka, Russia

S. Z. Rogovina
Semenov Institute of Chemical Physics, Russian Academy of Sciences, Moscow, Russia. E-mail: s.rogovina@mail.ru

A. V. Roshchin
Institute of Radio Engineering and Electronics, Russian Academy of Sciences (Fryazino Branch), pl. Vvedenskogo 1, Fryazino, Moscow 141120, Russia

M. Rukhadze
Department of Biology, Faculty of Exact and Natural Sciences, Iv. Javakhishvili Tbilisi State University, Tbilisi, Georgia

S. N. Rusanova
Kazan National Research Technological University, K. Marx Street, 68, Kazan, Tatarstan 420015, Russia

L. Rusishvili
Department of Biology, Faculty of Exact and Natural Sciences, Iv. Javakhishvili Tbilisi State University, Tbilisi, Georgia

V. T. Shashkova
Semenov Institute of Chemical Physics, Russian Academy of Sciences, Kosygina 4, Moscow 119991, Russia

A. A. Shaulov
Semenov Institute of Chemical Physics, Russian Academy of Sciences, Moscow, Russia.
E-mail: ajushaulov@yandex.ru

A. N. Shchegolikhin
Emanuel Institute of Biochemical Physics, Russian Academy of Sciences, Moscow, Russia

S. Yu. Sofina
Kazan National Research Technological University, K. Marx Street, 68, Kazan, Tatarstan 420015, Russia

D. V. Solomatin
Semenov Institute of Chemical Physics, Russian Academy of Sciences, Moscow, Russia

A. B. Solovieva
Semenov Institute of Chemical Physics, Russian Academy of Sciences, Moscow, Russia.
E-mail: ann.solovieva@gmail.com

A. O. Stankevich
Semenov Institute of Chemical Physics, Russian Academy of Sciences, Kosygina 4, Moscow 119991, Russia

O. V. Stoyanov
Kazan National Research Technological University, K. Marx street, 68, Kazan, Tatarstan 420015, Russia. E-mail: ov_stoyanov@mail.ru

T. Tatrishvili
Department of Chemistry, Iv. Javakhishvili Tbilisi State University, I. Chavchavadze Ave. 3, 0179 Tbilisi, Georgia; Institute of Macromolecular Chemistry and Polymeric Materials, I. Javakhishvili Tbilisi State University, I. Chavchavadze Ave. 13, 0179 Tbilisi, Georgia

E. Tavdishvili
Department of Biology, Faculty of Exact and Natural Sciences, Iv. Javakhishvili Tbilisi State University, Tbilisi, Georgia

N. E. Temnikova
Kazan National Research Technological University, K. Marx Street, 68, Kazan, Tatarstan 420015, Russia

N. L. Zaichenko
Semenov Institute of Chemical Physics, Russian Academy of Sciences, Kosygina 4, Moscow 119991, Russia. E-mail: zaina@polymer.chph.ras.ru

G. E. Zaikov
N.M Emanuel Institute of Biochemical Physics, Russian Academy of Sciences, Kosygina 4, Moscow 119991, Russia. E-mail: chembio@sky.chph.ras.ru

B. I. Zapadinskii
Semenov Institute of Chemical Physics, Russian Academy of Sciences, Kosygina 4, Moscow 119991, Russia

LIST OF ABBREVIATIONS

AlkXh-Co	cobalt alkylxhantogenate
AOC	aluminum organic compounds
APs	amphiphilic polymers
BA	boric acid
CCU	Center of Collective Use
CL	caprolactam
COA	carbooligoarylenes
COCs	cycloolefin copolymers
CP	cross polarization
4,4'-DADPS	4,4'-diaminodiphenylsulfone
4,4'-DADPM	4,4'-diaminodiphenylmethane
4,4'-DATPO	4,4'-diaminotriphenyloxide
DEAC	diethylaluminumchloride
DEDTC-Co	cobalt-diethyldithiocarbamate
DSC	differential scanning calorimetry
EADC	ethylaluminumdichloride
EDG	elastic-deformation grinding
EG	extrusion grinding
ENB	ethylidene norbornene
EO	ethylene oxide
EPDM	ethylene-propylene-diene rubbers
ESP	elastic-strain powdering
ETS	ethyl silicate
EVA	ethylene with vinyl acetate
EVAMA	modified ethylene copolymers with vinyl acetate
GS	(3-glycidoxypropyl)trimethoxysilane
GPC	gel permeation chromatography
GRT	ground rubber tire
HDPE	high-density polyethylene
HIC	hydrophobic interaction chromatography
HPHTS	high-pressure and high-temperature sintering
HTSD	high-temperature shear deformation

IPO	inorganic polyoxides
IPP	isotactic polypropylene
iPrXh-Co	cobalt iso-propylxhantogenate
ISAC&S	intensive stress action compression and shear
KW	Kruskal–Wallis
LDPE	low-density polyethylene
MAO	methylaluminoxane
MFI	melt flow index
MFR	melt flow rate
mp	melting point
mPCTPO	4-protio-3-carbamoyl-2,2,5,5-tetraperdeuteromethyl-3-pyrroline-1-yloxy
MPDA	methylphenyldiamine
MTHFA	metatetrahydrophtalanhydride
MTT	methyltetrazolium blue
MW	Mann–Whitney
MV	Mooney viscosity
OBOs	oligomer boron oxides
o-DCB	o-dichlorobenzene
PA	polyanthracene
PBS	phosphate-buffered saline
PD	photoditazine
PDT	photodynamic therapy
PEG	polyethylene glycol
PEI	polyetherimide
PEO	polyethyleneoxide
PEPA	polyethylenepolyimine
PLA	polylactide
PMHS	polymethylhydrosiloxanes
PMMA	polymethyl methacrylate
PN	polynaphtalene
PO	propylene oxide
PPh	polyphenilenes
PPS	porphyrin photosensitizers
PRB	plasmon resonance band
PS	photosensitization
PVA	polyvinyl alcohol
PVP	polyvinylpyrrolidone

RP	rubber powder
SAW	surface acoustic wave technology
SDV	styrene-divinylbenzene
SEC	size-exclusion chromatographic
SPR	surface plasmon resonance
SSSE	solid-state shear extrusion
SV	solution viscosity
τind	induction time
TAA	trialkylaluminum
TEA	triethylaluminum
TFHQ	tetrafluorohydroquinone
Tg	glass transition temperature
TGA	thermogravimetric analysis
THF	tetrahydrofuran
TMA	thermomechanical analysis
TPEs	thermoplastic elastomers
TPP	meso-tetraphenylporphyrine
TPVs	thermoplastic vulcanizates
TSPC	thermostable protein complex
UCPS	upper critical point of solubility
VCN	very cold neutron
MTHFA	metatetrahydrophtalanhydride

FOREWORD

The volume presents the results of investigation made in the past years at the Department of Polymers and Composite Materials of Institute of Chemical Physics of the Russian Academy of Sciences.

The Institute of Chemical Physics of the Russian Academy of Sciences is one of the main chemical institutes of the Russian Academy of Sciences. It was established in 1931. The first head of the Institute was Nobel Prize winner academician N. N. Semenov.

The Institute was established with the aim of the "introduction of physical theories and methods into chemistry, chemical industry, and other branches of economics." More than 80 years of the activity of the Institute has shown that this problem was solved. However, the main achievement of the Institute's team and its Head N. N. Semenov was the establishment of a new branch of natural sciences, chemical physics. N. N. Semenov defined the chemical physics as a "science describing the fundamentals of chemical transformations and the associated problems of substance structure." This idea indicates that chemical physics has a comprehensive character, and it is not surprising that its ideas and concepts are being used in all areas of natural sciences including biology and medicine.

Initially, the basic line of investigation in the Institute was the theory of chain reactions, processes of combustion and explosion, later chain reactions of nuclear fission, and polymerization reactions. At present, in the Institute, studies of kinetics and mechanism of heterogeneous chemical reactions and catalysis, liquid-phase oxidation reactions, and kinetics of chemical reactions in biological systems are being developed.

The current Semenov Institute of Chemical Physics consists of six large departments, each with its own "face", namely,

- i) Department of Kinetics and Catalysis
- ii) Department of Polymers and Composite Materials
- iii) Department of Combustion and Explosion
- iv) Department of Substance Structure
- v) Department of Dynamics of Chemical and Biological Processes
- vi) Department of Chemical and Biological Safety

The Department of Polymers and Composite Materials of Semenov Institute, traditionally one of the world leaders in the area of synthesis and investigation of the polymers and polymer composite materials, had been organized in 1958. The academician N. S. Enikolopov—a famous scientist in the field of polymers and polymer composites—was the head of the department from 1963 to 1990. For the past 25 years, academician A. A. Berlin has beens the head of this department. The team of the department has experience in the successful development of high-strength composite materials based on filled polymer composite and nanocomposite materials with the given functional properties, multicomponent polymer and elastomeric compositions, noncombustible materials based on inorganic and hybrid polymers and composites, and biodegradable composite materials based on synthetic and natural polymers. Based on the works currently performed in the Department of Polymers and Composite Materials in this field, one can distinguish the following priority areas:

1. Development of high-strength composite materials on the basis of thermoset and thermoplastic matrices reinforced by glass, carbon, organic, boron, and other high-modulus fibers.

The study of the influence of numerous factors on the physicomechanical properties of composites allows one to synthesize the composite materials with increased strength.

The main directions of investigations are related to the study of composites fracture mechanisms: fiber adhesion to binders; development and modification of polymer binders; use for composite reinforcing of new types of glass fibers, produced from rocks (basalt, tuff, etc.); and creation of scientific foundations of technology of prepreg and composite production.

2. Development of polymer-filled composite and nanocomposite materials of new generation based on large-tonnage industrial thermoplasts – polyolefins.

Synthesis of composites promotes the improvement of the traditional methods of introduction of fillers into polymer matrix via mixing of components in polymer melt and progress of the polymerization filling technology proposed first in the Department of Polymers and Composite Materials and successfully tested in experimental-industrial scale. The application of the polymerization filling technology allows one to maximally realize the properties of the polymer nanocomposites at low and ultrahigh-filling degrees, especially at the synthesis of composites based on ultra-high-molecular-weight polyethylene (UHMWPE). These works are developed in the Department of Polymers and Composite Materials in the following promising area:

Development of constructional and functional nanocomposite materials of new generation based on polyolefins (polyethylene, polypropylene, ethylene, and propylene copolymers with vinyl comonomers) and nano-dispersed fillers (carbon, layered silicates, etc.) with the set of improved operational characteristics (increased rigidity, thermal stability, barrier properties, decreased combustibility, controlled electric and radar absorbent properties, and improved tribological characteristics) and technology of their production.

Synthesis of new ultra-high-filled polymer composite materials (up to 80–85 vol.%) combining heat conducting and dielectric properties and plasticizing capacity based on binary (micro/nano) heat-conducting fillers and UHMWPE and polymerization technology of their production.

3. Development of nanofilled compositions cured by UV radiation.

Nanofilled compositions cured by UV based on acrylic, methacrylic, and urethane monomers, oligomers, photoinitiators, and filler nanoparticles are developed. These compositions are used to produce nanoparticles of dyes, luminophores, and other photosensitive compounds for optical and electronic industries and filled by TiO_2, SiO_2, or Al_2O_3 – for protective and abrasion-resistant coatings featuring improved quality.

4. Blending and chemical modification of polymers under conditions of shear deformation using extruders and other mixers. Multicomponent polymer and elastomer compositions.

For production of polymer mixtures, a fundamentally new method of polymer blending based on joint action of high pressure and shear deformations realized in extruders and some kinds of mixers was developed. It was shown that under these conditions, the significant structural transformation of the compounds and changes in their reactivity take place, so some chemical reactions can be carried out in solid phase in the absence of solvents.

Nowadays, the development of new approaches in the regulation of structure of the polymer composite materials gains special attention, as a result, gaining set of new properties required for practical application in modern articles and constructions. So, among multicomponent polymer systems, the composite materials related to class of thermoplastic elastomers are widespread. The production of such materials using method of blending under conditions of shear deformation gives the possibility to combine two important characteristics peculiar to each of the components: preservation of the mechanical properties of the initial polymers at operation with the processability into articles by a technology applied for linear thermoplastic polymers. A great advantage of the thermoplastic elastomeric materials is also their ability for multiple processing without substantial decrease in properties that allows one to almost completely use the wastes of production and used articles and gives additional economic effect.

5. Noncombustible materials based on inorganic and hybrid polymers and composites.

New priority scientific area developed in the Department of Polymers and Composite Materials is a production of composite materials based on one of the most widespread classes of natural polymers – inorganic thermoplastic polyoxides with increased elastic and strength characteristics. The binders of organic–inorganic composites is proposed to use the polyoxides based on boron, fluorinated borophosphates of different metals, blends of polyoxides with different thermoplastic polymers, and hybrid polymers of boron polyoxides with monomeric and oligomeric nitrogen-containing compounds. In the presence of inorganic thermoplastic components, it is possible to perform molding and chemical modification and synthesize hybrid organic–inorganic polymers and blends with polyolefins noncombustible nonflammable materials. Besides, these materials are characterized by the absence or substantial decrease of the toxic product amount released at thermal action and inflammation. The production of the materials based on hybrid polymer composites assumes the application of ecologically clean technologies in the absence of solvents.

6. Synthesis of biodegradable composite materials based on blends of synthetic and natural polymers.

The synthesis of biodegradable composite materials based on the blends of synthetic and natural polymers is one of the priority areas of modern polymer chemistry. Among such compositions, the blends based on industrial large-tonnage thermoplastic polymers and renewable polysaccharides characterized by easy biodegradability are of special interest.

With the aim to improve the production methods of biodegradable polymer compositions and operational properties of the materials based on them, a method of mixing of the above-mentioned classes of polymers under action of high-temperature shear deformations in the absence of solvents was used. The distinctive feature of the method proposed is a possibility to obtain under particular conditions the powders with increased homogeneity and uniform distribution of the components even in the case when one of them is infusible polymer (polysaccharides). The application of the shear deformation method for the production of the powder polymer blends based on infusible polysaccharides and synthetic polymers is original approach to simultaneous utilization and processing of the polymer wastes of both synthetic and natural origin. Thus, the developing problem is not only scientific, but promotes the solution of the ecological problems.

7. Theory and computer simulation.

Special attention is given to problems of nonlinear dynamics, structural transitions, and kinetics of solid-state reactions (soliton regimes).

The mechanical properties and fracture of fibrous and filled composites as well as the structure, mechanical and relaxation properties of polymeric and inorganic glasses, crystals, and liquids are also of scientific interest.

In 2014, the Institute, hosted a conference dedicated to the 90th anniversary of academician Enikolopov, where the main achievements of Department were presented.

The papers presented in this book concern most of the above-mentioned directions of Department's investigations and the editors hope that this book will be very interesting for readers.

Alexandr A. Berlin
Director of Semenov Institute of Chemical Physics
Russian Academy of Sciences
4 Kosygin str., 119991 Moscow, Russia
berlin@chph.ras.ru
Svetlana Z. Rogovina
Leader Scientific Researcher
Semenov Institute of Chemical Physics
Russian Academy of Sciences
4 Kosygin str., 119991 Moscow, Russia
s.rogovina@mail.ru
Gennady E. Zaikov
Head of Polymer Division
N.M. Emanuel Institute of Biochemical Physics
Russian Academy of Sciences
4 Kosygin str., 119991 Moscow, Russia
chembio@sky.chph.ras.ru

PREFACE

Additives are selected depending on the type of polymers to which they will be added or the application for which they will be used. Appropriate selection of additives helps develop value-added plastics with improved durability, as well as other advantages. This research book provides a range of modern techniques and new research on the use of additives in a variety of applications.

The methods and instrumentation described represent modern analytical techniques useful for researchers, product development specialists, and quality control experts in polymer synthesis and manufacturing. Engineers, polymer scientists, and technicians will find this volume useful in selecting approaches and techniques applicable to characterizing molecular, compositional, rheological, and thermodynamic properties of elastomers and plastics.

CHAPTER 1

5-ETHYLIDENE-2-NORBORENE-OLEFIN COPOLYMERS: SYNTHESIS, STRUCTURE, MORPHOLOGY, PROPERTIES, AND FUNCTIONALIZATION OF COPOLYMERS

I. N. MESHKOVA[1], A. N. SHCHEGOLIKHIN[2],
L. N. RASPOPOV[3], E. V. KISELEVA[1], S. P. KUZNETSOV[4],
V. G. GRINEV[1], and L. A. NOVOKSHONOVA[1]

[1]Semenov Institute of Chemical Physics, Russian Academy of Sciences, Moscow, Russia

[2]Emanuel Institute of Biochemical Physics, Russian Academy of Sciences, Moscow, Russia

[3]Institute of Problems of Chemical Physics, Russian Academy of Sciences, Chernogolovka, Russia

[4]Lebedev Physical Institute, Russian Academy of Sciences, Moscow, Russia

CONTENTS

ABSTRACT

Copolymers of ethylene (propylene) and 5-ethylidene-2-norbornene (ENB) containing 5–65 mol% of the cyclic comonomer units (ENB) in the main chain were synthesized by using homogeneous *ansa*-zirconocene catalysts of C_2 symmetry. They are *rac*-Et(Ind)$_2$ZrCl$_2$ (**1**), *rac*-Et(H$_4$Ind)$_2$ZrCl$_2$ (**2**), *rac*-Me$_2$Si(Ind)$_2$ZrCl$_2$ (**3**), Me$_2$C(Ind)$_2$ZrCl$_2$ (**4**), *rac*-Me$_2$C(*t*-BuInd)$_2$ZrCl$_2$ (**5**), and methylaluminoxane (MAO). It is shown that the ethylidene groups of ENB do not take part in the copolymerization and the very cyclic comonomer unit bears the pendent unsaturated group. According to the obtained kinetic data (on copolymer output, the specific rate of ethylene consumption, and conversion of ENB), catalyst **1** shows the highest activity in the copolymerization of ethylene with ENB in comparison with the other catalysts used. The incorporation of ENB into copolymer chain proceeds with greater difficulty than with unsubstituted norbornene. The values of r_1 are higher and the values of r_2 are lower for ENB than those reported for the copolymerization of ethylene with norbornene. The introduction of *t*-Bu substituent into Ind-ligand of *ansa*-zirconocene (catalyst **5**) considerably complicates the ENB insertion into the copolymer. The copolymers were characterized by FTIR, DSC, NMR, and X-ray spectroscopy as well as by the original technique of scattering with very cold neutrons having energies between 2×10^{-7} and 5×10^{-4} eV. With the growth of ENB content in the copolymer, its crystallinity, melting enthalpy, and melting temperature progressively decrease. Embedding of ENB units into the polyethylene main chain leads to a considerable increase of the T_g value and better optical transparency of the resulting material. The copolymer with 30 mol% of ENB, prepared by catalyst **1**, was optically transparent in the UV and visible parts of the spectrum and had the highest T_g (83°C) and density ($\rho > 1,000$ kg/m^3). Ozonolysis under the mild reaction conditions permits the conversion of the ethylidene groups of poly(ethylene-*co*-ENB)s into polar carbonyl, aliphatic carboxylic, and aldehydes side groups.

INTRODUCTION

Cycloolefin copolymers (COCs) are an interesting and practically important family of copolymers synthesized by metallocene catalysts. Copolymers of ethylene[1–3] and propylene[4–6] with norbornene have unique thermal and optical properties. On this basis, the firms Mitsui and Ticona manufacture industrial-grade Apel and Topas polymer materials with the very high glass transition temperature and transparency.[7] To improve compatibility with other materials and for better adhesive properties, the COC functionalization is carried out by one-step copolymerization of ethylene with cyclic monomers containing polar groups[8-10] or by the copolymerization of ethylene or propylene and cyclic monomers with double bonds, which are capable of taking part in post-polymerization polymer-analog transformations, such as 5-vinyl-2-norbornene,[11-13] 5-ethylidene-2-norbornene (ENB),[14-17] 4-vinyl-1-cyclohexene,[18] and dicyclopentadiene.[19]

It is known that the relatively cheap substituted norbornene ENB is used as termonomer in the ethylene propylene diene terpolymer (EPDM) synthesis with *ansa*-zirconcene catalysts.[20-23] The maximum content of ENB in EPDM is 15 (12) mol%. In ethylene–ENB copolymerization by different types of catalysts (with *ansa*-zirconocenes of C_2 symmetry,[14-16] constrained-geometry catalysts[17]), it is possible to obtain copolymer products containing 50–60 mol% of ENB.

In this chapter, the kinetic parameters of the ethylene–ENB copolymerization process and the copolymer composition depending on the nature of C_2 symmetry *ansa*-zirconocene catalyst (bridge, ligands in Ind of *ansa*-zirconocene) are considered. For the synthesis of the polymer materials that combine high glass transition temperature, high density, and optical transparency with the capability for postpolymerization polymer-analog transformation, the morphology, density, as well as thermal and optical properties of synthesized poly(ethylene-*co*-ENB)s are studied. The possibility of modifying the copolymers by ozonation of their side groups >C=CH–CH$_3$ is also investigated.

EXPERIMENTAL PROCEDURE

MATERIALS

The catalyst components *rac*-Et[Ind]$_2$ZrCl$_2$ (1), *rac*-Et[H$_4$Ind]$_2$ZrCl$_2$ (2), *rac*-Me$_2$Si[Ind]$_2$ZrCl$_2$ (3) (Aldrich), and methylaluminoxane (Aldrich, 10 wt% in toluene) were used as received. *Rac*-Me$_2$C[Ind]$_2$ZrCl$_2$ (4) and *rac*-Me$_2$C[(t-Bu)Ind]$_2$ZrCl$_2$ (5) were synthesized in the laboratory of Prof. I. E. Nifantyev (Moscow State University).

Spectral-grade toluene (Aldrich) was purified by refluxing over sodium wire with subsequent distillation under nitrogen. Ethylene and propylene were of polymerization grade. ENB (Aldrich, 99%; a mixture of exo and endo isomers) was distilled over Na under vacuum at room temperature.

ETHYLENE (PROPYLENE)–ENB COPOLYMERIZATION

The formation of the catalytic complex and the following olefin copo-lymerization were carried out in a 0.5 dm³ glass reactor equipped with units for feeding the solvent, catalyst, comonomers and a stirrer to mix the reaction mixture. The reagents (toluene, methylaluminoxane [MAO], zirconocene, and ENB) were introduced into reactor at room temperature. After heating the reaction mixture up to the copolymerization tempera-ture, toluene in reactor was saturated by the comonomer. The saturation time was 1.5–2 min. During copolymerization, the ethylene (propylene) pressure, the temperature, and the stirring rate of reactive mass were maintained constant. The cyclic comonomer conversion led to a change in the ENB:olefin molar ratio in the copolymerization. The copolymer-ization conditions are presented in Table 1.1. The concentration of ethyl-ene or propylene in toluene was calculated using the experimental values of Henry (for ethylene $k_H = 0.9 \times 10^{-3}Q^{2990/RT}$mol/L·atm; for propylene $k_H = 2.1 \times 10^{-3}Q^{3340/RT}$mol/L·atm. The catalyst activity (A) and the rate of ethylene (propylene) incorporation into the copolymer chain were charac-terized by the consumption of olefin during the process (R_p, kg of PE(PP)/ mol of Zr mol comonomer h).

COPOLYMER CHARACTERIZATION

The content of ENB in the samples was estimated via calculations from the yield of the copolymer products and the consumption of ethylene (pro-pylene) in copolymerization, as well as by the method of ozonation of the side double bonds of copolymer,[24] which is given the next paragraph.

The composition of some ethylene-ENB copolymers was also deter-mined by ^{13}C NMR (Bruker/XWIN-NMR, solvent TCE, 103°C). To cal-culate the ENB content in the copolymer chain, the formula applied for the determination of copolymer composition with unsubstituted norbornene[25] was used:

mol ENB% = $\{[2I_{C7} + I_{C1-C4} + I_{C2-C3}]/3I_{CH2}\} \times 100$,

where I_{CH2}, I_{C7}, I_{C1-C4}, and I_{C2-C3} are the total peak area in the ethylene-ENB spectrum in the ranges 27.73–28.73, 30.90–31.86, 37.57–40.10, and 45.07–48.85 ppm, respectively.

TABLE 1.1 The Influence of the Catalyst Nature and Conditions[a] of the Homo- and olefin–ENB Copolymerization on the Activity of *ansa*-Zr-cene Catalysts and Copolymer Compositions

Run	Catalyst	The Initial Mol. ratio ENB: Olefin	[Zr], mmol/L	Mol Ratio [Al]: [Zr]	t_{pol}, min	Yield, g	A^b	Conversion of ENB, %	[ENB] in Copolymer mol%
1	1	Ethylene	0.065	3,000	60	0.6	15,000	–	0
2	1	Ethylene	0.26	1,000	70	5.6	4,400	–	0
3	1	2	0.026	2,550	60	4.6	14,600	100	5.3
4	1	6.5	0.244	1,000	87	5.55	1,760	78	9.5
5	1	8.6	0.106	1,875	85	5.64	2,020	74	28
6	1	7.7	0.035	2,510	110	9.5	2,080	24	30.5
7	1	16	0.143	1,500	160	1.29	1,645	62.7	39
8	1	22.5	0.115	1,850	180	0.98	1,000	9.7	43
9	1	ENB	3.6	120	300	0.25	–	3.5	100
10	2	Ethylene	0.023	4,160	90	2	10,400	–	0
11	2	11.8	0.06	1,800	190	1.89	1,170	24	62
12	3	Ethylene	0.132	1,370	65	4.2	5,800	–	0
13	3	Ethylene	0.4	1,060	53	3.75	5,000	–	0
14	3	2.2	0.084	2,230	50	2.9	4,870	100	9.4
15	3	4.4	0.087	2,030	100	4.7	1,900	100	20
16	3	6.5	0.083	2,010	290	3	1,000	37	31
17	3	6.5	0.31	1,000	90	4.22	1,110	79.4	12.6
18	3	10.4	0.02	2,800	170	0.4	800	23.8	37
19	4	Ethylene	0.107	900	120	3.5	2,040	–	0
20	4	6.5	0.264	1,000	129	6.57	1,820	100	17
21	5	Ethylene	0.245	1,000	30	3	9,700	–	0
22	5	6.5	0.245	1,000	65	2.9	5,580	16.6	3.3
23	1	Propylene	0.056	1,700	20	1	3,000	–	–
24	1	1.56	0.056	1,300	190	1	120	100	50.4
25	1	1.65	0.56	1,700	200	1.2	150	120	52

Notes: [a] In run 1, the ethylene concentration is 0.024 mol/L, T_{pol} = 33°C. In runs 2–8 and 10–22, the ethylene concentration is 0.058 mol/L, T_{pol} = 40°C. In run 9, the ENB concentration is 0.99 mol/L; T_{pol} = 40°C.
[b] The ethylene (propylene) consumption in homo- and copolymerization with ENB, kg of PE (PP)/mol Zr mol $C_2H_4 \cdot (C_3H_6)$ h.

The obtained results coincide with data of other works. Thus, the ENB content in sample run 4 (Table 1.1), according to ^{13}C NMR, is 9.02 mol%; from the copolymer yield and the consumption of ethylene 9.50 mol%; and by ozonation 9.66 mol%.

The molecular mass of the copolymers was characterized by viscometry in decalin at135°C.

The density of the synthesized samples of homo- and copolymers of the ethylene with ENB was determined by the flotation method (ASTM D1505) at 20°C using ethanol/water mixture as working fluid. The measurement accuracy was ±0.1%.

The Fourier transform infrared (FTIR) spectra of the copolymers were obtained by a Perkin-Elmer FTIR-1720 spectrometer.

X-ray studies of the melt-molded ethylene–ENB copolymer samples with different contents of ENB were performed on a DRON-2.0 general-purpose diffractometer (Cu Kα radition, 20 mA, 40 kV, and 14°–28° range). The degree of crystallinity of copolymer samples was estimated according to the standard Hermans–Weidinger method,[26] which was modified as described by Raspopov and Belov[27] .The size of the copolymer crystallites was determined from the half-width of the reflections [110] and [200] and calculated by the Scherrer formula.[28] As a reference, we used the [200] line of graphite with reflection angle $2\theta = 24.8°$, which is close to the angles of reflections of copolymers under examination.

The supramolecular structure of the ethylene–ENB copolymers was also tested by the original technique of very cold neutron (VCN) scattering with energies 2×10^{-7} to 5×10^{-4} eV.[29] The neutron wavelengths were in the range 1–100 nm. In the order of magnitude, these values coincide with the characteristic sizes of the supramolecular structure of the polymers. VCN experiments were performed with a time-of-flight spectrometer in the wavelength range 4–60 nm. The measurements were conducted at 300 K using 150–250 µm copolymer samples with a measured surface area of 2.5×3.5 cm^2.

The transparency of ethylene–ENB copolymers in the near-UV and visible range was investigated using an instrument "Specol UV-vis" (the spectral width was 2 nm).

Differential scanning calorimetry (DSC) measurements of the copolymer products (for T_m, ΔH, and T_g) were carried out on Perkin-Elmer DSC-7 calorimeter at a heating rate of 10 K/min.

The samples of homopolymers and E–ENB copolymers for determination of their properties were prepared by compression molding.

Functionalization of ethylene–ENB copolymers by ozonolysis reaction

The ozonation procedure is based on the ability of unsaturated compounds to attach ozone to the double bonds at a high rate (one ozone molecule is attached to one double bond). The bimolecular rate constant of ozone addition to C=C varies over the range 10^5–10^6 L/(mol·s).[24] Measurements were performed on a double-bond analyzer ADS-5.[30] The copolymer solution in an organic solvent (carbon tetrachloride or *o*-dichlorobenzene, 1.5–2.5 g of the copolymer per liter) was loaded into the reactor, and an ozone–oxygen mixture with controlled concentration of ozone (1×10^{-6} mol/L) was passed through the reactor at 0°C. The amount of absorbed ozone was proportional to the content of double bonds in the copolymer. To avoid the degradation of polymer products, the concentration of ozone in the ozone–oxygen mixture did not exceed 1×10^{-6} mol/L. The ozonation method has high sensitivity (threshold sensitivity is of 10^{-11} moles per double bond), accuracy of ±1%, and high performance (0.5–5 min per measurement).

RESULTS AND DISCUSSION

The copolymerization of ethylene and ENB, catalyzed by the *ansa*-zirconocene systems, as well as the ethylene[11] or propylene[13] copolymerization with 5-vinyl-2-norbornene, proceeds by virtue of regioselective insertion of an endocyclic double bond of the cyclic comonomer into the main chain (Scheme 1.1). The ethylidene double bond of the ENB does not participate in the copolymerization.[15]

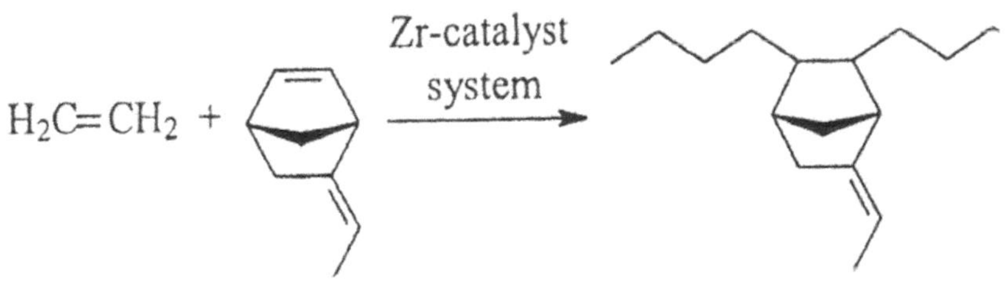

SCHEME 1.1 Main Chain.

FTIR spectra of the ethylene–ENB copolymers show absorption bands at 3,038, 1,687, and 808 cm⁻¹, which are characteristic for the ethylidene unsaturation.[20] Any absorption in the vicinity of 712 cm⁻¹, which is normally assignable to the endocyclic double bond of the ENB the monomer molecule, is absent in the spectra of the copolymers (Fig. 1.1).

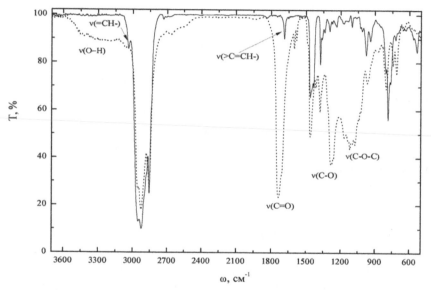

FIGURE 1.1   FTIR Spectra of Ethylene–ENB Copolymers Before (—) and After (···) Ozonolysis.

The compositions of the prepared ethylene–ENB copolymers, calculated from the ethylene consumption and yield of copolymer product, correlate with the results of ozonation of the ethylidene double bonds (Table 1.2). This means that the *ansa*-zirconocene catalyst systems used produce copolymers in which every cyclic comonomer unit bears the pendent unsaturated group.[15]

TABLE 1.2 Content of ENB and Ethylidene Double Bonds in the Ethylene–ENB Copolymers

Run	Olefin	Catalyst	Initial Mol.ratio ENB:Olefin	$[ENB] \times 10^3$ in Copolymer, mol/g	$[C=C] \times 10^3$ in Copolymer, mol/g
4	Ethylene	1	6.5	2.56	2.62
5		1	8.6	5.27	5.5
7		1	16	6.16	6.02
11		2	11.8	6.95	6.75
14		3	2.2	2.56	2.79
15		3	4.4	4.26	4.38
18		3	10.4	6	6.7
24	Propylene	1	1.56	6.2	5.7
25		1	1.65	7.5	8.2

KINETICS OF ETHYLENE–ENB COPOLYMERIZATION BY C_2 SYMMETRIC ANSA-ZIRCONOCENE SYSTEMS

The consumption of ethylene in the homo- and copolymerization and conversion of ENB in copolymerization are given in Table 1.1. Catalyst **1** shows the highest activity (on copolymer output, the specific rate of ethylene consumption, and conversion of ENB) in the polymerization of ethylene and in the copolymerization of ethylene with ENB in comparison to the other catalysts used.

At constant ethylene concentration in solvent during copolymerization, the ENB:ethylene ratio decreases with time due to the ENB consumption, depending on the catalyst activity and the relative reactivity of the comonomers.

The ethylene–ENB copolymerization with the catalysts used is a nonstationary process. Figure 1.2 shows the kinetic curves of ethylene consumption in ethylene–ENB copolymerization for different metallocenes. The increase of ethylene consumption rate observed for catalyst **1, 3,** and **4** can be associated with the gradual reduction of ENB concentration in reaction zone (decrease of ENB:ethylene ratio) with time as a result of ENB consumption in the copolymerization and with the consequent increase of the fraction of ethylene centers. The effect appears more visible for the more active catalyst **1**. At low ENB insertion into the copolymer chain, which was obtained with catalyst **5**, the ENB:ethylene ratio in the reaction zone changes very little. The reduction of ethylene consumption rate with time is mainly due to the lowering of activity of the catalyst itself. The decrease of ethylene consumption rate observed after the rate increase for catalyst **1, 3**, and **4** can have the same reason.

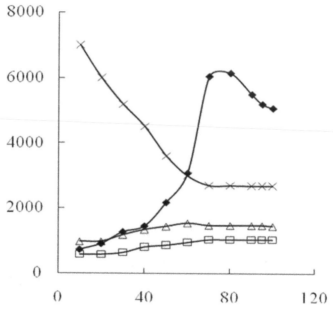

FIGURE 1.2 Change of Ethylene Consumption Rate During Ethylene–ENB Copolymerization With different *ansa*-Zirconocene Catalysts. Catalyst: ♦ – **1** (Table 1.1, no. 4), Δ – **3** (no. 17), □ – **4** (no. 20), and × – **5** (no. 22).

Ethylene–ENB copolymers with the cyclic comonomer content from 10 to 65 mol% were obtained. Based on the composition of the copolymers, which were prepared under conditions where the conversion of the cyclic comonomer did not exceed 25 mol%, the reactivity ratios of comonomers were calculated via the Finemann–Ross and Bohm methods. As seen from Table 1.3, the insertion of ENB into the copolymer chain in the presence of the tested *ansa*-zirconocene catalysts proceeds with more difficulty than with the unsubstituted norbornene. The values of r_1 are higher and the values of r_2 are lower for ENB than those reported for the copolymerization of ethylene with norbornene. The steric volume and the flexibility of a bridge in the *ansa*-zirconocene and the presence of substituents in the Ind-ligand affect the ability of ENB to be inserted into the copolymer chain as well as in the case of the unsubstituted norbornene.[31]

TABLE 1.3 Parameters of Copolymerization of Ethylene with ENB and with Norbornene (N) According to Fineman–Ross Equation[32] and BOHM Method[33]

Catalyst	Cyclic Comonomer	T_{pol}, °C	r_1	r_2	$r_1 r_2$	References
1	ENB	40	10.5	0.004	0.042	15
1	N	40	3	0.03	0.09	34
1	N	30	1.9	0.03	0.057	35
1	N	30	2.33	0.031	0.072	36
2	ENB	40	7.20[a]	—	—	
2	N	25	2.20[a]	—	—	37
2	N	50	3.20[a]	—	—	1
3	ENB	40	8.7	0.007	0.061	15
3	N	30	2.71	0.0525	0.143	36
5	ENB	40	~190[a]	—	—	—

Notes: [a]Calculated on Bohm method, $m_2/m_1 = 1/r_1 \cdot M_2/M_1$.

Under identical conditions of copolymerization, the ENB content in copolymers was 9.5 mol% for catalyst **1** with Et-bridge (Table 1.1, no. 4), 12.6 mol% for catalyst **3** with Me$_2$Si-bridge (no. 17), and 14 mol% for catalyst **4** with Me$_2$C-bridge (no. 20).The introduction of *t*-Bu substituent into Ind-ligand of *ansa*-zirconocene (catalyst **5**) leads to a decrease in ENB insertion into the copolymer chain up to 3.3 mol % ENB (no. 22) compared to 14 mol% ENB in copolymer obtained by catalyst **4** (no. 20).

For catalysts **1** and **3**, the product $r_1 r_2$ is equal to 0.042 and 0.061, respectively. This means that the structure of the produced ethylene–ENB copolymers is intermediate between block and alternating. According to the obtained copolymerization constants, the tendency to alternate of the comonomer molecules in the copolymer macrochain is higher for catalyst **1** than for catalyst **3**.

The molecular weight (intrinsic viscosity) of the ethylene–ENB copolymers decreases with increase of the cyclic monomer content (Table 1.4).

TABLE 1.4 Intrinsic Viscosity of Ethylene–ENB Copolymers Obtained Catalyst 1 Depending on ENB Content (Decalin, 135°C)

Run	[ENB] in Copolymer, mol%	[η], dL/g
1	0 (PE)	2.30
5	28.6	0.24
7	39.0	0.18
8	43.0	0.15

Supramolecular structure, transparency, and density of PE and ethylene–ENB copolymers

The results of the study of the ethylene–ENB copolymer supramolecular structure by X-ray diffraction and VCN scattering methods as well as data on density and transparency are summarized in Table 1.5. It can be seen that as the content of ENB units in the partly crystalline ethylene–ENB copolymers increases, the overall degree of crystallinity as well as the crystallite dimension decreases, and eventually the copolymer becomes completely amorphous. In the diffractograms for a series of ethylene–ENB copolymers (Fig. 1.3), the peaks belonging to a crystalline PE phase can be seen only for the copolymers having ethylene mass content above 12 mass% (Fig. 1.3, traces 3 and 4). The partly crystalline copolymer having the ENB content of 50 mass% has shown the amorphous halo peak at $2\Theta = 18.9°$, while the amorphous copolymer with the ENB content of 87 mass% exhibits the maximum at $2\Theta = 17.4°$. The diffractogram of an ENB-homopolymer (Fig. 1.3, trace 1) reveals the amorphous halo with two maxima at 16.3° and 22°, which is similar to that observed for the unsubstituted norbornene homopolymer.

It is worth noting that the X-ray diffraction and the VCN scattering methods may give, to some extent, controversial results. Specifically, the mean dimensions of the supramolecular formations in several copolymers evaluated by VCN scattering turned out to be higher than those found by X-ray diffraction (Table 1.5, runs 2 and 11). Further, according to the X-ray data, the samples 4 and 12 are devoid of any crystalline phase, while the VCN scattering data for the same samples indicate the presence of nonuniformities with mean diameters 10.0 and 8.7 nm, the volume fractions of these being equal to 0.1 and 0.2, respectively. Probably the observed discrepancies are related to the fact that, in contrast to X-ray diffraction, which is mainly sensitive to only strictly ordered crystalline domains, VCN scattering is capable of detecting some almost amorphous or highly defective structures as well.

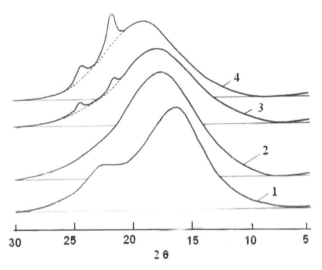

FIGURE 1.3   Wide-Angle X-ray Diffraction Patterns of poly-ENB (1) and the Ethylene–ENB Copolymers. Intensity is Normalized (10^4 pulse/s).[ENB] in Copolymer in Mass%: 100 (1), 87 (2), 66 (3), and 50 (4).

Notably, the possibility of the presence of the latter structures in the studied ethylene–ENB copolymers has been supported also by the observed lower optical transparency of the samples 4 and 12 (Table 1.5) as compared, for example, to sample 13, which is categorized as perfectly amorphous material in terms of both the X-ray and the VCN scattering analyses data. The general optical characteristics of the prepared copolymers in the near-UV and the visible parts of the spectrum have been studied using optical transmission measurements in the range of $l = 200$–750 nm. The data relevant to the optical properties of both PE and the ethylene–ENB copolymers are presented in Table 1.5. The transmittance of the copolymers increases with increasing content of the ENB units in the copolymer main chain. For the copolymer samples containing up to 20 mol% of ENB units, gradual formation of zones of transparency in the UV and visible parts of the spectrum has been noted, while at higher partial contents of ENB in the copolymer main chain the materials reveal fairly good transparency ($T > 50\%$ at $l > 350$ nm).

TABLE 1.5   Morphology, Optic Transparency, and Density of PE and Ethylene–ENB Copolymers

Run	Catalyst	[ENB] in Copolymer, mol%	X-ray Studies		Very Cold Neutron Scattering		Transparency ($\lambda = 750$ nm), %[b]	Density, kg/m³
			χ	$(L_{200} + L_{110})/2$, nm	φ	l, nm		
1	1	0 (PE)	0.77	12	n.d.[a]	n.d.	56	949
3	1	5.3	0.47	8.5	0.5	13	n.d.	955
6	1	30.5	amorphous	0	0.1	10	77.2	1,024
10	2	0 (PE)	0.64	24	n.d.	n.d.	36	956
11	2	62	amorphous	0	transparent	0	n.d.	1,200
14	3	9.4	0.31	8	0.3	12	n.d.	963
15	3	20	amorphous	0	0.2	8.7	73.5	986
16	3	31	amorphous	0	transparent	0	84.7	1,000

Notes: χ, degree of crystallinity; L, dimensions of crystallites by X-ray; φ, volume; l, dimensions of nonuniformity fractions by VCN. [a]Not determined. [b]The light transmission through the support = 89.9%.

Density is one of the most important parameters controlling the properties of homo- and copolymers. This parameter is determined by the polymer composition and the conditions of its preparation. From Table 1.5, it is seen that the density of the ethylene–ENB copolymers increases with increase in ENB content in copolymer and depends only insignificantly on the nature of the used catalyst as well as the density of the PE synthesized.

THERMAL PROPERTIES OF ETHYLENE–ENB COPOLYMERS

The thermal properties of the prepared copolymers show a clear dependence on their composition. The thermal characteristics of the ethylene-ENB copolymers in comparison with those of homo-PE and homo-poly-ENB samples were studied by DSC and are given in Table 1.6. According to the DSC data, and in good correlation with the above-mentioned X-ray and VCN scattering data, the partly crystalline ethylene–ENB copolymers show a sharp decrease of their crystallinity degree with increasing ENB content in the copolymer. Consequently, the crystal phase melting enthalpy and the melting temperature progressively decrease. Further increase of the ENB content in the ethylene–ENB copolymers leads to a fully amorphous state of the polymers. As is well known, the glass transition of the homo-PE takes place at temperatures (T_g) below 0°C. Embedding of 5-ethylidene-2-norbornene units into the polyethylene main chain leads to a considerable increase of T_g value of the resulting material. Among the copolymers studied in this work, the highest T_g value of 83°C was observed for a copolymer containing 30 mol% ENB and prepared using catalyst 1. This T_g value is typical for ethylene-substituted norbornene copolymers of similar composition prepared by the same catalyst.[34,38–40]

TABLE 1.6   Results of Investigation of PE and Ethylene–ENB Copolymer by DSC[a]

No.	Catalyst	Composition of Polymer Product	ΔH, J/g	Crystallinity,[b] %	T_m, °C	T_g, °C
1	1	0 (PE)	174.42	0.59	133.2	Below 0
3	1	5.3	92.56	0.33	124.7	Below 0
5	1	28	–	Amorphous	136.4	81.2
6	1	30.5	–	Amorphous	131	83.1
9	1	100.0 (poly-ENB)	–	Amorphous	121.6	57.6
10	2	0 (PE)	–	0.66	100.2	Below 0
11	2	62	175.85	Amorphous	–	69
12	3	0 (PE)	78.76	0.6	–	Below 0
14	3	9.4	11.3	0.27	131.7	Below 0
15	3	20	–	0.03	119.3	36.5
16	3	31	–	Amorphous	134.6	60
17	3	37	226.4	Amorphous	134	72.6
18	4	0 (PE)	36.6	77.3		Below 0
19	4	14	199.2	12		66
20	5	0 (PE)	190.4	68		Below 0
21	5	3.3		65		Below 0

Notes: [a]Second melting. [b]Calculated in view of ΔH of the PE monocrystal = 294 J/g.

FUNCTIONALIZATION OF ETHYLENE–ENB COPOLYMERS

The ethylene–ENB copolymers, which intrinsically show a good set of thermal and optical properties, can further benefit from the availability in their chains of side double bonds which, as has been mentioned previously, are able to participate in the ozonolysis reaction. The highest content of the ethylidene $>C=CH-CH_3$ groups in the copolymers prepared in this work was 6.9×10^{-3} mol/g. For comparison, the content of the same groups in a homopolymer of 5-ethylidene-2-norbornene was 8.3×10^{-3} mol/g, while the content of terminal C=C groups in a homo-PE having, for example, $M_n = 100,000$ is only ~5×10^{-5} mol/g.

Ozonation of the ethylene–ENB copolymers is a convenient way of modifying PE-based polymers by generating polar groups pendent to the main chain. The rate constants of ozone interaction with double bonds by several orders exceed those characteristic for reactions of O_3 with other groups.[24] Ozonides of organic compounds, such as many other derivatives from the peroxides class, readily decompose during storage or under heating, the main decomposition products being the corresponding aldehydes and carbonic acids.[41]

According to our FTIR data (Fig. 1.1), ozonolysis of the $>C=CH-CH_3$ groups pendent to the main chain of the copolymers under study seems to be capable of generating plenty of derivative functional groups. For instance, a series of absorption bands exist in the vicinity of 1,700–1,725, 1,740, and 1,750 cm^{-1}, which obviously originate from the vibrations of the carbonyl moiety in aliphatic carboxylic acids, ketones, and aldehydes. One more absorption characteristic for the ozonated material appears in the range 1,150–1,040 cm^{-1}, which, most probably, belongs to the stretching vibration of the $>C-O-$ group. However, the asymmetric stretching of aliphatic ethers bonds ($>C-O-C<$) is normally observed in the same frequency range (1,140–1,085 cm^{-1}), and one cannot exclude the possibility of aliphatic ether formation as a result of ozonolysis of such copolymers.

Thus, the obtained results confirm the feasibility of functionalization of the main chain of PE through copolymerization of ethylene with a cyclic ENB comonomer, followed by ozonolysis of pendent ethylidene groups.

CONCLUSION

The main factors controlling the kinetics of the ethylene–ENB copolymerization by C_2-symmetric *ansa*-zirconocene catalysts, namely, the composition, morphology, and properties of the obtained ethylene–ENB copolymers, depend on the nature of the *ansa*-zirconocene catalyst (type of bridge, the presence of substituent in the indenyl ring) and the content of the cyclic comonomer in the reaction copolymerization mixture.

The modification of crystalline as well as amorphous phases takes place with the growth of the ENB content in the copolymer. The presence of the highly defective nano-sized structures in the ethylene–ENB copolymers detected by VCN scattering method decreases the transparency of the amorphous ethylene–ENB copolymer materials.

The ethylene–ENB copolymer chain with pendent reactive sites is amenable to further polymer-analog transformations, such as the ozonolysis reaction. By ozonolysis under the mild reaction conditions of ethylidene double bonds, the carbonyl, carboxy, and ester polar groups can be introduced into ethylene–ENB copolymer. For the first time, we applied the method of ozonation of ethylidene double bonds for detecting the molecular compositions and functionalization of the ethylene–ENB copolymers.

ACKNOWLEDGMENTS

We are grateful to Dr. Maria Carmela Sacchi for measurements of the copolymers' ^{13}C NMR spectra, and Prof. I.E. Nifantyev for *ansa*-zirconocenes.

KEYWORDS

- **5-ethylidene-2-norbornene**
- **olefin**
- **homogeneous *ansa*-zirconocenes catalysts**
- **copolymerization, ozonolysis**

REFERENCES

1. Kaminsky, W.; Spiel, R. *Makromol. Chem.* 1989, 190, 515–526.
2. Dieter, R.; Fink, G. *Macromolecules*, 1998, 31, 4669–4673, 4674–4680, 4681–4683, 4684–4686.
3. Tritto, I.; Boggioni, L.; Sacchi, M. C.; Locatelli, P.; Ferro, D. R. *Macromol. Symp.* 2004, 213, 109–121.
4. Hensche, O.; Koller, F. *Macromol. Rapid Commun.* 1997, 18, 617.
5. Carbone, P.; Ragazzi, M.; Tritto, I.; Boggioni, L.; Ferro, D. R. *Macromolecules*. 2003, 36, 891–899.
6. Hasan, T.; Ikeda, T.; Shiono, T. *Macromolecules*. 2005, 38, 1071–1074.
7. *Modern Plastics Internationa*, 1995, 72 (9), 137.
8. Benedikt, G.; Ed Elce, M.; Goodall, B. L.; Kalamarides, H. A.; McIntosh, L. H.; Rhodes, L. F.; Selvy, K. T. *Macromolecules*. 2002, 35, 8978–8978.
9. Diamanti, S. J.; Khanna, V.; Hotta, A.; Yamakawa, D.; Shimizu, F.; Kramer, E. J.; Fredrickson, G. H.; Bazan, G. C. *J. Am. Chem. Soc.* 2004, 126, 10528–10529.
10. Sujith, S.; Joe, D. J.; Na Sung, J.; Park, Y.-W.; Choi, C. U.; Lee, B. Y. *Macromolecules*. 2005, 38, 10027–10033.
11. Marathe, S.; Siviram, S. *Macromolecules*. 1994, 27, 1083–1086.
12. Lasarov, H.; Pakkanen, T. T. *Macromol. Chem. Phys.* 2000, 201, 1780–1786.
13. Yazin, S.; Fink, G.; Hauschild, K.; Bachman, M. *J. Polym. Sci., Part A: Polym. Chem.* 2002, 40, 1454.
14. Naga, N.; Tsuchiya, G.; Toyota, Y. *Polymer*. 2006, 47, 520–526.
15. Meshkova, I. N.; Grinev, V. G.; Kiseleva, E. V.; Raspopov, L. N.; Kuznetsov, S. P.; Udovenko, A. I.; Shchegolikhin, A. N.; Ladygina, T. A.; Novokshonova, L. A. *Polym. Sci., Ser. A*. 2007, 49, 1165–1172.
16. Hasan, T.; Ikeda, T.; Shiono, T. *J. Polym. Sci., Part A: Polym. Chem.* 2007, 45, 4581–4587.
17. Li, H.; Li, J.; Mu, F. J. *Polymer*. 2008, 49, 2939–2944.
18. Kaminsky, W.; Arrowsmith, D.; Winkelbach, H. R. *Polym. Bull.* 1996, 36, 577–584.
19. Li, X. F.; Hou, Z. M. *Macromolecules*. 2005, 38, 6767–6769.
20. Kaminsky, W.; Miri, M. *J. Polym. Sci., Part A: Polym. Chem.* 1985, 23, 2151.
21. Zhengtian, Y. U.; Marques, M.; Raush, M. D.; Chien, J. C. W. *J. Polym. Sci., Part A: Polym. Chem.* 1995, 33, 2795–2802.
22. Chien, J. C. W.; Xu, B. P. *Makromol. Chem., Rapid Commun.* 1993, 14, 109.
23. Malberg, A.; Lofgrein, B. *Appl. Polym. Sci.* 1997, 66, 35–44.
24. Lisizyn, D. M.; Poznyak, T. I.; Razumovsky, C. D. *Kinet. Catal.* 1976, 17, 1049.
25. Tritto, I.; Marestin, C.; Boggioni, L.; Sacchi, M. C.; Brintzinger, H.-H.; Ferro, D. R. *Macromolecules*. 2001, 34, 5770–5777.
26. Hermans, P. H.; Weidinger, A. *Makromol. Chem.* 1961, 50, 98.
27. Raspopov, L. N.; Belov, G. P. *Acta Polym.* 1990, 41, 215.
28. Martynov, MA; Vylegzhanina, K. A. *X-Ray Diffraction Analysis of Polymers.* "Khimia", ("Science" in Rus.); Publishing House: Leningrad, 1972; p 96.

29. Kuznetsov, S. P.; Lapushkin, Y. U. A.; Mitrofanov, A. V.; Shestov, S. V.; Udovenko, A. I.; Shelagin, A. V.; Meshkova, I. N.; Grinev, V. G.; Kiseleva, E. V.; Raspopov, L. N.; Shchegolikhin, A. N.; Novokshonova, L. A. *Crystallography*. 2007, 52(3), 712.
30. Double Bonds Analyser Reduses Analysis Time and Increases Sensitivity. Soviet Instruments and Scientific Deices Editorials/advertisements. 1979, 3, 16, p.42.
31. Kaminsky, W.; Engehausen, R.; Korf, I. *Angew. Chem. Int. Ed. Engl.* 1995, 34, 2273.
32. Fineman, M.; Ross, S. D. *J. Polym. Sci.* 1950, 5(2), 259.
33. Bohm, L. L. *J. Appl. Polym. Sci.* 1984, 29, 282.
34. Yong, K.; Seung, C.; Park, Y.; Han, J. J.; Kebede, B. In *Future Technology for Polyolefin and Olefin Polymerization Catalysis*; Terano, M., Shiono, T., Eds.; Technology and Education Publishers: Tokyo, 2002; p 74.
35. McKnignt, A. L.; Waymoth, R. M. *Macromolecules*. 1999, 32, 2816–2825.
36. Tritto, I.; Roggioni, J. C.; Thorshang, K.; Sacci, M. C.; Ferro, D. R. *Macromolecules*. 2002, 35(3), 616.
37. Kaminsky, W.; Bark, A. *Polym. Intern.* 1992, 28, 251.
38. Kaminsky, W. *Polymery*. 1997, 42(10), 587.
39. Kaminsky, W. *Macromol. Symp.* 2004, 213, 101–108.
40. Begstrom, C. H.; Vaananen Taito, L. J. *J. Appl. Polym. Sci.* 1997, 63(8), 1071.
41. Razumovsky, C. D.; Lisizyn, D. M. *Polym. Sci., Ser. A*. 2008, 50, 1165–1172.s

CHAPTER 2

INSIGHT INTO NEW APPLICATION ASPECTS FOR PHOTOPOLYMERIZABLE ACRYLIC COMPOSITIONS

N. L. ZAICHENKO[1], B. I. ZAPADINSKII[1], A. V. KOTOVA[1],
I. A. MATVEEVA[1], V. T. SHASHKOVA[1], L. A. PEVTSOVA[1],
A. O. STANKEVICH[1], R. G. KRYSHTAL[2], A. V. MEDVED[2],
and A. V. ROSHCHIN[1]

[1]Semenov Institute of Chemical Physics, Russian Academy of Sciences, Moscow, Russia
[2]Institute of Radio Engineering and Electronics, Russian Academy of Sciences (Fryazino Branch), Moscow, Russia
Email: zaina@polymer.chph.ras.ru

CONTENTS

ABSTRACT

Aspects of two new applications of photopolymerizable acrylic compositions, namely as a registering medium for optical data recording and a
polymer sensor element for gas analysis, are discussed. The first application is based on the possibility of generation of surface plasmon resonance
by gold nanoparticles formed in a 3D network matrix produced by irradiating a liquid acrylic composition containing a dissolved gold salt with a
mercury lamp. It was established that the efficiency of formation of gold
nanoparticles is determined by the parameters of the polymer network and
by the chemical structure of the oligomer block, with the most suitable
matrixes being those having a loose network and containing groups capable of participating in complexation. The obtained matrixes may serve
as registering media for optical data recording because the agglomeration
of gold nanoparticles is initiated by subsequent external effects (heating
and/or irradiation), and the plasmon resonance band of gold nanoparticles
arises in the visible spectral region (500–600 nm) at the point of external effect application. The second application deals with polymer films
containing microcavities with selective absorption centers, which were
obtained from oligoester acrylates. Experimental studies of absorption–
desorption of morpholine as an analyte by a molecularly imprinted polymer film (obtained with morpholine as a template) using surface acoustic waves demonstrated significantly increased sensitivity of the polymer
sensor element to morpholine in comparison with other analytes. A new
method to measure the response of the sensitive element based on the molecularly imprinted polymer, which takes into account the size of output
signal and its relaxation time simultaneously, is proposed.

INTRODUCTION

Application of photopolymerizable oligomers has opened up an important
stage in polymer technology based on the direct transition from the liquid
(oligomer) to the solid state (polymeric product) without the difficult and
labor-intensive stages of synthesis, secretion, purification, granulation,
and processing of high-molecular-weight compounds. Moreover, they are
of low toxicity and volatility, and their transformation into the polymer is

characterized by small shrinkage and takes place without high temperatures and pressures.

All the above-mentioned properties of photopolymerizable oligomers permit their use in the production of enamels and lacquers for wood and metal, binders for composite materials, glues, optical devices, contact lenses, anaerobic sealants, electrical insulating materials, typographic printing forms, etc. But recently some new applications of photopolymerizable oligomers have been proposed, such as the creation of nanocomposite materials with unique optical, electronic, magnetic, and catalytic properties and the development of molecularly imprinted polymers. These materials can be used as a new generation of sensitive elements for gas and liquid detectors and as recording media for optical data storage.

EXPERIMENTAL PROCEDURE

Hydrogen tetrachloroaurate(III) tetrahydrate (trihydrate) ($HAuCl_4 \cdot 4(3)$ H_2O) (Merck, analytical grade) with a gold content of 49% and ethylene glycol monomethacrylate (MEG, Aldrich) with a main component content of 98% were used as supplied. The methods for synthesizing the oligomers used have been described previously.[1,2]

METHOD OF PREPARATION OF OLIGOMERIC COMPOSITIONS CONTAINING GOLD(III)

1. A 0.03 g (0.073 mmol) portion of $HAuCl_4 \cdot 4H_2O$ was dissolved in 0.22 g (0.38 mmol) of MEG in vitro nigro to obtain a solution with a salt concentration of 12 wt% (as recalculated to $HAuCl_4$). The solution was stored in the dark at 8–10°C and then used to prepare the acrylic compositions to be tested.

2. A mixture 0.74 g of acrylate and 0.01 g of Darocur 4265 photoinitiator (a 1:1 mixture of diphenyl(2,4,6-trimethylbenzoyl)phosphine oxide and 2-hydroxy-2-methylpropiophenon) was placed in vitro nigro, after which 0.25 g of $HAuCl_4$ solution in MEG was added under intense stirring; that is, the composition contained 3.0 wt% $HAuCl_4$. In a similar way, compounds **1–9** and two $HAuCl_4$-free reference OUM-1 and BDM-based compositions (**10** and **11**) were

        prepared from the following acrylates: OUM-1 OUM-2, OUM-3, OUM-4, OCM-2, DGDA, BDM, EDM, and OTFDM (Table 2.1).

3.   To obtain the compositions **23–27**, HAuCl$_4$ (3 wt %) was dissolved directly in a mixture of an acrylic oligomer and the photoinitiator (1 wt%) , while the composition **33**, based on the MEG-incompatible perfluorinated acrylate, was prepared with the use of oligomer OCM-2 as a cosolvent.

PHOTOINITIATED POLYMERIZATION OF OLIGOMERIC COMPOSITIONS CONTAINING GOLD(III)

A 1 g portion was applied onto a quartz or silicate glass plate 75 × 35 mm in size, with a joint sealing insert 0.09 mm in thickness (2 mm for electron microscopy analysis) and covered with an identical plate, after which the assembly was irradiated with a DRT_1000 mercury lamp for 2 min from either side of the sample. The oligomers were analyzed by gel permeation chromatography (GPC) on a Breeze chromatograph (Waters) equipped with 100, 500, and 1000 E Ultrastyrogel columns; the eluent, tetrahydrofuran, was passed at a flow rate of 1 mL/min.

        The electronic absorption spectra of the initial gold(III)-containing compositions and their time evolution during the reaction were recorded on a Specord UV–vis spectrophotometer. The absorption spectra of the test films (on a glass support or between glass plates) after irradiation were recorded on a Shimadzu vis–NIR 3101PC scanning spectrophotometer over the wavelength range 200–2500 nm. The reflection spectra were recorded using a Shimadzu MPC-310 unit. The size of gold particles in the polymer matrix were determined by transmission electron microscopy (TEM) on an LEO912 AB OMEGA instrument (Carl Zeiss, Germany) with the energy filter integrated into the optical system of the microscope (resolution of 0.2–0.34 nm). The samples were slices of the indicated polymers 50–200 nm in thickness, which were prepared at room temperature by using a Leica Ultracut UCT microtome (Leica, Germany) with a diamond cutter. The slices were fixed at 300-mesh copper TEM grids (Ted Pella, Canada).

RESULTS AND DISCUSSION

UV-RADIATION-INDUCED FORMATION OF GOLD NANOPARTICLES IN A 3D POLYMER MATRIX

Among the studies on the development of nanocomposite materials with unique optical, electronic, magnetic, and catalytic properties, a promising direction is the synthesis of nanomaterials by means of dispersing a metal, in particular gold or silver, throughout an olygomer matrix.[3,4] The process of formation of nanosized gold particles has been intensely investigated in recent years, especially in connection with the discovery of an absorption band in the visible spectrum associated with localized surface plasmon resonance (SPR) of metal nanoparticles.[5,6] According to the Mie electromagnetic theory, the position of the plasmon resonance band (PRB) depends on the aspect ratio, particle size, and dielectric properties of the medium.[7] Many authors have assigned the absorption band in the 500–600 nm range to gold particles of 30–90 nm in diameter.[8,9] It was demonstrated that a number of organic compounds, for example, sulfur-containing compounds,[10] specifically influence the size and stability of the nanoparticles formed.

The main field of applications of the unique properties of gold nanoparticles is the development of sensors for detecting DNA or proteins by measuring the change in the local refractive index caused by the adsorption of an analyte molecule on the metal surface.[11] Gold nanoparticles dispersed in a polymer matrix have been suggested for use as elements of optical memory devices[12] or as IR-reflecting coatings on organic glasses: eyeglass lenses or glazing of premises.[13] Among the methods for introducing gold nanoparticles into the matrix of a linear polymer, typically poly(methyl methacrylate)(PMMA), the most effective one is the photoreduction of soluble gold-containing salts dispersed throughout the matrix.[14]

The aim of this work was to examine whether plasmon resonance can arise in gold nanoparticles dispersed in the 3D network of polyacrylates and how the parameters of the network influence the kinetics of formation of nanoparticles and the spectral properties of the synthesized material. Such matrices have not been studied previously. Gold nanoparticles were formed during the photoreduction of gold salts dissolved in an acrylic oligomer composition, the 3D polymerization of which was induced by photochemical initiation.

The structural formulas of the oligomers used (Table 2.1) are given below:

$CH_2=C(CH_3)COO–R–OOCC(CH_3)=CH_2$ (I)

where R = $(CH_2)_4$ for BDM, R = $(CH_2)_{20}$ for EDM, and R = $(CH_2CH_2CH_2CH_2O)_m(CH_2)_4$ for OTFDM (m = 15–20); oligo(urethane acrylates)

R-OOCNH—⟨CH₃⟩—NHCOO$(CH_2CH_2CH_2CH_2O)_n$CONH—⟨CH₃⟩—NHCOOR'

Here, n = 15–20 and R = R' = CH_2CH_2OMet for OUM-1; R = CH_2CH_2OMet and R' = $CH(CH_3)_2$ (66 : 33) for OUM-2; R = CH_2CH_2OMet and R' = $CH(CH_3)_2$ (33 : 66) for OUM-3; R = R' = $CH(CH_3)_2$ for OUM-4; R = R' =$[CH_2CH(CH_3)O]_{5-6}Met$ for OUM-5; R = R' = $[CH_2CH(CH_3)O]_{6-7}$ Met for OUM-6; R = R' = $(CH_2CH_2O)_{6-8}Met$ for OUM-7; R = R' = CH_2CH_2OAcr for OUA-1; R = CH_2CH_2OAcr and R' = $CH(CH_3)2$ (66 : 33) for OUA-2; R = R' = $(CH_2CH_2O)_{6-9}Acr$ for OUA-3; and R = R' = $CH_2CH(CH_3)OAcr$ for OUA-4.

Here, n = 15–20 and R = $[CH_2CH_2CH_2CH_2O]_n$ for OUAIPF.

Here, R = CH_2CH_2 for OUAIE and R = $CH_2CH(CH_3)$ for OUAIP.

$$ROOCNH-\overset{\overset{\displaystyle CH_3}{|}}{\underset{}{\bigcirc}}-NHCOO-(CH_2CH(CH_3)O)_nCO-\overset{\overset{\displaystyle CH_3}{|}}{\underset{}{\bigcirc}}-NHCOOR$$

Here, $n = 30$–35 and R = MetOCH$_2$CH$_2$ for OUM-2000T.[15]

MetOCH$_2$CH$_2$OOCO-R-OCOOCH$_2$CH$_2$OMet

Here, R = CH$_2$CH$_2$OCH$_2$CH$_2$ for OCM-2,[16]

R = (CH$_2$)$_2$O(CH$_2$)$_2$OCOO(CH$_2$)$_2$O(CH$_2$)$_2$OCOO(CH$_2$)$_2$O(CH$_2$)$_2$ for OCM-2-1,[17]

R = CH$_2$C(CH$_2$Br)$_2$CH$_2$ for OCMN,

$$O[CH_2CH_2OCOOCH_2\text{-}C\overset{}{\diagdown}\diagup C\text{-}CH_2]_2$$
$$B10H10 \quad \text{for OCMC}$$

$$\text{for OCMP}$$

(structure for OCMP with $-CH_2$, CH$_2$-O-CH$_2$, CH$_3-$, BrCH$_2$, CH$_2$Br BrCH$_2$, CH$_2$Br)

CH$_2$=CHCOO(CH$_2$CHCH$_2$O)$_n$CHCH$_2$O-⟨⟩-$\overset{\overset{\displaystyle CH_3}{|}}{\underset{CH_3}{|}}$-⟨⟩-OCH$_2$CH(OCH$_2$CHCH$_2$)$_n$OOCCH=CH$_2$
(with OH, OH ... OH, OH)

DGDA, $n = 5$–8

We had established that crystal hydrates of HAuCl$_4$ are highly soluble in hydroxyl-containing acrylic oligomers.[13] For example, the solubility of HAuCl$_4$ in MEG is 30 wt%; for this reason, this monomer was selected as an active solvent for HAuCl$_4$ salts. From the experimental point of view, the optimum concentration of the Au salt in MEG is 12 wt%; so this solution was used as a stock solution in preparing the acrylic compositions. When composition **1** was irradiated between quartz plates by unfiltered light of the mercury lamp, HAuCl$_4$ rapidly disappeared (within 3–4 min), as evidenced by the vanishing of its absorption band at

TABLE 2.1 Structure of Initial Acrylates, Their Basic Properties and Notation of Their Compositions with a 25 wt% Solution of HAuCl$_4$ in meg

Comp. Number	Notation	Name, Block Structure	Viscosity, cP	d^{20}_4, g m^{-3}	n_D^{20}
–	MEG	Ethylene glycol mono-methacrylate	4.0	1071	1.4520
1	OUM-1	Oligourethane methacrylate (II) R=R'= – (CH$_2$)$_2$OOCC(CH$_3$)=CH$_2$	T_m = 29–34°C	1095	1.5000
2	OUM-2	Oligourethane methacrylate (II) R=– (CH$_2$)$_2$OOCC(CH$_3$)=CH$_2$ R'=–CH(CH$_3$)$_2$ (66: 33)	>20,000	1086	1.5012
3	OUM-3	Oligourethane methacrylate (II) R=– (CH$_2$)$_2$OOCC(CH$_3$)=CH$_2$ R'=–CH(CH$_3$)$_2$ (33: 66)	>20,000	1078	1.4992
4	OUM-4	Oligourethane methacrylate (II) R=R'= –(CH$_2$CH(CH$_3$) O)$_6$OC-(CH$_3$)=CH$_2$	>20,000	1065	1.4861
5	OCM-2	Oligocarbonate methacrylate	160	1196	1.4660
6	DGDA	Diphenilopropane glycerolate diacrylate	20 000	1180	1.5590
7	BDM	Butanediol di(meth)acrylate [I R=(CH$_2$)$_4$]	4.7	–	–
8	EDM	1,20-Eicosanediol di(meth)acrylate [I R=(CH$_2$)$_{20}$]	T_m = 115°C	–	–
9	OTFDM	Oligotetrahydrofuran-α,ω-diol di(meth)acrylate [I R=(CH$_2$)$_4$O)$_m$ (CH$_2$)$_4$]	T_m = 25–28°C	998	1.4640 at 30°
10	OUM-1	Reference sample, without Au	–	–	–
11	BDM	Reference sample, without Au	–	–	–

320 nm; that is, Au^{3+} was reduced to Au^0 without the formation of products capable of absorbing in visible and near-UV spectral regions (Darocur 4265 photoinitiator has an extinction coefficient of 8.2×10^{-3} M^{-1} cm^{-1} at 320 nm) (Fig. 2.1a). Within this time interval, the photopolymerization of the acrylic components resulted in the formation of a transparent film.

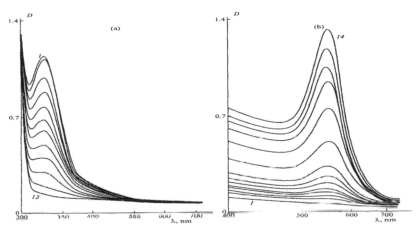

FIGURE 2.1 Absorption Spectra of a $HAuCl_4$ Solution in Acrylic Composition 1 in the Course of Irradiation: (a) $HAuCl_4$ Consumption: curves 1–12 Correspond to Irradiation Times from 0 to 5.3 min; (b) PRB Intensity Growth: Curves 1–14 Correspond to Irradiation Times from 23 to 113 min.

In a number of cases, for example, in OUM-based compositions (No. 1–4 in Table 2.1), 3–5 min after complete consumption of $HAuCl_4$, an absorption band near 340 nm appeared, the intensity of which increased for 15–20 min. Only 20–25 min after the beginning of irradiation, an absorption in the visible spectrum emerged near 550nm (Fig. 2.1b), which is assigned to a PRB of gold nanoparticles, and its intensity increased for 60–90 min. Clearly, the formation of gold nanoparticles in the film and their growth and agglomeration occur at a lower rate compared to the rate of the reduction of the salt. As a result of these photoinduced processes, a rose-violet film was formed. When the sample was irradiated between silicate glass plates (in contrast to quartz plates), the picture was somewhat different. The 340 nm absorption band was not observed as an individual peak, whereas the PRB was shifted to 580 nm and exhibited a shoulder near 640 nm. The above data were obtained for films 20–25 μm in thickness. Note, however, that, under similar conditions of photoinitiated polymerization, it is possible to obtain transparent, colorless, gold-containing films, 100 μm and more in thickness, which, when irradiated, feature a PRB in the visible spectrum. It was demonstrated[12] that, upon $HAuCl_4$ photoreduction in a PMMA matrix, the formation of nanoparticles and PRB are observed only at elevated temperatures and not during irradiation. It is likely that heating up to the glass transition temperature and above is needed for the diffusion-controlled growth of the reduced gold phase, which is due to the thermodynamic incompatibility of PMMA and Au.

Since irradiation with a 1000 W UV lamp is accompanied by significant heating due to the irradiation, we studied the effect of temperature on PRB generation. Table 2.2 lists the results of irradiation (hv) and/or heating of composition 1, fixed between quartz and silicate glass plates. To exclude the effect of temperature during irradiation, the assembly with the test film was cooled with a water flow (10–15°C). In all the above-mentioned cases, the compositions fixed between quartz plates yielded a PRB near 550 nm upon irradiation, whereas the 340 nm band manifested itself as an individual peak only in irradiated oligourethane methacrylates.

TABLE 2.2 Effect of the Conditions on the Formation of Gold Particles after UV Irradiation of Composition

Conditions	Quartz		Glass	
	PRD Wavelength, nm	Color	PRD Wavelength, nm	Color
90°C	565	Violet	Absent	Golden
90°C + hv	540	Rose-violet	580	Violet
Hv	540	Rose-violet	Absent	Colorless

The photoreaction in the respective sample fixed between silicate glass plates occurred in a different manner. When heated, it produced no PRB; instead the film turned golden yellow due to the formation of sample up to 70–90°C depending on the duration of formation of gold particles 1–20 μm in size. When the sample was irradiated without heating, no spectral changes in the film were observed, and the sample remained transparent and colorless. Only the joint action of light and heat produced a PRB at 580 nm and the sample turned violet.

When a liquid composition was introduced between the silicate glass plates and from the very beginning was subjected to the joint action of

UV irradiation and heat (90°C + hv), fairly larges Au particles formed in 100–200 μm films even at early stages of the process, giving rise to golden color and opalescence. The composition that is homogeneous before curing becomes optically nonuniform. As double bonds continue to disappear, gold nanoparticles with a new absorption band near 600 nm are formed. It was revealed that the polymerization of MEG and its mixture with acrylic oligomers are markedly inhibited by the gold salt or by the products of its decomposition. In comparison with the reference ($HAuCl_4$-free) sample, the rate of the polymerization of composition 5 (Table 2.1) decreased by 2–3-fold. An increase in the salt concentration by 3–13% leads to an increase in the time it takes to cure a film prepared from composition 1 from 12 to 38 min and in the size and concentration of reduced gold particle that make the sample turn golden yellow. Thus, the photopolymerization is largely inhibited due to the participation of $HAuCl_4$ in free-radical reactions. Inhibition can also occur due to the reflection of initiation radiation by the gold particles formed.

As the photoinitiator concentration was increased from 3.2 to 5 wt%, the inhibiting action of $HAuCl_4$ was overridden and the process of particle formation was retarded; as a result, the sample remained colorless. When heated, the cured samples turned violet-golden.

Thus, the photoinduced processes, namely the reduction of gold, polymerization of acrylates, and the growth and agglomeration of gold particles, are interrelated. In most cases, the polymer matrix formed exhibited good mechanical properties.

The chemical structure of the acrylic components and the density of the polymer network markedly affect the photoreduction of $HAuCl_4$ and emergence of PRB near 580 nm. Figure 2.2 shows the kinetic curves for the consumption of the gold salt in compositions 1 and 4–6 (numbers at the curves corresponds to the numbers of the compositions in Tables 2.1 and 2.3). Table 2.3 lists the rate constants for this reaction, whereas Figure 2.3 displays the time evolution of the absorption at 580 nm in the films of the acrylic compositions, which has been shown to reflect the formation of Au^0 particles ~40 nm in size.[14] As can be seen, the interchain length of the network virtually does not affect the rate of $HAuCl_4$ photoreduction, but essentially influences the maximum intensity of the PRB. Under a given condition, the maximum rate of the formation of nanoparticles is observed for OUM-based low-density networks, whereas, in rigidly cross-linked OCM-2, BDM, and trimethylolpropane triacrylate matrixes this process occurs much more slowly. Under the same conditions, an increase in the rigidity of the oligomer block (composition **6**) causes a decrease in the rate of gold reduction and growth of gold particles (Table 2.3).

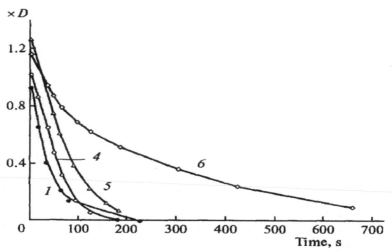

FIGURE 2.2   Consumption of HAuCl$_4$ During Photoreduction in Acrylic Matrices (λ = 320 nm; Here and Below, the Curve Number Corresponds to the Number of the Respective Composition in Tables 2.1 and 2.3).

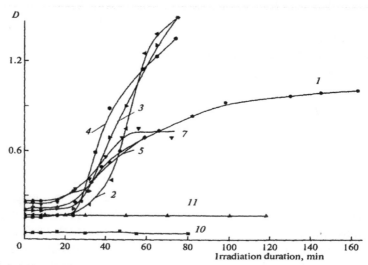

FUGURE 2.3   Time evolution of the intensity of the PRB near 580 nm for the acrylic compositions tested.

TABLE 2.3 Gold-Containing Acrylic Compositions and Some of the Properties of the Films Formed in the Course of Photochemical Reactions

Composition Number	Oligomer	$k \times 10^2$, s^{-1}	λ, nm	I_m (500–2100 nm)	D/λ
1	OUM-1	2.53	580/640	1.583	949.6
1a	OUM-1 after 14 days	–	–	1.223	736.4
1b	+ 5% GDMA	–	625	0.423	253.4
2	OUM-2	–	570	0.563	339.1
3	OUM-3	–	560	0.576	345.3
4	OUM-4	2.51	580/620	0.45	267.3
5	OCM-2	1.66	560	–	–
5a	OCM-2 after 7 days	–	–	0.27	161.8
5b	5% HAuCl$_4$	–	–	0.22	130.7
6	DGDA	0.37	–	–	–
7	BDM	–	580/620	1.13	676.1
8	EDM	–	–	0.313	187.1
9	DMPF	–	600	0.953	570,0

Figures 2.4 and 2.5 show the absorption spectra of 0.2 mm thick films prepared from gold-containing acrylic matrixes after 20 min of irradiation between silicate glass plates under the same conditions ($90°C + h\nu$). One can claim that the network internodal length in the polymer matrix is the parameter that governs the formation of gold nanoparticles and responsible for the appearance of a PRB. Of the systems studied, those with OUM-based sparse network are most suitable for the formation of gold nanoparticles capable of generating plasmon resonance: in the case of OUM-2, OUM-3, and OUM-4, the absorption spectrum of the composites features only one narrow peak, namely the PRB. This means that, in these matrixes, gold nanoparticles of approximately the same size, that is, 30–90 nm, are formed. The spectrum of the polymer prepared from composition **1** (Fig. 2.4) differs from the spectra of the other OUM-based polymers, which suggests that this somewhat more rigid polymer provides the minimum size cell required for the formation of gold nanoparticles exhibiting a PRB.

On the other hand, under the same condition, the spectra of densely cross-linked polymers (prepared from compositions **5, 5a, 5b**, and **6**; Table

2.3) feature virtually no absorption in the region of PRB, except for thin films from composition **5**, which exhibits a low-intensity PRB. It appears that the cell size of these polymers prevents them from acting as reactors for the processes under study, that is, the accommodation of bulky $HAuCl_4$ molecules and photoreduction to the growth and agglomeration of gold particles. In these cases, the formation of gold particles most likely occurs in the defective areas of the polymer network, and the absence of the PRB shows that, to form 30–90 nm gold particles, the cell walls should act as a stabilizing factor. Gold-containing, densely cross-linked networks exhibit a strong absorption over the entire 320–2100 nm range (Figs. 2.4 and 2.5).

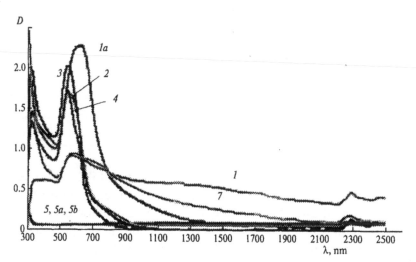

FIGURE 2.4   Absorption Spectra of the Films of the Acrylic Nanocompositions after the Completion of the Photochemical Processes.

The chemical structure of the oligomer block also substantially influences the formation of gold nanoparticles. Compounds containing groups capable of complexation with the metal or its ions (e.g., those involving $CH_2–O–CH_2$ ether bonds) promote the formation of nanoparticles. In the first approximation, the overall absorption within 500–2100 nm (the PRB and the long-wave shoulder belonging to smaller particles[18]) can be considered a measure of the concentration of nanoparticles that give rise to the PRB. Table 2.3 lists comparative data on the absorption of gold-containing films, which show that the films prepared from OUM-1 or OTFDM (compositions **1** and **9**) contain 3–5 times more nanoparticles than the films prepared from the purely hydrocarbon oligomer with a $(CH_2)_{20}$ block (composition **8**). Differences in the solvating ability of the acrylates also affect the position of the PRB maximum, which is in the range 560–640 nm. Note that solutions of $HAuCl_4$ in the acrylates are unstable especially when exposed to light. For example, composition **1** irradiated immediately after the preparation features a broad band over the entire range covered (Fig. 2.4, spectrum 1), whereas after being kept in the dark for 14 days and photoirradiation, the sample exhibits a narrow PRB (spectrum 1a). In the course of the photoreduction of the completely cured films, the ratio between the intensity of the bands in the range 500–2100 nm can change, and consequently the color of the film can also change; but after the completion of the photochemical reactions, the samples are stable, and the spectra of the films displayed in Figures 2.4 and 2.5 remain unchanged at room temperature for 70 days.

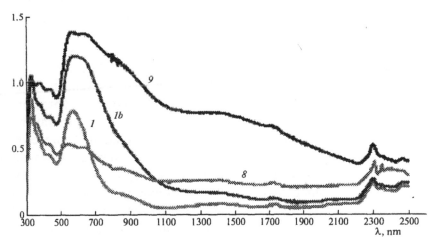

FIGURE 2.5   Effect of the Structure of the Oligomer and Conditions of the Photoreactions on the Generation of the PRB.

The chemical structure of the oligomer block also substantially influences the formation of gold nanoparticles. Compounds containing groups capable of complexation with the metal or its ions (e.g., those involving $CH_2–O–CH_2$ ether bonds) promote the formation of nanoparticles. In the first approximation, the overall absorption within 500–2100 nm (the PRB and the long-wave shoulder belonging to smaller particles[18]) can be considered a measure of the concentration of nanoparticles that give rise to the

PRB. Table 2.3 lists comparative data on the absorption of gold-containing films, which show that the films prepared from OUM-1 or OTFDM (compositions **1** and **9**) contain 3–5 times more nanoparticles than the films prepared from the purely hydrocarbon oligomer with a $(CH_2)_{20}$ block (composition **8**). Differences in the solvating ability of the acrylates also affect the position of the PRB maximum, which is in the range 560–640 nm. Note that solutions of $HAuCl_4$ in the acrylates are unstable especially when exposed to light. For example, composition **1** irradiated immediately after the preparation features a broad band over the entire range covered (Fig. 2.4, spectrum 1), whereas after being kept in the dark for 14 days and photoirradiation, the sample exhibits a narrow PRB (spectrum 1a). In the course of the photoreduction of the completely cured films, the ratio between the intensity of the bands in the range 500–2100 nm can change, and consequently the color of the film can also change; but after the completion of the photochemical reactions, the samples are stable, and the spectra of the films displayed in Figures 2.4 and 2.5 remain unchanged at room temperature for 70 days.

In the course of the photoreduction of $HAuCl_4$ to Au^0, chlorine-containing products are formed, which can play a significant role in the formation and agglomeration of nanoparticles. It makes sense to assume that, along with solvating intermediate products of gold reduction, groups containing ether bonds in the oligomer block bind chlorine molecules (atoms or ions), thereby impeding the reverse reaction of dissolution of reduced gold in the chlorine-containing solution, and thereby promoting the growth of gold particles. The probability of occurrence of the reversible reaction of formation of Au^0 nanoparticles and their dissolution in "chlorinated water" with predominant dissolution of nanosized particles due to their higher surface energy is confirmed by the fast formation of gold nanoparticles and the emergence of a PRB upon introduction a reducer, for example, glycidyl methacrylate or oxalic acid (Figure 2.5, spectrum 1b).

It is commonly assumed[19] that the ability of gold-containing nanocomposites to reflect IR radiation is associated with the presence of ~20 nm gold particles (such particles give rise to a PRB). Our results show, however, that films prepared from loose polymers (optimal for the emergence of a PRB) virtually does not reflect IR radiation (Figure 2.6, spectra 2, 3, and 4), whereas the densely cross-linked composites exhibit the best reflectivity. According to Swanson and Billard,[18] the long-wave shoulder of the PRB in the absorption spectrum is associated with the presence of ~5 nm particles; it seems that such nanoparticles are responsible for the ability of the material to reflect IR radiation.

The maximum efficiency of IR radiation reflection by the composites under study, 25–30%, is observed for composites **1** and **7**, which are characterized by the most intense absorption within 600–2100 nm. Note that a marked reflectivity is demonstrated by OUM-2-based matrixes (compositions **5**, **5a**, and **5b** in Table 2.3), although they exhibit no absorption in the visible and near-IR spectrum. One can expect that by selecting a proper acrylic matrix and conditions for the photoprocess, it is possible to synthesize nanocomposites with ultrafine gold particles and, consequently, substantially enhance their IR reflectivity.

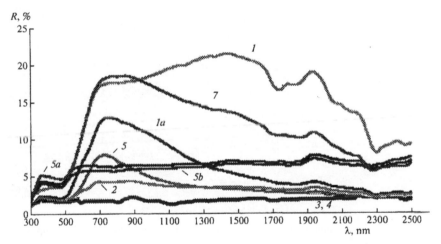

FIGURE 2.6   Reflection Spectra of the Films of the Acrylic Nanocomposites.

Electron microscopy analysis showed that the cured acrylic matrixes contain nanosized gold particles (Fig. 2.7). The concentration of nanoparticles increases after thermal treatment. At temperatures above the glass transition temperature, the process of secondary formation of fine gold particles begins.

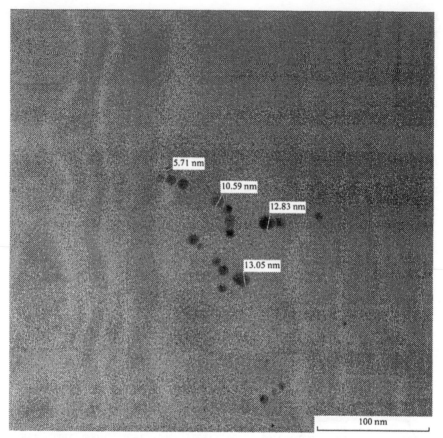

FIGURE 2.7   Gold Nanoparticles Formed upon UV Irradiation of an Acrylic Composition Containing HAuCl$_4$.

The next step of this work was to investigate in detail the effects of the structure of acrylic oligomers and the parameters of the network of a cross-linked polymer – the product of their photopolymerization – on the formation of gold nanoparticles during the photoreduction of hydrogen tetrachloroaurate(III) in solutions of liquid acrylic compounds and to study the kinetics of consumption of the salt and the kinetics of the formation of gold nanoparticles characterized by the PRB in the range 500–600 nm.[20]

We began with the study of the photoinitiator's influence. An increase in the photoinitiator concentration from 1 to 3.63 wt% leads, under identical conditions, to a gain in the rate of photoreduction of $HAuCl_4$ by a factor of ~4 up to $k_d = 0.12$ s^{-1} and in the rate of PRB generation, but not in the maximum optical density ($D = 1.18$ and 0.97, respectively). Recently, an efficient method was developed to form gold nanoparticles via the photolysis of $HAuCl_4$ in aqueous media; the typical polymerization initiator $HO(CH_2)_2OC_6H_4COC(CH_3)_2OH$ (Irgacure 2959) generating •$C(CH_3)_2OH$ radicals was used as reducing agent.[21] The latter reduces Au^{3+} to Au^{2+} and then to Au^0. It is possible that, in our case, one of the photoinitiator components, 2-hydroxy-2-methyl propiophenone [$C_6H_5COC(CH_3)_2OH$], participates in the reduction of the gold salt and the formation of nanoparticles. In this case, the consumption of the initiator for the reduction of the gold salt may determine the inhibiting effect of $HAuCl_4$ on the photopolymerization of acrylates. Thus, the proper choice of the initiator's structure and its concentration in an acrylic matrix offers additional possibilities to control processes leading to the formation of nanoparticles of the desired shape, as was shown for the reaction in aqueous media.

A much stronger influence on the photolysis rate is related to the spectral composition of the irradiating light; a silicate glass filter sharply decreases the rate of photolysis and the rate of acrylate polymerization (Fig 2.8). Note that, during irradiation of composition 28 placed between two pieces of transparent KBr glass, the decomposition of the gold salt primarily proceeds in a liquid medium or in a very weakly cross-linked gel. (For diacrylates, the gel point is reached at a double-bond conversion of 1–5%.) Slower photoreactions simultaneously occur between two pieces of silicate glass, and the gold salt is mostly reduced in a network polymer matrix.

On the basis of our data on the suitability of gold nanoparticles in loose matrixes,[13] the effect of 3D network parameters (internodal and interchain distances) on the generation of the PRB then was studied for oligo(urethane acrylate) (OUM) compositions with similar structures of the oligomeric block. The network density in the homo- and copolymers of OUM was determined by using the Cluff and Gladding method for studying the swelling of network polymers from the number-average molecular mass of internodal subchains (cross-links) of 3D networks.[22]

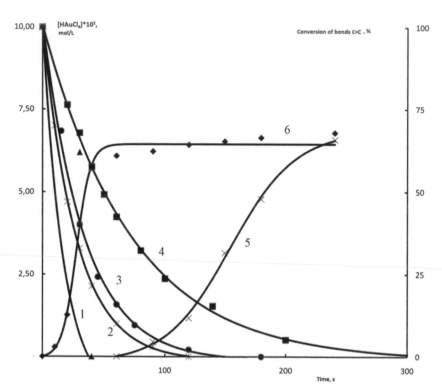

FIGURE 2.8 Consumption of the Gold Salt in the Media of (1, 2) DMPF and (3, 4) OUM_1 During Irradiation of the Composition Through (2, 4) Silicate Glass, (1) KBr glass, and (3) Quartz and the Kinetics of DMPF Photopolymerization During Irradiation Through (5) Silicate Glass and (6) KBr glass.

Figure 2.9 shows the kinetic curves illustrating variation in the optical density of the PRB (hereafter, D at the wavelength of the PRB maximum) in OUM and OUA matrixes with different oligomeric block lengths (OUM-1, OUA-1, OUA-4, OUM-5, OUM-6, OUM-7, OUA-3, and OUM-2000T) and internodal distances (OUM-2, OUM-3, and OUA-2) with the content of functional groups (a partial replacement of acrylic groups with isopropyl groups; Table 2.4). Note, above all, that in the case of the oligomers under consideration containing the oxyalkylene block, the network parameters affect the rate of PRB generation, but not the limiting optical density (D_{max}).

TABLE 2.4 Spectral and Kinetic Characteristics of $HAuCl_4$ Photolysis in Acrylic Compositions

Composition	Oligomer	λ_{max}, nm $HAuCl_4$	k_p 10^2, s^{-1}	λ_{max}^b, nm	D_{max} (PRB)	τ_{ind}^c, min
Oligomer + MEG + 3% $HAuCl_4$ + 1% photoinitiator						
1	OUM-1	322[a]	4.39	532/532	1.18	15.7
2	OUM-2	325[a]	4.37	538/532	1.22	15.7
3	OUM-3	325[a]	4.15	535/532	1.15	5.7
4	OUM-4	325[a]	3.22	562/532	1.05	5.7
5	OUM-5	323[a]	4.05	533/530	1.17	15.7
6	OUM-6	323[a]	5.10	536/533	1.25	5.7
7	OUM-7	326[a]	4.98	533/530	1.16	5.7
8	OUM-2000T	323[a]	5.24	542/536	1.22	15.7
9	OUM-4F	320	6.14	535/545	0.45	25.7
10	OUA-1	323[a]	4.58	536/533	1.15	15.7
11	OUA-2	323[a]	4.80	539/536	1.15	5.7
12	OUA-3	323[a]	5.16	530/530	1.17	5.7
13	OUA-4	323[a]	5.23	539/533	1.10	5.7
14	OUAIE	323	5.74	–/521	0.20	35.7
15	OUAIP	323	5.47	533/524	0.20	35.7
16	OUAIPF	320	11.71	540/530	1.22	3.7
17	OCM-2	320	10.81	533/528	1.07	5.7
18	OCM-2-1	320	6.93	542/535	0.87	5.7
19	OCMC	320	5.12	560/545	1.04	5.7
20	OCMN	320	6.30	554/575	1.10	5.7
21	OCMP	323	7.22	554/566	0.63	0.7
22	ДМPF	323	13.62	560/536	0.91	1.7
Oligomer + 3% $HAuCl_4$ + 1% photoinitiator						
23	OCM-2	320	8.21	545/535	1.06	5.7
24	DMTG	325	2.57	–	–	–
25	DMEG	323	6.03	539/536	0.70	20.7

TABLE 2.4 *(Continued)*

Composition	Oligomer	λ_{max}, nm HAuCl$_4$	$k_p\ 10^2$, s^{-1}	λ_{max}[b], nm	D_{max} (PRB)	τ_{ind}[c], min
26	DAHG	323	10.31	542/536	0.94	5.7
27	DAEHG	326	11.57	536/530	0.99	5.7
28	FHAC OCM-2 MEG	320	9.07	550/535	0.62	5.7
Reference samples						
29	MEG	320	10.11	525	1.08	3.7
30	MPG	320	9.58	525	0.99	3.7

Notes: [a]The band at 340 nm appears. [b]The values of λ_{max} at the appearance of the PRB and the final position of the PRB are in the numerator and denominator, respectively. [c]The induction period (time from the onset of irradiation to the appearance of PRB.

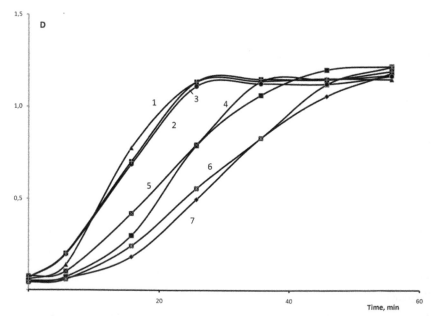

FIGURE 2.9 Effect of the Length of the Oligomer block in Diacrylates on the Kinetics of PRB Accumulation: (1) OUM-3, (2) OUA-3, (3) OUA1, (4) OUM-2, (5) OUM-1, (6) OUM2000-T, and (7) OUM-5.

It follows from comparison of the rate of accumulation of the PRB in acrylic and methacrylic matrixes having identical oligomeric block structures (Table 2.4) that gold nanoparticles are formed in acrylates at a noticeably higher rate than that in methacrylates (Fig. 2. 9, curves 3 and 5 or 2 and 7). This result is unexpected because the polymerization of acrylates occurs faster than that of methacrylates, and for polyfunctional acrylic oligomers the contribution of slower processes generating nanoparticles in a network matrix is greater than that in the case of methacrylates. In contrast, the majority of nanoparticles responsible for the PRB are formed even in the network matrix, and the difference in the rate of accumulation of the PRB in OUM and OUA matrixes may be determined by the network density; polyacrylates are more elastic than their methacrylic analogs.

The intensity of the PRB depends on the concentration of the gold salt in the original composition: at $[HAuCl_4]_0 = 0.085$ and 0.12 M for OUM-1, $D_{maxPRB} = 1.2$ and 1.63, respectively. These D values are close to those for the initial concentration of the salt (0.979 and 1.66). It is known[23] that the PRB of metal nanoparticles is characterized by a large and easily tunable optical absorption cross section. Thus,[24] gold nanoparticles with a diameter of ~40 nm have a calculated absorption cross section of 2.93×10^{-15} m^2, corresponding to a molar extinction coefficient of $\varepsilon = 7.66 \times 10^9$ L/(mol·cm) at the PRB wavelength, a value that is four orders of magnitude higher than that for $HAuCl_4$ (7.3×10^5 L/(mol cm) at 320 nm). With the use of common relationships, it may be assumed that the concentration of nanoparticles is four orders of magnitude lower than the initial concentration of the salt; that is, each nanoparticle contains ~10^4 Au atoms.

The value of D attained for the PRB of nanoparticles in the above loose OUA networks is close to the maximum value for polymer matrixes under the given experimental conditions: the photolysis of $HAuCl_4$ in solutions of monomers, that is, MEG or oligo(1,2-propylene glycol) methacrylate, with $n = 5$–6 (compositions **35** and **36**) leads to a maximum optical density of the PRB in the matrix of the linear polymer that is close to the above values for networks: ~1.0.

For oligourethanes with a long oligomeric block, the substitution of aliphatic isocyanates for aromatic isocyanates leads to a significant increase in the rate of formation of gold nanoparticles responsible for the PRB: OUAIPF versus aromatic OUA-1 (Fig. 2.10, curves 1, 2).

In the case of OUAIPF, the value of D_{max} (1.22) is the highest among all studied compositions with close interchain distances, while the highest rate of a gain in D is typical for the oligo(ester acrylate) DMPF (Fig. 2.10), although the attained value of D_{max} is noticeably lower than that for other compositions based on polyfurit.

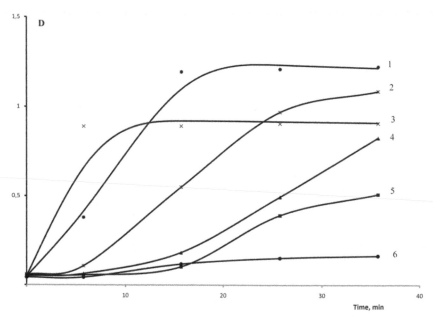

FIGURE 2.10 Kinetics of PRB Accumulation in Matrices Containing no m-Toluene Diisocyanate: (1) OUAIPF, (2) OUA-1 (for comparison), (3) DMPF, (4) OUM-4F, (5) OUAIE, and (6) OUAIP.

At the same time, for "rigid" acrylate networks with short oligomeric blocks, that is, OUM-4F, OUAIE, and OUAIP (Fig. 2.10, Table 2.4, compositions **9**, **14**, **15**), the photolysis of the gold salt occurs in the same manner, but almost no absorption is observed in the visible region.

In summary of the above-described data on the effect of the parameters of the acrylic polymer network on the generation of gold nanoparticles, it should be emphasized that, to a certain extent, this effect is leveled off through the method of introduction of $HAuCl_4$ via its dissolution in MEG or incomplete consumption of double bonds during 3D polymerization (Fig. 2.8). The use of MEG as a monomer results in an increase in the internodal distance, while incomplete polymerization causes an overall decrease in the network density. In the present work, acrylic oligomer networks with glass transition temperatures T_g from −15°C (oligomer OUM-

3) to >80°C (aromatic acrylates) were studied, although most experiments were performed at a temperature above T_g of the matrix because the temperature of the sample attained 90°C when the samples were irradiated with the nonfiltered light of the mercury lamp at a power of 1000 W.

The next stage of our study was related to gaining insight into the effect of groups introduced into the oligomeric block – specifically in terms of their electronic and steric properties – on the formation of gold nanoparticles (Fig. 2.11). In order to reveal the structure of the block, no MEG, which is an active solvent for the gold salt, was added to the compositions. It turned out that, even for the short-chain oligo(carbonate methacrylate) OCM-2 and its oligomer homologue OCM-2-1, gold nanoparticles may be generated with approximately the same efficiency as in the case of OUM and OUA with long oligomeric blocks (Fig. 2.11, curve 1).

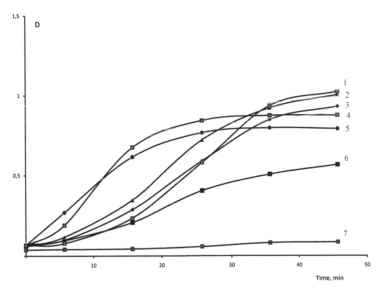

FIGURE 2.11 Effect of the Structure of the Oligomer Block in Diacrylates on the Kinetics of PRB Generation: (1) OCM-2, (2) OCMN, (3) OCMC, (4) OCM-2-1, (5) OCM-2 + FHAC, (6) DMEG, and (7) DMTG.

This extremely unexpected and efficient formation of the PRB in the matrix of OCM-2 is probably due to the participation of carbonate groups in photoprocesses (OCM-2 is one of the best substrates for radiation-induced or photoinitiated polymerization) or an increase in the network elasticity due to the "free rotor" effect of the –C(O)–O– group. At the same time, halogenated, perfluorinated, and bulky carborane groups in the oligomeric block insignificantly affect the generation of the PRB (Fig. 2.11, curves 2, 3, 5). However, in a very "rigid" network based on DMEG, the formation of nanoparticles is less efficient (Fig. 2.11, curve 6).

A specific behavior is demonstrated by compounds containing sulfide sulfur: In the DMTG-based composition **24**, the gold salt is consumed at the normal rate, but there is no absorption in the visible region (Fig. 2.11, curve 7). It may be supposed that 4 nm gold nanoparticles consisting of ~140 Au atoms (according to the mass spectrometry data) are formed in this case of absorption in the UV region,[25] as in the case of DMTG, or a bimodal distribution of nanoparticles is realized with two PRB bands at 560 nm and a shoulder at 370 nm.[26]

It turns out that only in OUA matrixes based on *m*-toluene diisocyanate does a band at 340 nm (along with the PRB near 530 nm) appear during generation of gold nanoparticles. The absorption at 340 nm occurs immediately after consumption of the gold salt before the appearance of the PRB, which then increases together with the PRB and does not disappear. Figure 2.12 shows the spectra of PRB development in OUMs based on oligotetrafuran-α,ω-diol and ethylene glycol monoacrylate, which differ only in the used diisocyanate: *m*-toluene diisocyanate or isophorone diisocyanate. The indicated band is not seen in the case of aliphatic isophorone diisocyanate. Moreover, the band at 340 nm is absent in the case of composition **9**, including aliphatic OUM_4F, which is *N*-phenylsubstituted at the urethane group. It is possible that in flexible-chain polyurethanes based on *m*-toluene diisocyanates, the same electronic interactions are realized as those in sulfur-containing matrixes,[26] which leads to the formation of gold particles with a bimodal distribution (30–40 and ~4 nm) and two absorption bands at 560 nm and a shoulder near 340 nm.

Earlier, the effect of regeneration of the gold salt was revealed when the reduced gold-containing film was heated in the PMMA matrix.[27] Figure 2.13 illustrates the kinetics of evolution in the intensity of the band of $HAuCl_4$ and of the PRB of gold nanoparticles in the OCM-2-based network at an initial salt concentration of 3 wt%. The photolysis of the gold salt was performed photochemically, and, after its complete decomposition, samples were kept in a thermostat at 70 and 110°C. Heating of the film at 70 and 110°C leads to the regeneration of the salt and its subsequent slow consumption (Fig. 2.13, curves 1, 2). However, this effect is not exhibited under our "standard" conditions during simultaneous UV irradiation and heating at 90°C, owing to the photoreduction of the salt.

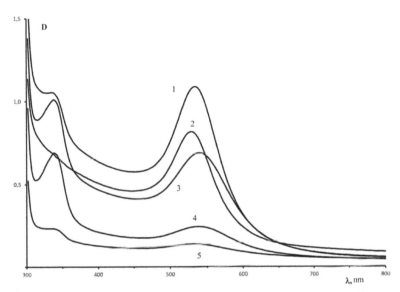

FIGURE 2.12 Absorption Spectra of OUM-1 After Irradiation for (5) 15, (3) 40, and (1) 60 min and of OUAIE Containing no *m*-Toluene Diisocyanate After Irradiation for (4) 20 and (2) 80 min.

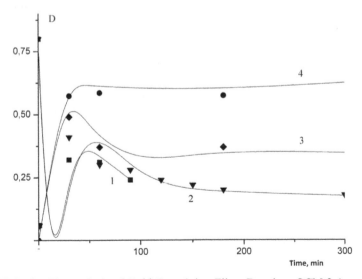

FIGURE 2.13 Thermolysis of Gold-Containing Films Based on OCM-2 (composition 5): (1, 2) Changes in the Optical Density of the $HAuCl_4$ Band at 70 and 110°C; (3, 4) the Same for the PRB at 110 and 70°C.

The concentration of PRB-responsible, "photostable" Au nanoparticles reaches a maximum during heating at 110°C but not at 70°C and then decreases; that is, under these conditions, nanoparticles participate in reverse oxidation reactions up to the regeneration of Au^{3+}. A drop in the intensity of the PRB is observed only at high $[HAuCl_4]_0$ (2–3 wt%) and is not observed at 70°C at any concentrations of the salt.

As was shown above, parallel reactions of acrylate photopolymerization and gold salt photoreduction give rise to a transparent, colorless material containing dissolved gold. Films based on this material are stable enough; after 25 months, the samples became only pale yellow without any PRB traces. The obtained matrix may serve as a registering medium for optical data recording because a further external effect – heating and/ or irradiation – initiates the agglomeration of gold nanoparticles, and the plasmon resonance band of gold nanoparticles arises in the visible spectral region (500–600 nm) at the point of external effect application.

A principal advantage of these registering media is their extremely high sensitivity. The molar extinction coefficient at the PRB wavelength is five orders of magnitude higher than that for a typical dye, indocyanine green ($\varepsilon = 1.08 \times 10^4$ L/(mol·cm) at 778 nm).[24]

The first experiments demonstrated that gold-containing polyacrylic films show promise as a registering medium during both thermal and photochemical external effects. In Figure 2.14a, the image obtained after 5 min UV irradiation of a film based on composition **1** through a metallic grid with a hole diameter of 200 μm is presented, while in Figure 2.14b an arbitrary image obtained on the surface of the transparent colorless film by means of a point heating element with a temperature of 150–160°C is shown. The colored regions absorb at the PRB wavelength.

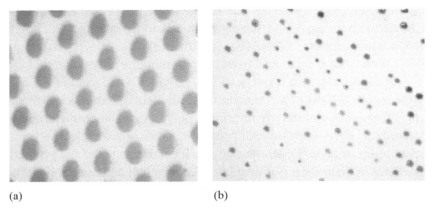

(a) (b)

FIGURE 2.14 Images obtained (a) during irradiation of the gold-containing transparent film based on composition 1 through a metallic grid with a hole diameter of 200 μm and (b) by means of a point heating element with a temperature of 150–160°C. Magnification ×45.

As a result, having no analogous highly sensitive polymeric media, a 3D network polymer containing nanoparticles of Au^0 for optical information recording was created.[28]

APPLICATION OF MOLECULARLY IMPRINTED POLYMERS FOR SURFACE ACOUSTIC WAVE SENSOR ELEMENTS WITH HIGH SENSITIVITY

Chemical sensors using surface acoustic wave technology (SAW) are characterized by high reliability and sensitivity, and are easy to produce and economical at mass production.[28] They beneficially differ from other sensors by their small size, weight, and compatibility with the format of modern digital devices. Moreover, they can be used in passive and wireless regimes.[29] The main principle of their action is that the change in the number of molecules on the acoustic line surface causes a change in the rates of spreading and decay of signal registered at the end of SAW line. However, the chemical selectivity of existing sensors based on SAW technologies is not sufficient to detect with high probability a specified compound with one detector.

In recent years, some new fast methods for detecting vapors of explosives and narcotics have been developed based on molecularly imprinted polymers.[30–32] In this technique, some polymerizable compound (mono-

mer, oligomer, or their mixture) that is able to create specific intermolecular bonds with an analyte (a substance that is to be detected) is chosen.
Then, a liquid mixture of these components is photopolymerized, and the
analyte molecules are removed from the polymer phase by washing, heating, evacuating, etc. As a result, a special polymer with "nanopores" having the size and form of the analyte molecule (solidified imprint of analyte) is formed. A molecularly imprinted polymer can be characterized as
a "sewed up" polymer phase with predominant absorption selectivity with
respect to the analyte. Depending on the special features of the synthetic
procedures, a distinction is drawn between covalent and noncovalent im-
printing.

The aim of this work was to develop sensitive elements for a SAW gas
detector based on highly sensitive polymeric films obtained from photopolymerizable acrylic compositions.

First, we studied the mechanism of diffusion and absorption processes
of the analyte molecules in polymer films containing microcavities with
selective absorption centers. A complete system of partial differential
equations with initial and boundary conditions that describe the absorption–diffusion processes was derived, and an algorithm for its solution
was developed.[34]

Then the experimental study of absorption–desorption of an analyte
by a molecularly imprinted polymer film was performed with the use of
SAWs. The dependence of the phase $\Delta\varphi$ of the SAW on the mass of the
substance absorbed in the film was derived.

Morpholine (C_4H_8ONH) was selected as the analyte for the absorption
experiments because it is the simplest structural simulator of explosives
with the structure of a cyclic amine, is soluble in most organic solvents,
and has a low boiling point (129°C).

Molecularly imprinted polymer films were obtained by photopolymerization of the diacrylate glicerolate bisphenole

$$CH_2=CHCOO(CH_2CHCH_2O)_n CHCH_2O-\!\!\!\bigcirc\!\!\!\underset{CH_3}{\overset{CH_3}{-C-}}\!\!\!\bigcirc\!\!\!-OCH_2CH(OCH_2CHCH_2)_nOOCCH=CH_2$$

on the surface of the sensitive element of the SAW sensor in the presence of morpholine as a template following the procedure described by
Binchack et al.[30] The film studied had a length of 3 mm, and a thickness
of 1.5 μm.

For experimental studies of the analyte absorption–desorption by a molecularly imprinted polymer film, we constructed a special installation unit, schematically shown in Figure 2.15.[33]

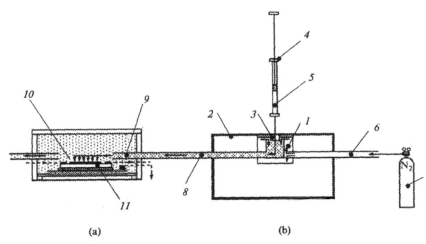

FIGURE 2.15 Units for experimental studies of analyte absorption-desorption by a molecularly imprinted polymer film with the use of surface acoustic waves. See text for notation.

It consists of a measuring chamber (a) containing a chemical sensor based on a SAW delay line and a gas sample preparation block (b). The sample preparation block is a vaporization chamber (1) placed in a thermostat (2). The membrane (3), made of an elastic silicone polymer material, is mounted in the upper part of the vaporization chamber. Through this membrane, an analyte sample is introduced with the use of a standard chromatographic microsyringe (4). The analyte is contained in container (5) in the liquid or gaseous state. The temperature of the vaporization chamber is maintained at a level higher than the boiling point of the analyte. The chamber is blown with a carrier gas supplied through the communication line (6) from a cylinder (7). A portion of liquid analyte exiting from syringe needle vaporizes, and its vapors are carried with a carrier gas flow into pipeline (8) and thence to the sensor through the entrance pipe (9). The analyte vapors are absorbed by molecularly imprinted polymer film (10) on the surface of the SAW sensor (11). They correspond to the film (16) on the surface of the SAW detector (3) shown in Figure 2.16.

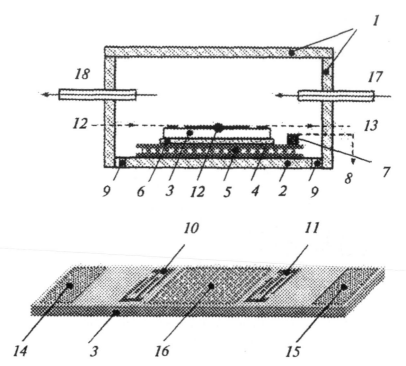

FIGURE 2.16 Measuring chamber and chemical detector based on a surface acoustic wave sensor. See text for notation.

The chemical sensor on the SAW sensor in measuring chamber (a) has been described by us earlier.[34,35] In Figure 2.16, the construction of the measuring chamber and sensor device used in this work are shown in more detail.

In this construction, the SAW sensor is situated in a hermetically sealed chamber (1) on the base (2). The sensor (3) is a delay line with an acoustic line from a piezoelectric material, e.g., ST quartz or $LiNbO_3$. The delay line is fastened to the "hot" surface (4) of a Peltier thermoelectric cell (5). In this variant, the sensor with delay line is fastened to the thermoelectric cell using a heat-conducting paste (6) usually employed in the production of radio-electronic equipment. The thermosensitive resistor (7) is placed on the working surface of the thermoelectric cell to measure its temperature. The resistor is connected to a temperature control block through contact line (8). A hermetically sealed, heat-insulating gasket (9) is situated between the base (2) and side walls of the chamber (1). Interdigital transducers (10) and (11) are situated on the surface of the delay line. Electric transmitting lines for the introduction (12) and outputting (13) of sinusoidal microwave signals are connected to the transducers. In the usual SAWs, acoustic absorbers (14) and (15) are placed between the ends of the piezoelectric plate and interdigital transducers for absorbing spurious SAWs reflected from acoustic line edges. The molecularly imprinted polymer film (16) whose absorption properties are studied is situated between the input (10) and the output (11) interdigital transducers. Gas pipes for the introduction (17) and removal (18) of the air flow to be analyzed pass through the hermetically sealed chamber case.

A SAW excited by the input interdigital transducer delay line propagates along the acoustic line in its surface layer with thickness on the order of the acoustic wavelength. For this reason, these waves are very sensitive to the state of the surface and the properties of the medium adjacent to the surface.

Changes in the physical properties of a molecularly imprinted polymer film adjacent to the surface of the acoustic line, for instance, in its mass or viscosity, caused by the absorption of an analyte result in changes in the parameters of the SAW propagation (the rate of propagation and damping coefficient). In practice, changes in the phase or amplitude of the electromagnetic signal passing through the delay line, caused by changes in the amplitude and phase of the SAW under the action of sorption processes, are measured.

In this work, we studied the absorption of analytes by molecularly imprinted polymer films synthesized from oligoester acrylates in the presence of one of the analytes as a template. Special attention was given to the thermal stabilization of delay lines on the SAWs. Commercial single-stage thermoelectric cells with a standard control circuit provided temperature stability of at least 0.003°C during 10 min and at least 0.01°C during 10 h over the temperature range 2–67°C at 20–40°C temperature of the medium around the measuring chamber.

During measurements, a flow of dry, chromatographically pure nitrogen is passed through the measuring chamber (dew point not higher than –60 C) at a controlled flow rate of 5–30 cm^3/min. The analyte vapors were introduced into the nitrogen flow using the microsyringe (4) through membrane (3) of vaporization chamber (1) (Fig. 2.15). The temperature of vaporization was 130°C, which was higher than the boiling points of all the analytes used in our experiments. The temperature of the thermostat

(2) was also 130°C, and the temperature of the SAW sensor (11) with the absorbing molecularly imprinted polymer film (10) was 40°C.

A portion of the analyte vapor formed in vaporization chamber (1) after its injection with the microsyringe passed through gas pipe (8) into the sensor device and was absorbed by the polymer film (10) on the surface of the SAW sensor (11). After the absorption was completed and analyte vapor removed from the sensor device with carrier gas flow, the desorption of the analyte from the film began.

It was found in the measurements of the amplitude–frequency characteristics of samples with polymer films having various thicknesses without an adsorbed analyte that, at film thicknesses no larger than 1.5 µm, studies at frequencies up to 80 MHz could be performed. Subsequently, all measurements with molecularly imprinted polymer films with such a thickness were performed at an even higher frequency of 120 MHz.

The experimental time dependences of the phase shift $\Delta\varphi$ of the SAW sensor signal are shown in Figure 2.17. Experiments were performed for a molecularly imprinted polymer with 10% sewn morpholine with sequentially injecting 0.4 µL acetone, 0.1 µL ethanol, and 0.1 µL morpholine into the carrier gas. Note that, first, we also studied SAW sensors with 5% sewn morpholine, but sensors with 10% morpholine were in all respects more preferable, and, subsequently, only sensors with such molecularly imprinted polymers were made. All the sensors were placed in a vacuum chamber prior to measurements and held at 40°C in 10^{-5} Pa vacuum for 3 h.

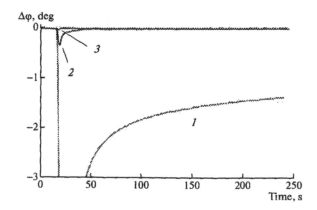

FIGURE 2.17   Time dependence of signal phase shift $\Delta\varphi$ after the injection of (1) 0.1 µL morpholine, (2) 0.1 µL ethanol, and (3) 0.4 µL acetone into carrier gas.

Changes in the output signal (response) phase when analyte vapors passed through the measuring chamber and after analyte vapor left the chamber, which occurred in approximately 10 s after the largest change in the response, are shown in Figure 2. 17. It follows from these curves that the SAW sensor with a molecularly imprinted polymer is several times more sensitive to morpholine than to acetone or ethanol. We also see that the phase changes after the removal of morpholine vapor return to the initial level much more slowly than with the other analytes. Complete sample restoration at room temperature and atmospheric pressure requires substantial time, on the order of several hours.

Changes in the phase of surface acoustic waves caused by the injection of aqueous solutions of morpholine with various concentrations into the vaporization chamber are shown in Figure 2.18. Each subsequent injection of the analyte was made without waiting for the complete restoration of the molecularly imprinted polymer, immediately after heating the sample for 10 min directly in the measuring chamber in a flow of pure nitrogen. The experimental conditions were $T_{SAW} = 42°C$, sample volume $= 0.2\ \mu L$, and polymer molecularly imprinted by 10%.

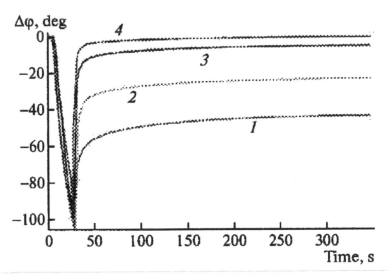

FIGURE 2.18 Time dependence of signal phase shift $\Delta\varphi$ for aqueous solutions of morpholine of various concentrations: (1) 10, (2) 5, (3) 1, and (4) 0%.

We see from Figure 2.18 that the phase shifts over the whole desorption region of the time dependence curve is higher when the initial morpholine concentration in water is higher. Accordingly, the time of restoration of the molecularly imprinted polymer sample increases. Crude estimates show that, in the region of slow morpholine desorption from the film, the signal phase changes are proportional to the concentration of morpholine during the whole time of desorption.

It was established that it is possible to carry out relative measurements without full restoration of the molecularly imprinted polymer sample after its treatment by the analyte. It was enough to get it in the nitrogen flow at 65°C for 10 min; after that, the sample properties practically do not change for some hours.

The experimental time dependences of phase shifts were approximated by theoretical curves calculated with the use of optimized kinetic parameter values. The kinetic parameter characterizing the selectivity of absorption was singled out.[33]

Moreover, we proposed to use a modified method to measure the response of the sensitive element based on the molecularly imprinted polymer, which takes into account the size of output signal and its relaxation time simultaneously. That is, we proposed to use an integral response – the dependence of phase (amplitude) changes on time. Using this method, we have tested the selectivity of the developed SAW sensor element to morpholine as compared with benzene, cyclohexane, and dioxane. The integral responses (in dB) on analytes containing these substances for integration time = 300 s are presented in Table 2.5.[36] It is evident that the integral response on the morpholine-containing solution is 10 dB higher than on other solutions.

TABLE 2.5 Integral Responses in dB for Different Analytes (10 volume % solutions in ethanol).

Analyte	Probe Volume, µl	Sensor Temperature, °C	Velocity of Carrier Gas, cm³/min.	Integral Response in dB
Benzene	0.4	28	6.1	−23

Cyclohexene	0.4	28	6.1	−15.1
Dioxane	0.4	28	6.1	−10.7
Morpholine	0.4	28	6.1	0

It should be noted that the described SAV sensitive element with morpholine-modified polymer film is only one example of the potential uses of molecularly imprinted polymer technology for the fabrication of sensitive gas detectors.

CONCLUSIONS

1. The possibility of generation of surface plasmon resonance by gold nanoparticles formed in a 3D network matrix produced by irradiating a liquid acrylic composition containing a dissolved gold salt with a mercury lamp is demonstrated for the first time; irradiation is needed to accomplish the photoreduction of Au^{3+} to Au^0 and photoinduced polymerization of the test acrylates.

2. The efficiency of the formation of gold nanoparticles is determined by the parameters of the polymer network and by the chemical structure of the oligomer block, with the most suitable matrixes being those having a loose network and containing groups capable of participating in complexation.

3. Chlorine-containing products of $HAuCl_4$ reduction can influence both the rate of reduction of the salt and the process of formation of gold nanoparticles. It has been shown that the SPR of gold

nanoparticles can be generated in a 3-D network matrix obtained during UV irradiation of a liquid acrylic composition containing dissolved gold salt as a result of Au^{3+} photoreduction accompanied by photoinitiated polymerization of acrylates. Decisive factors determining the accumulation rate of nanoparticles and their maximum concentration are the density of the acrylic network and the complexing ability of the oligomeric block. These parameters affect the rate of generation of gold nanoparticles but not the limiting optical density of the plasmon resonance band of nanoparticles.

4. The kinetic data on the consumption of $HAuCl_4$ show that, in oligomeric matrixes with a long oligomeric block, the rates of salt consumption are close. This process proceeds in the gel and is completed long before the end of network formation. The gold salt inhibits the polymerization of acrylates. Nevertheless, in all the studied cases, we managed to obtain high-quality films in which the PRB was generated in the visible spectral region after subsequent irradiation.

5. It has been found that only in oligo(urethane(meth)acrylate) matrixes based on m-toluene diisocyanate does the band at 340 nm arise (it increases and does not disappear), along with the PRB near 560 nm, during the generation of gold nanoparticles.

6. The obtained matrixes may serve as registering media for optical data recording because the agglomeration of gold nanoparticles is initiated by subsequent external effects – heating and/or irradiation – and the plasmon resonance band of gold nanoparticles arises in the visible spectral region (500–600 nm) at the point of external application.

7. Polymer films containing microcavities with selective absorption centers were obtained from oligoester acrylates.

8. It was found in measurements of the amplitude–frequency characteristics of samples with polymer films having various thicknesses without a sewed analyte that, at film thicknesses no larger than 1.5 µm, studies at frequencies up to 80 MHz could be performed.

9. Experimental studies of absorption–desorption of morpholine as an analyte by a molecularly imprinted polymer film (obtained with morpholine as a template) using the SAWs demonstrated significantly increased sensitivity of the polymer sensor element to morpholine in comparison with other analytes.

10. A new method to measure the response of the sensitive element based on the molecularly imprinted polymer, which takes into account a size of output signal and its relaxation time simultaneously, has been proposed.

KEYWORDS

- **Oligomeric composition**
- **oligourethane acrylate**
- **oligocarbonate methacrylate**
- **gold nanoparticles**
- **surface plasmon resonance**
- **molecularly imprinted polymer**
- **acoustic wave sensor**

REFERENCES

1. Zapadinskii, B. I.; Kotova, A. V.; Matveeva, I. A.; Pevtsova, L. A.; Stankevich, O. A.; Shashkova, V. T.; Barachevskyii, V. A.; Dunaev, V. A.; Timashev, P. S.; Bagratashvili, V. N. UV-radiation-induced formation of gold nanoparticles in a three-dimensional polymer matrix. *Rus. J. Chem. Phys. B.* 2010, 4, 864.
2. Link, S.; El-Sayed, M. Spectral properties and relaxation dynamics of surface plasmon electronic oscillations in gold and silver nanodots and nanorods. *J. Phys. Chem.* 1999, 103, 8410.
3. Danie, M. C.; Austruc, D. Gold nanoparticles: assembly, supramolecular chemistry, quantum-size-related properties, and applications toward biology, catalysis, and nanotechnology. *Chem. Rev.* 2004, 104, 293.
4. Buchachenko, A. L. Nanochemistry: a direct route to high technologies of the new century. *Rus. Chem. Rev.* 2003, 72, 375.
5. Mulvaney, P. Surface plasmon spectroscopy of nanosized metal particles. *Langmuir.* 1996, 12, 788.
6. Link, S.; Mohamed, M. B.; El Sayed, M. A. Simulation of the optical absorption spectra of gold nanorods as a function of their aspect ratio and the effect of the medium dielectric constant. *J. Phys. Chem. B.* 1999, 103, 3073.
7. Maier, S. *Plasmonics: Fundamentals and Applications*; Springer: New York, 2007; Vol. XXVI, p 224.
8. Njoki, P. N.; Lim, I. S.; Mott, D.; Park, H. I.; Khan, B.; Mishra, S.; Sujakumar, R.; Luo, J. Size correlation of optical and spectroscopic properties for gold nanoparticles. *J. Phys. Chem. C.* 2007, 111, 14664.

9. Hatter, E.; Fendler, J. H. Heterodyne apertureless near-field scanning optical microscopy on periodic gold nanowells. *Adv. Mater.* 2004, 16, 1685.

10. Esumi, K.; Suzuki, S.; Yamahira, A.; Torigoe, K. Interactions between alkanethiols and gold-dendrimer nanocomposites. *Langmuir.* 2001, 17, 6860.

11. Smith, E. A.; Corn, R. M. Surface plasmon resonance imaging as a tool to monitor biomolecular interactions in an array based format. *Appl. Spectrosc.* 2003, 57, 320.

12. Tanaka, T.; Yamaguchi, K.; Yamamoto, S. Fabrication of multilayered photochromic memory media using pressure-sensitive adhesives. *Opt. Commun.* 2002, 212, 45.

13. Xu, X.; Gibbons, T. H.; Cortie, M. B. Gold nanoparticles for physics, chemistry and biology. *Gold Bull.* 2006, 39(4), 156.

14. Smirnova, L. A.; Aleksandrov, A. P.; Yakimovich, N. O.; Sapogova, N.; Kirsanov, A. V.; Soustov, L. V. UV-induced formation of gold nanoparticles in a poly(methyl methacrylate) matrix. *The Reports of Russian Academy of Sciences (in Rus).* 2005, 400, 779.

15. Berlin, A. A.; Korolev, G. V.; Kefeli, TYa; Sivergin, YuM. *Acrylic Oligomers and Related Materials "Khimiya" ("Chemistry", in Rus.)*; Publishing House Moscow: Moscow, 1983.

16. Shashkova, V. T.; Zelenetskaya, T. V.; Zapadinskii, B. I. Effect of the structure of a network matrix on the formation of gold nanoparticles during photoreduction of its salts in solutions of acrylic oligomers. *Polym. Sci. Ser. A.* 1995, 37, 1123.

17. Radugina, A. A.; Shashkova, V. T.; Kuzaev, A. I.; Brikenshtein, A. A.; Kefeli, T. Effect of the structure of a network matrix on the formation of gold nanoparticles during photoreduction of its salts in solutions of acrylic oligomers. *Polym. Sci., Ser. A.* 1984, 26, 1695.

18. Swanson, N. L.; Billard, B. D. Optimization of extinction from surface plasmon resonances of gold nanoparticles. *Nanotecnology.* 2003, 14, 353.

19. Kreibig, U.; Fauth, K.; Quinten, I.; Schonauer, D. Z. Many-cluster-systems: models of inhomogeneous. *Matter. Phys. D.* 1989, 12, 505.

20. Zapadinskii, B. I.; Shashkova, V. T.; Pevtsova, L. A.; Kotova, A. V.; Matveeva, I. A.; Stankevich, O. A. Effect of the structure of a network matrix on the formation of gold nanoparticles during photoreduction of its salts in solutions of acrylic oligomers. *Polym. Sci. Ser. B.* 2013, 55, 31.

21. Ahmed, M.; Narain, R. Photochemical reactions of ketones to synthesize gold nanorods. *Langmuir.* 2010, 26, 18392.

22. Kotova, A. V.; Glagolev, N. N.; Matveeva, I. A.; Cherkasova, A. V.; Shashkova, V. T.; Pevtsova, L. A.; Zapadinskii, B. I.; Solov'eva, A. B.; Bagratashvili, V. N. Effect of parameters of a crosslinked polyacrylic matrix on the impregnation of photoactive compounds in supercritical fluids. *Polym. Sci. Ser. A.* 2010, 52, 522.

23. Hglund, C.; Zdch, M.; Kasemo, B. Enhanced photocurrent in crystalline silicon solar cells by hybrid plasmonic antireflection coatings. *Appl. Phys. Lett.* 2008, 92, 013113.

24. Qun, H.; Jianhua, Z.; Hui, C. Polymer composites having highly dispersed carbon nanotubes. Patent Appl. No. 20080004364, USA, 2008.

25. Thomas, J.; Anija, M.; Cyriac, J.; Pradeep, T.; Philip, R. Optical nonlinearity in glass-embedded silver nanoclusters under ultrafast laser excitation. *Chem. Phys. Lett.* 2005, 403, 308.

26. Taubert, A.; Wiesler, U.-M.; Mellen, K. Scattering of surface electromagnetic waves by Sn nanoparticles. *J. Mater. Chem.* 2003, 13, 1090.
27. Agareva, N. A.; Aleksandrov, A. P.; Smirnova, L. A.; Bityurin, N. M. Laser generation of nanostructures on the surface and in the bulk of solids. *Perspekt. Mater.* 2009, 5. (in Rus.).
28. Zapadinskii, B. I.; Kotova, A. V.; Matveeva, I. A.; Pevtsova, L. A.; Stankevich, O. A.; Shashkova, V. T.; Bagratashvili, V. N.; Sarkisov, O. M.; Berlin, A. A.; Timashev, P. S. Polymeric media for recording of optical information and method of its production. Patent Appl. No 2429256, RF, 2011.
29. Wohltjen, H.; Dessy, R. E. Surface acoustic probe for chemical analysis I. Introduction and instrument description. *Anal. Chem.* 1979, 51, 1458.
30. Binchack, W.; Buf, W.; Klent, S.; Hamsch, M.; Ehrenpfort, J. A combination of SAW-Resonators and Conventional Sensing Elements for Wireless Passive Remote Sensing, Proceedings of the IEEE 2000 Ultrasonic Symposium: San Juan, Puerto Rico; pp 495–498, 2000.
31. Gendrikson, O. D.; Zherdev, A. V.; Dzantiev, B. B. Opportunities to improve selectivity of determining oxidoreductase substrate using artificial receptors. *Biol. Chem. Rev.* 2006, 46, 149. (in Rus).
32. Paredes, J. I.; Martinez-Alonso, A.; Tascon, J. M. D. The effect of selective absorption sites in the diffusion processes in polymer films. *Microporous Mesoporous Mater.* 2003, 65, 93.
33. Hilal, N.; Kochkodan, V.; Khatib, L.; El Busca, G. Biomimetic sensors for toxic pesticides and inorganics based on optoelectronic/electrochemical transducers-an overview. *Surf. Interf. Anal.* 2002, 33, 672.
34. Losev, V. V.; Medved', A. V.; Roshchin, A. V.; Kryshtal, R. G.; Zapadinskii, B. I.; Epinat'ev, D.; Kumpanenko, I. An acoustic study of the selective absorption of vapors by microporous polymer films. *Russ. J. Phys. Chem. B.* 2009, 3(6), 990.
35. Kryshtal, R. G.; Kundin, A. P.; Medved', A. V.; Shemet, V. V. Impulse characteristic as a response of a liquid sensor based on shear-horizontal surface acoustic waves. *Tech. Phys.* 2002, 47, 1316.
36. Medved', A. V.; Kryshtal, R. G.; Bogdasarov, O. E. An acoustic study of the selective absorption of vapors by microporous polymer films. *J. Commun. Technol. Electron.* 2005, 50, 651.

CHAPTER 3

ON THE SPECIFIC REACTIVITY OF CARBOOLIGOARYLENES: REPRESENTATIVES OF OLIGOMERS WITH CONJUGATE SYSTEM

V. A. GRIGOROVSKAYA

Semenov Institute of Chemical Physics, Russian Academy of Sciences, Moscow, Russia; Email: valgrig3@ mail ru

CONTENTS

ABSTRACT

Features of the reactivity of soluble carbooligoarylenes (COA) based on the condensed aromatic compounds such as naphthalene, anthracene, and their mixtures with benzene in different chemical processes have been studied. It has been shown that their increased reactivity in various chemical processes is due to the presence of long conjugation system in their structure. This feature leads to the polymeranalogical transformations of carbooligoarylenes, which are not possible for corresponding monomeric and dimeric compounds. Because of the characteristics of dissolving in organic solvents and softening on heating, oligoaryilenes represent a base for the synthesis of a series of products of different functions, particularly, polymer materials resistant to thermal, thermooxydative, thermochemical, and radiation–thermal influences. It was established that effective stabilizers of thermal-oxidative degradation for industrial polymers, flame retardants, water-soluble oligomeric cation exchange resins, and antifriction additives can be obtained by using soluble COA as the base. The totality of this data suggests that the range of applications of soluble COA can be much greater.

INTRODUCTION

Being representatives of the polymers with conjugate system, carbooligoarylenes (COA) are of great interest due to the increasing demand for highly resistant polymer materials in different industrial fields. The aim of this research is to obtain soluble COA by one-stage synthesis from condensed aromatic compounds such as naphthalene, anthracene, and their mixtures with benzene. We have developed two methods for the synthesis of COA: a thermal dehydropolycondensation [1,2] and the cationic oxidative dehydropolycondensation in the presence of a Friedel–Crafts catalyzer and an oxidizer, proposed initially for the synthesis of polyphenylene.[3] The application of the cationic oxidative dehydropolycondensation method to a condensed aromatic compounds, first with naphthalene, then with anthracene, as well as their mixtures with benzene, made it possible to obtain oligomers that are soluble in organic solvents and that soften on heating.[4,5] We modified this method of synthesis of oligomers by using copper chloride as an oxidant instead of nitrobenzene, thereby improving the technological parameters of synthesis.[6-8] This research focuses on the systematic study of chemical properties of COA and their dependence on the length of oligomeric chains, that is, the value of molecular weight (M_n).

EXPERIMENTAL PART

The aim of this study is to obtain soluble COAs by one-stage synthesis from the condensed aromatic compounds naphthalene and anthracene using the method thermal dehydropolycondensation [1,2] or cationic oxidation.[4-8] The fractionation of COA was performed by fractional precipitation from benzene–methanol mixture or by exclusion chromatography (to study the enthalpy of dissolution of COA). The column was filled with styrene-divinylbenzene (SDV)-gel; tetrahydrofuran was used as an eluent. The M_n of COA and of their fractions were measured in chloroform by measuring the condensation heat effects.[9] IR spectra were examined using UR-20 instrument, electronic absorption spectra-SF-4 spectrometer. The thermomechanical analysis (TMA) curves were recorded under a specific load of 0.8 kg/cm and a rate of temperature increase of 1.5 deg/min. Thermal stability was estimated by thermogravimetry on an automatic thermo-balance ATB-2 or Perkin-Elmer. Stabilizing action of COA was studied in reactions of non-initiated oxidation of ceresin at 170°C and 5% concentration of COA (above the critical). Content of SO_3H groups in sulfonated COA was measured by means of potentiometric titration. Enthalpies of dissolution for oligoarylene on the base of naphthalene and its fractions in benzene, toluene, and 1,2,4-trichlorobenzene were determined on a Calvet calorimeter with margin error not more than 1%.

RESULTS AND DISCUSSION

THERMAL PROPERTIES OF COA

According to TMA, soluble COA softens on heating, but the repeated heating after pressing of samples causes increase in their softening temperature resulting in the formation of insoluble and non-fusible products.[10] Thus, the insoluble and non-fusible products are the evidence of proceedings of various condensation and polycondensation processes during heat treatment of COA.

The study of COA by thermogravimetric analysis (TGA) method has shown their property of high thermal stability both in an inert environment and in the air curing. The temperatures at which intensive weight loss begin on the air for soluble COA, based on naphthalene, anthracene, and mixtures of them with benzene, are >450°C–500°C, according to the TGA data.[11] It has been established that the insoluble fractions of the COA have less resistance to thermooxidative degradation but greater stability and greater amount of stable balance in inert atmosphere in comparison with soluble fractions. TGA curves on the air for all soluble COA have autocatalytic character that indicates the principle distinction of their destruction mechanism from that of saturated polymers.[12] According to IR spectroscopy and elemental analysis data, reactions of dehydropolycondensation similar to that observed at TMA and leading to the products with less solubility and less resistant to thermooxydative destruction occur during the induction period. After accumulating a certain concentration of these structures, an intensive process of thermal oxidation begins. These views are confirmed by the results of the absorption of oxygen by the soluble and insoluble fractions of COA on static installation. Analysis of TGA curves of soluble polymer-homological fractions of COA, differing only in M_n values, established that in an inert atmosphere, the thermal stability increases monotonically with increase in M_n and on air, its dependence on M_n has more complex character: in the beginning, the thermostability increases due to thermal destruction of COA on low-molecular fragments, and then, after reaching a certain value of M_n at its subsequent increase, thermostability on the air begins to decrease.[13] This feature is connected with the proceeding of two kinds of competing processes during heat treatment on the air: thermal destruction with the formation of low-molecular-weight fragments and thermooxidative destruction of insoluble fractions, resulting in dehydropolycondensation reactions. Analysis of the products formed during the heating of the COA samples with different values of M_n up to the beginning of the intensive mass loss (~350°C) showed that when the value of M_n of the original oligoarylene was higher, the content of insoluble fractions obtained was more and M_n of remaining soluble parts was lower.[14] These results indicate that most high molecular-weight fractions of COA participate in the reactions of polycondensation in the first instant, which confirms their high reactivity. Thus, these data suggest that the reactivity in the polycondensation and oxidation processes of both original COA and their fractions increase with increase in the M_n values.

STABILIZING ACTION OF COA

Previously, it was found that conjugated compounds are capable of show-ing a stabilizing action in the processes of thermal-oxidative degradation of various polymer systems.[15] So it seemed important to study the influ-ence of M_n value of soluble COA on their stabilizing activity. Polymer-homological fractions of polyanthracene (PA) and polynaphtalene (PN) with various values of M_n, obtained directly after synthesis were studied as COA. Their stabilizing activity in the reactions of non-initiated oxida-tion of ceresin at 170°C was studied.[16] Fractions of the PA and PN with a variety of M_n were identical among themselves, according to the data of IR and electronic spectra, as well as the results of element analysis. Charac-teristics of fractions of the PA, as well as 9,9'-bianthryl, which can be con-sidered as the first member of the homologous series of oligoanthracenes, are shown in Table 3.1.

TABLE 3.1 The Characteristics of the Fractions of Polyanthracene and Induction Time (τ_{ind}) of the Ceresin Oxidation at 170°C in the Presence of 5% PA

Samples		M_n	Polymer Matrix Composite Con-centration, $\times 10^{-17}$, spin/g	Tem-perature of Softening, °C	τ_{ind}, min
	1	2230	67	380	20
	2	1620	31	320	23
	3	1080	8.7	270	44
Fractions of polyanthra-cene	4	920	3.3	240	225
	5	780	1.6	210	2450
	6	680	1.5	185	2060
	7	–	1.5	165	1250
9,9'-bianthryl		354	$<10^{-3}$	308	~650

During the study of stabilizing action of PA fractions, it was found that they can be both weak and strong inhibitors depending on their M_n values. Strong inhibitors are characterized by occurrence of the critical concentra-tions. The data from Table 3.1 show that at the inhibitor concentration of 5% (above the critical), the dependence of the induction period (τ_{ind}) on M_n

value has a sharply defined extreme character with the maximum value of τ_{ind} (2450 min) at $M_n = 780$, which corresponds to 4–5 monomeric units in the chain of the molecule. This value is more than 100 times greater than the relevant values for PA with $M_n = 2230$ and 1620. It is interesting to note that at M_n increase for PA from 550 to 780, the induction period is increased from 1250 to 2450 min, while the concentration of PMC in the inhibitor practically does not change (Table 3.1). It is also evident from Table 3.1 that 9,9'-biantryl, characterized by the absence of paramagnetism, provides induction period of 650 min. These results suggest that M_n value is the determining factor for the stabilizing activity in the PA studied. This assumption is confirmed by the data for PN, according to which when the M_n value of PN is increased from 1100 to 1250, τ_{ind}. also increases from 21 to 124 min. By comparing the results of the PA and PN, it can be observed that the stabilizing activity of COA depends significantly on the structure of the basic unit, evidently on the conjugation extent in the segment. The extreme character of τ_{ind} dependence on the M_n value of inhibitor (COA) can be explained, as in the case of thermal-oxidative stability of COA, by proceedings of two types of competitive processes. On the one side, the increase of length of conjugate system causes decrease of excitation energy of polymeric inhibitor molecules that increases the effective rate constant of interaction between inhibitor and peroxidate radicals and, hence, a speed of deactivation of these radicals. The last depresses the critical concentration of inhibitor and increases τ_{ind} during thermooxidation. On the other hand, during the process of increase in M_n, the probability of oligomeric inhibitor interaction with air oxygen also increases. The first consequence of that is the reduction of the effective concentration of the inhibitor, and the second consequence is the appearance of the initiating action of ex-inhibitor. At certain M_n value, the second factor starts to prevail and an extremum appears on the curve of τ_{ind} dependence on M_n. A position of the last depends on the structure of the oligomeric inhibitor chain.

SULFONATION OF COA

Interesting and significant results on the influence of M_n value on the reactivity of COA were obtained during the study of COA sulfonation reaction, which is characteristic of aromatic compounds.[17] Sulfonation reaction was examined for PN, main structural unit of which is 1,4–disubstituted naphtylene link.[18] The analysis of the IR spectra of formed oligomeric products has shown that they are sulfonic acids of oligoarylenes and joining of sulfuric groups occurs, first of all, in 1,4-disubstituted rings of naphtylene segments.[19] Sulfonic acids of OA are readily soluble in water and insoluble in organic solvents and are not softened or sublimated at high temperatures. In the air, sulfonic acids of OA are stable up to 250°C.

It was noticed that polyphenilenes (PPh) show much less reactivity in sulfonation process than PN. It is connected, apparently, with the difference in the degree of conjugation of PPh and PN links. At the same time, reactivity of COA is largely determined by the total length of oligomeric chain, that is, M_n value, which can be observed from the data of Table 3.2. It is evident from Table 3.2 that not all initial PNs participate in sulfonation reaction at temperatures below 100°C. As one can expect, in less time and lower temperature, the smaller part of the initial product reacts. It should be noted that in all cases, the degree of sulfonation increases with the decreasing of reacted PN part. The detected peculiarity is connected, obviously, with unequal reactivity polymer-homological fractions, which are parts of initial OA fractions being characterized by highest molecular weight and possessing the highest reactivity that participate in the reaction of substitution for the first time leading to the formation of products with the highest degree of substitution of aromatic rings. If the reaction conditions are hard, the most part of initial OA would enter the reaction of substitution, but the average degree of sulfonation of resulting oligomeric product decreases. At the same time, the tendency of decrease in the M_n value of unreacted oligomer is observed (Table 3.2).

TABLE 3.2 Influence of Temperature and Time of Sulfonation of PN with $M_n = 870$ on the Yield and Degree of Sulfonation of Oligomeric Products

Temperature (°C)	Time (h)	Share of Reacting Oligomer[a]	Content of SO₃H Groups on a Link in a Reaction Product	M_n of Unreacted Oligomer[a,b]
40	0.5	0.10	5.20	790
60	0.5	0.23	1.70	740
60	4.0	0.61	1.3	720
80	0.50	0.73	1.20	720
80	4.0	0.93	0.81	700

100	0.5	1.00	0.64	–
100	19.0	1.00	0.74	–
140	0.5	1.00	2.21	–

Notes: [a]It is defined as the difference between the unit (1) and a share of unreacted PN, insoluble in water, but soluble in chloroform; [b]The variability in all cases does not exceed ±5%.

The conclusion about increase in reactivity of COA with the increase in M_n value is confirmed also in the conditions when COA participates completely in the sulfonation reaction. So, sulfonation of PN with $M_n = 870$ at 100°C for 19 h results in a product containing 0.74 SO_3H groups on a link, which is contrary to a result of processing for PN with $M_n = 1890$, where SO_3H group contents in the final product reaches only 1.7 on a link. Thus, from the totality of the above data, it can be concluded that the reactivity of COA in sulfonation reaction increases with the increase in M_n value. It is interesting to note that although it is impossible to reach such high degrees of sulfonation, this can be obtained in the soft conditions for PN for the monomer naphthalene even under the most severe conditions.[17]

So, the data presented proves that as for the original COA and their fractions with increasing value of M_n, their reactivity increases in different processes such as polycondensation, thermal oxidation, interaction with peroxidate radicals, and sulfonation.

COMPLEXING OF COA

It is known that polymers and oligomers with conjugate system, possessing high donor-acceptor activity, tend to form complexes before the chemical reaction. Moreover, during different chemical transformations of COA in organic solvent, the last may also complex with oligoarylene. In this respect, the results of the influence of M_n value on the enthalpy of dissolution of COA in solvents of different polarity are interesting. PN and its fractions with various M_n values, obtained by exclusion chromatography method, were the focus of this study. Dissolution enthalpies in benzene, toluene, and 1,2,4-trichlorbenzene were determined on a Calvet calorimeter with a margin error not more than 1%.[20] Polymer homology of studied fractions is confirmed by the data of electronic and IR spectra, as well as monotonous character of T softening dependency on M_n value. Besides, according to the data of exclusion chromatography, all the studied fractions are very homogeneous in composition and close to each other. Enthalpy of dissolution of PN and its fractions in solvents of different polarity are presented in Table 3.3.

TABLE 3.3   Thermal Effects of Dissolution of PN and its Fractions

Oligoarylene	M_n	ΔH comm (cal/g)		
		In Benzene	In Toluene	In 1,2,4-trichlor-benzene
PN Original	790	+ (3.3 ± 0.1)	+(4.1 ± 0.2)	+(6.4 ± 0.1)
	450	−(3.5 ± 0.4)	−(3 ± 0.1)	−(0.7 ± 0.1)
	490	−	−(1.72 ± 0.14)	−
	550	−(2.7 ± 0.7)	−	+(0.4 ± 0.2)
	640	−(2.1 ± 0.2)	−(1.5 ± 0.2)	−
Fractions of PN Obtained by Method of Exclusion Chromatography	710	−	−(0.49 ± 0.1)	−
	780	−	+(2.1 ± 0.2)	+(5.0 ± 0.1)
	850	−	+(5.8 ± 0.1)	−
	1010	+(6.5 ± 0.1)	+(6.5 ± 0.9)	+(8.1 ± 0.7)
	1260	−	+(7.8 ± 0.7)	+(10.4 ± 0.2)
	1320	−	+(11.4 ± 0.1)	−

From Table 3.3, the following conclusions can be made:
i)   A thermal effect of dissolution of initial unfractionated PN is higher than that of narrow PN fraction with similar M_n value.
ii) For each OA, the thermal effect of dissolution increases with the increase in solvent polarity.
iii) The thermal effects of dissolution always increase with an increase in fraction M_n.

The obtained results can be explained on the basis of the following equation:

$$-\Delta H_{comm} = \Delta H_{s-s} + \Delta H_{p-p} - \Delta H_{p-s},$$

where ΔH_{comm} is common enthalpy of dissolution of the polymer. ΔH_{s-s} is enthalpy of disconnection of solvent molecules, ΔH_{p-p} is enthalpy of disconnection of polymer molecules, and ΔH_{p-s} is enthalpy interaction of polymer with the solvent.

Thus, the increase in the thermal effect of dissolution, observed with the increase in solvent polarity, can be explained by a significant increase of energy of the interaction of COA with the solvent. In order to compare the dissolution results for various PN fractions in the same solvent, the observed increase of the heat effect with the increase in M_n should also be connected with a significant increase in the energy of the interaction of PN with the solvent. The dependence of the enthalpies of COA dissolution on M_n value is, obviously, connected with the features of COA electronic structure, in particular, with the presence of conjugated double bonds in a chain.

It should be noted that the dependence of soluble COA reactivity on M_n value is not trivial. Indeed, according to the results of the construction of molecular models of the COA, as well as the data from NMR of wide lines, the structure of COA molecules is not coplanar, and the angle of rotation between neighboring links increases from PPh to PA: if for PPh, it is ~20°, then for PA, it is ≥45°.[21] Nevertheless, the totality of the obtained data let us to declare unequivocally that the effect of the conjugation is sufficiently pronounced for soluble COA, owing to what a clear dependence of reactivity of the last on chain length, that is, values of M_n, takes place. This feature of COA reactivity makes it possible to achieve on their basis a variety of polymeranalogic transformations, including that which is basically impossible for corresponding monomeric and dimeric compounds. Reaction products of interaction of soluble COA with *n*-diethynylbenzene,[22,23] polyethyleneglycol, and oligomers on its basis,[24] and epoxy compounds such as epichlorohydrin, glycidol, glycidilmethacrilat,[25–28] and polyoximethylenes [29–32] after hardening allow to obtain various polymeric materials with the desired set of properties, particularly, high stability to thermal, thermooxidizing, thermochemical, and radiation-thermal influences. It should be noted that the analysis of fractionated samples of oligomeric products of COA transformations has shown that the higher is the fraction M_n, the larger is the degree of substitution in the aromatic link.[23,25] The specific reactivity of soluble COA allows to synthesize effective stabilizers of thermooxidative degradation of industrial polymers,[33,34] fire retardants,[35] oligomeric sulfoacids, including water soluble,[36] antifrictional additives [37,38] as their base

The analysis of literature data allows us to assume that the dependence of reactivity on chain length has enough general character for polymers with a system of conjugated bonds contrary to polymers with saturated bonds in chains. So, Hutareva et al[39] and Jandarova et al[40] reported that at decarboxylation of polypropiolic acid, the increase in chain length essentially influences the process course. A change of a rate and a direction of polyoxiarylenes reactions in comparison with model compounds were also obtained by Liogon et al.[41]

In conclusion, we note that results obtained in the present study exhibit the clear dependency of reactivity of soluble COA on their molecular weight value. This property allows us to synthesize different products with soluble COA as a base. Finally, it can be observed that these results give the grounds to believe that the range of use of soluble COA and products of their polymeranalogic transformations can be expanded in future.

KEYWORDS

- Carbooligoarylene
- conjugate system
- benzene
- naphthalene
- anthracene
- dehydropolycondensation
- thermooxydative
- thermochemical
- thermoradiative stability
- solubility
- molecular weight sulfonation
- polymer
- homolog
- oligomer

REFERENCES

1. Berlin, A. A.; Grigorovskaya, V. A.; Skurat, V. E. Influence of polymeric fraction of thermolyzed anthracene containing paramagnetic centers on the course of law-temperature thermolysis of anthracene and bianthryl. *Polym. Sci. (in Rus.).* 1966, 8(11), 1976.

2. Berlin, A. A.; Grigorovskaya, V. A.; Kushnerev, M. Ya; Skachkova, V. K. Structures and certain characteristics of polymer fractions produced during the low-temperature pyrolysis of anthracene. *Polym. Sci. A (in Rus.).* 1968, 10, 2687.

3. Kovacic, P.; Kyriakis, A. Polymerization of benzene to p-polyphenyl. *Tetrahedron Lett.* 1962, 8, 467.

4. Berlin, A. A.; Grigorovskaya, V. A.; Parini, V. P.; Chernikova, S. A. Production of aromatic polymers in the presence of aluminium chloride and influemce of additives containing paramagnetic centers. *Polym. Sci. B (in Rus.).* 1967, 9, 423.

5. Berlin, A. A.; Grigorovskaya, V. A.; Skachkova, V. K. Method of Producing of Oligoarylenes. USSR Inventor's Certificate №. 298613, B. I. 11, 1971.

6. Astrakhantseva, N. I.; Brikenstein, A. A.; Berlin, A. A.; Grigorovskaya, V. A.; Skachkova, V. K. Influence of synthesis conditions on the yield, molecular weight and structure of oligoarelenes obtained in the presence of aluminium chloride and cupric chloride. *Polym. Sci. A" (in Rus.).* 1973, 15, 54.

7. Astrahanceva, N. I.; Brikenstein, A. A.; Berlin, A. A.; Grigorovskaya, V. A.; Shvetsova, S. I. Method of Producing of Oligoarylenes. USSR Inventor's Certificate, № 358325, B.I .№ 34, 1972.

8. Kostenkova, L. S.; Iotkovskaya, L. A.; Kuznetsov, V. V.; Siling, M. I.; Grigorovskaya, V. A. Improved technology of oligoarelene synthesis. *Plastics (in Rus.).* 1986, 5, 10.

9. Bekhli, E. Yu; Novikov, D. D.; Entelis, S. G. Thermomechanic study of oligomeric arylenes. *Polym. Sci. A (in Rus.).* 1970, 12, 2754.

10. Berlin, A. A.; Belova, G. V.; Grigorovskaya, V. A.; Ochorsin, B. A. Thermomechanic study of oligomeric arylenes. *Polym. Sci. B (in Rus.).* 1970, 12(11), 793.

11. Berlin, A. A.; Belova, G. V.; Grigorovskaya, V. A. Thermostability of of oligomeric arylenes. *Polym. Sci. A (in Rus.).* 1970, 12(10), 2351.

12. Berlin, A. A.; Grigorovskaya, V. A.; Selskaja, O. G. Thermostability of fractionationed samples of some oligoarylenes. *Polym. Sci. A (in Rus.).* 1970, 12(11), 2541.

13. Grigorovskaya, V. A.; Selskaja, O. G.; Khakimova, D. K.; Berlin, A. A. Thermal transformations of oligoarylenes. *J. Therm. Anal., (in Rus.).* 1975, 8, 431.

14. Grigorovskaya, W. A.; Selskaya, O. G.; Astrachanzeva, N. I.; Berlin, A. A. Influence of some factors on the thermostability of oligoarikebes. *Plastic Kautchuk, (in Rus.).* 1974, 21, 897.

15. Berlin, A. A.; Bass, S. I. *Ageing and stabilization of polymers, Moscow, "Khimia" ("Chemistry", in Rus)*; Publishing House: Moscow, 1966; p 129.

16. Firsov, A. P.; Grigorovskaja, V. A.; Pazhitnova, N. V.; Ivanov, A. A.; Berlin, A. A. Some factors influence on the stabilizing activity of oligoarylenes. *Polym. Sci. A (in Rus.).* 1973, 15, 1881.

17. Vorozhtsov, N. N. *Bases of Intermediate Products and Dyes. Izdatel'stvo nauchno-texnicheskoi literaturi, ("Science-Technology", in Rus.)*; Publishing House: Moscow, 1955; p 98.

18. Berlin, A. A.; Grigorovskaya, V. A.; Kissin, J. V. Study of oligoarilenes structure using IR- spectroscopy. *Polym. Sci. A (in Rus.)*. 1970, 12, 1497.

19. Grigorovskaya, V. A.; Davydova, G. I.; Berlin, A. A. Some peculiarities of sulfonation reaction of oligomeric arylenes. *Polym. Sci. B (in Rus.)*. 1980, 22, 315.

20. Grigorovskaya, V. A.; Miroshnichenko, E. A.; Kuzaev, A. I.; Vorobjeva, V. P.; Lebedev, Y. A. On the influence of average moleculae nass on dissolving enthalpy of fractioned oligoarylenes. *Reports of Russian Academy of Sciences. Ser. Chem. (in Rus.,)*. 1985, 284, 647.

21. Berlin, A. A.; Grigorovskaya, V. A.; Kushnerev, M. Ja; Urman, Ja. G. Structure of oligomers on the base of benzene, napthaline and anthracene. *Proc. Acad. Sci. USSR, Ser. Chem. (in Rus.)*. 1969, 11, 2568.

22. Berlin, A. A.; Skachkova, V. K.; Grigorovskaya, V. A.; Khoroshilova, I. P. et al. Method of producing of compositions. USSR Inventor's Certificate №473727, BI №22, 1975.

23. Berlin, A. A.; Skachkova, V. K.; Grigorovskaya, V. A.; Kuzaev, A. I.; et al. Interaction of oligoarylenes with p- diethylbenzene. *Polym. Sci. A" (in Rus.)*. 1979, 21.

24. Masenkis, M. A.; Khoroshilova, I. P.; Avrasin, J. D.; Berlin, A. A.; Grigorovskaya, V. A.; Skachkova, V. K. et al. About compositions preparation. USSR Inventor's Certificate №77639, 1974.

25. Grigorovskaya, V. A.; Mordvinova, N. M.; Begun, B. A.; Yartseva, I. V. Interaction of oligoarylenes and epichlorohydrin. *Polym. Sci. B" (in Rus.)*. 1992, 33, 29.

26. Grigorovskaya, V. A.; Mordvinova, N. M.; Begun, B. A.; Yartseva, I. V.; Kuznetsov, Y. L. Synthesis and properties of chlorohydrine-substituted oligoarylenes. *Plastics (in Rus.)*. 1993, 6, 8.

27. Grigorovskaya, V. A.; Yartseva, I. V. Polymers based on fluoro(meth)acrylates and fluropolyimides. *Polym. Sci. B, (in Rus.)*. 2000, 42, 534.

28. Grigorovskaya, V. A.; Begun, B. A.; Yartseva, I. V.; Tikhomirova, I. V.; Tugov, I. I. Interaction of oligoarylenes with epichlorhydrin. *Plastics (in Rus.)*. 1991, 2, 51.

29. Grigorovskaya, V. A.; Lazareva, O. L.; Selskaya, O. G. Oligomethylolnaphthylenes for thermostable materials and method of their production. USSR Inventor's Certificate № 860479, 1981.

30. Aseeva, R. M.; Grigorovskaya, V. A. Modified carbooligoarylenes as a base for thermostable and combustible materials, Proceedings of International Conference on "Composite-2004", Moscow, 2004, p 157.

31. Aseeva, R. M.; Grigorovskaya, V. A. Thermochemical properties of cured carbooligoarylenes, Proceedings of 6th International Conference on "Polymer materials of reduction flammability", Vologda, Russia, 2011, p 90.

32. Grigorovskaya, V. A.; Zaichenko, N. L.; Aseeva, R. M. About interaction of oligoarylenes with polyoxymethylenes. *Plastics (in Rus.)*. 2005, 9.

33. Zelenetskaya, T. V.; Grigorovskaya, V. A.; Aseeva, R. M.; Berlin, A. A. Stabilization of three dimensional polyestermethacrylates on the base of α, ω-dimethacrylate diethyleneglicole. *Plastics (in Rus.)*. 1974, 1, 60.

34. Berlin, A. A.; Shutov, F. A.; Aseeva, R. M.; Grigorovskaya, V. A. Method of stabilization of polymeric materials. USSR Inventor's Certificate №. 328725, 1972.

35. Aseeva, R. M.; Grigorovskaya, V. A.; Ruban, L. V.; Zaikov, G. E. Oligobromoarylenes as fusible and soluble product for production of polymer materials with law combustibility. USSR Inventor's Certificate № 1413930, 1988.

36. Grigorovskaya, V. A.; Pikalov, A. E.; Davydova, G. I. et al. Sulfonic acids of oligoarylenes as catalysts of ion-exchanging type. USSR Inventor's Certificate № 603647, B.I. №15, 1977.

37. Denisova, O. V.; Rutman, P. A.; Lobanzova, V. S.; Fanstein, I. V.; Grigorovskaya, V. A.; Kostenkova, L. S. Antifrictional lubricant for mechanical treatment of metals. USSR Inventor's Certificate № 1778164, B.I. № 44, 1992.

38. Grigorovskaya, V. A.; Lobancova, V. S.; Rutman, P. A. Study of different carbooligoarylenes as antifriction compounds in solid lubricating compositions, Proceedings of 6th Conference of Chemistry and Physics-chemistry of Oligomers, Kazan, Russia, 1997, p 74.

39. Hutareva, G. V.; Jandarova, M. L.; Shishkina, M. V. *Polym. Sci. B (in Rus.).* 1970, 12(7), 515.

40. Jandarova, M. L.; Gejderih, M. A.; Krenzel, B. A. Study of decarboxylation reaction of polymer acids with conjugation systems. *Proc. Acad. Sci., USSR (in Rus.), Ser. Chem.,* 1983, 78.

41. Liogon'kij, B. I.; Alexanyan, R. Z. On the specific of interaction of polyoxyarylenes with electrophylic reagents on the example of aryldiazonium salts. *Reports of Academy of Sciences of USSR" (in Rus.).* 1983, 270(2), 364.

CHAPTER 4

PORPHYRIN–POLYMER COMPLEXES IN MODEL PHOTOSENSITIZED PROCESSES AND IN PHOTODYNAMIC THERAPY

A. B. SOLOVIEVA and N. A. AKSENOVA

Semenov Institute of Chemical Physics of Russian Academy of Sciences, Moscow, Russia
E-mail: ann.solovieva@gmail.com, naksenova@mail.ru

CONTENTS

ABSTRACT

We have studied the effect of amphiphilic polymers (APs) with different structures (polyvinylpyrrolidone, polyethyleneoxide, and a triblock copolymers of ethylene- and propyleneoxide – Pluronics) on the photoactivity of porphyrin photosensitizers, disodium salt of protoporphyrin (dimegin), and diglutamic salt of chlorine e6 (photoditazin), in the processes of singlet oxygen photogeneration. The photooxidation of tryptophan and histidine were chosen as model systems. Also, we studied the activity of APs and photoditazin complexes in photodynamic treatment of cancer cell culture and purulent wounds in rats. We showed that APs, especially block copolymers of ethylene oxide and propylene oxide, can increase the photocatalytic activity of porphyrin photosensitizers (PPS) – dimegin and photoditazine – an N-methylglucamine salt of chlorine e_6 both in model experiments in water solution and in biological systems. In fact, AP may enhance the phototoxicity of photoditazin in cell cultures and wound healing by photodynamic therapy. We attribute the observed effect of the APs on the photoactivity of dimegin to the presence of polymer–porphyrin interactions resulting in the porphyrin disaggregation in aqueous phase. Using ^{1}H NMR spectroscopy, we have found that dimegin binds to the polymers via the PPS interaction mainly with the hydrophobic fragments of polymeric macromolecules. Among the studied polymers, polyvinylpyrrolidone appeared to have the most significant influence onto the photoactivity of dimegin in model systems and by photodynamic treatment of purulent wounds in rats.

INTRODUCTION

Photodynamic therapy (PDT) is a quickly emerging treatment modality employing the photochemical interaction of three components: light, photosensitizer, and oxygen. Tremendous progress has been made in the past two decades in new technical development of all components, as well as understanding of the biophysical mechanism of PDT.[1] This technique permanently entered the clinical practice for the treatment of superficial tumors and festering wounds.[1,2]

Porphyrins and its analogues – chlorines – are commonly used as photosensitizers due to their high activity and nontoxicity. The studies conducted so far have shown that excited triplet state molecules of porphyrin photosensitizers (^{3}PPS*) are capable of initiating photochemical reactions of two types.[3,4] First, a direct ^{3}PPS* interaction with biomolecules is possible (detachment of an electron or hydrogen atoms), leading to the formation of free radicals. As a result of interactions of free radicals with molecular oxygen, reactive oxygen species are created. In the reactions of the second type, an energy transfer from ^{3}PPS* to oxygen molecules takes place, and singlet oxygen $^1O_2$, an active oxidant, is generated. Besides, $^1O_2$ is capable of accepting electrons from a substrate leading to the generation of superoxide anion-radicals. At the final stage of the PDT action, both types of photochemical reactions result in the destructive processes in the vital cell structures and their death.[5] The treatment with the PPS application produces positive results, but sometimes, it causes side effects, such as temperature elevation, muscle weakness, and prolonged skin photosensitivity.[6]

Therefore, the studies on the reduction of PDT side effects, primarily by lowering the dose of the applied photosensitizers and reducing the duration of illumination, are of significant importance. The evolution of the PDT techniques is associated with the creation of new photosensitizer drug formulations of improved bioavailability and reduced nonspecific toxicity. In particular, to attain these objectives, it has been suggested that complexes composed of photosensitizers and amphiphilic polymers (APs) differing in nature, such as polyoxyethylene, polyvinylpyrrolidone, Pluronics, and polyvinyl alcohol, should be used for the PDT purposes. In experiments with other polymers forming complexes with photosensitizers, positive biological effects were also noted. To illustrate, when using a chlorine e6–polyvinylpyrrolidone complex, Chin et al[4,7,8] observed an increased selectivity of accumulation of this complex in the cancerous tissues and its accelerated removal from the normal tissues. In addition, the binding of photosensitizers to polymers frequently allows the photophysical properties of the complexes obtained to be substantially improved as compared with the original photosensitizers.[9–11] Of special interest is the possibility of using Poloxamers (also known as Pluronics) – triblock copolymers based on ethylene oxide (EO) and propylene oxide (PO) – for PDT applications. Pluronics, which are tissue compatible emulgators, were earlier demonstrated to have properties of immune adjuvant, show antithrombotic activity, and be capable of modifying the activity of neutrophils. One of the most promising approaches is the development of complexes based on a bis-N-methyl-glucamine salt of chlorine e6. It features a broad absorption band, low toxicity, water solubility, amphiphilic properties, and substantial photodynamic activity, even in small doses. We have recently demonstrated a substantial enhancement of the photocatalytic activity of photoditazine (PD) when combined with Pluronics and also the dependence of this property on the hydrophilicity of the polymer used. Obviously, the emergence of new photosensitizer formulations has a number of merits as applied to the PDT of infected, festering, and persistent wounds, burns, and trophic ulcers but requires, at the same time, that an insight be gained into the mechanisms governing the action of such preparations and that adequate PDT regimens be developed. However, despite the growing interest in the issue and the appearance of first positive clinical experience reports, there is practically no morphological and pathophysiological substantiation for the PDT of wounds and burns using photosensitizers modified with APs.

Recently, it has been shown that a number of APs can improve therapeutic properties of different photosensitizers. It has been shown that non-covalent[8,9] or chemical[10] attachment of a photosensitizer to hydrophilic polymers or APs improves its photophysical properties. Covalent attachment of Zn-protoporphyrin to polyethylene oxide led to the increase in the triplet state lifetime in cell suspension by an order of magnitude.[11] Encapsulation of protoporphyrin IX into methoxy poly (ethylene glycol)-b-poly (caprolactone) diblock copolymers micelles markedly increased photocytotoxicity over that of free protoporphyrin IX, by nearly an order of magnitude at the highest light dose used.[12] Moreover, singlet oxygen quantum yield of such conjugate also increased from 0.05 up to 0.17. Obviously, it means that poly(ethylene oxide) coils impede collision of the porphyrin with surrounding molecules thus hampering idle losses of triplet state energy. Formulation of chlorine e_6 with polyvinylpyrrolidone (PVP) was proved to increase photosensitizer tumor tissue selectivity, obviously due to the facilitation of its clearance from normal tissues.[13] Tremendous (nearly 10-fold) increase in the triplet state lifetimes were observed when hematoporphyrin was adsorbed on the surface of polysaccharide Percoll® microparticles. In contrast to this, attachment of Sn-chlorine e6 to dextran and further conjugation of this polymer to anti-melanoma antibodies led to the decrease in the triplet state lifetime.[14] These facts show that the lifetime of the sensitizer triplet state can be modulated by the properties of its microenvironment, and optimal pairs "photosensitizer-polymer carrier" should be selected in each case.

Among biologically compatible polymers, a large family of polyalkylene oxides is of particular importance due to a large variety of properties of these polymers, their commercial availability, and simplicity of their synthesis. EO and PO block copolymers (also referred to as Pluronics®, Fig. 4.1) are nowadays used in pharmacy and medicine as immunoadjuvants[15] and biologically compatible emulsifiers.[16] They were shown to inhibit thrombosis[17] and modulate neutrophil activity.[18] Pluronics micelles have already been studied as possible vehicles for hydrophobic photosensitizers. In this regard, Pluronic F127 micelles were studied for their ability to facilitate transcytosis of highly hydrophobic meso-tetraphenyl porphyrin through Caco-2 cells monolayer.[19] The same Pluronic was found to assist transdermal penetration of a hydrophilic prodrug aminolevulinic acid.[20] More hydrophobic Pluronic P123 was found to favor delivery of a hydrophobic photosensitizer verteporfin to plasma lipoproteins.[21]

Our own studies[22–32] revealed that application of PPS in the complexes with Pluronics could increase the activity of photosensitizers upon their photodynamic action onto tumor cells and tumors of laboratory animals, thus reducing the applied PPS doses by 10–20 times[22–24] and also in the PDT of extended purulent wounds.[25–27] In the previous work, we also demonstrated that Pluronics augmented the activity of dimegin and PD in the model photooxidation processes in aqueous phase.[28–32] These phenomena are likely caused by the enhancement of singlet oxygen generation quantum yield upon porphyrins solubilization in Pluronic micelles. To verify the above-mentioned statement, we have studied the influence of Pluronic F127 (the least toxic one of polymeric surfactants) and other APs on the activity of a hydrophobic tetraphenylporphyrin (TPP) and water-soluble dimegin and PD in the photogeneration of singlet oxygen and photooxidation of histidine and tryptophan in water and D_2O. In addition, we investigated the activity of complexes PPS–AP in malignant cells by photoirradiation and in PDT of infected wounds in rats.

EXPERIMENTAL PART

MATERIALS

Meso-TPP (Fig. 4.1b), L-tryptophan, and L-histidine were purchased from Sigma-Aldrich Corp. (USA). Disodium salt of 2,7,12,18-tetrame-thyl-3,8-di(1-methoxyethyl)-13,17-di(2-oxycarbon-ylethyl) porphyrin, (Dimegin,

DMG; Fig. 4.1b) was synthesized and kindly gifted by Prof. G. Ponomarev. N-methyl-bisglucamine salt of chlorine e6 (trade name: Photoditazine®, PD, Fig. 4.1c) was generous gift of Veta-Grand Corp. (Russia).We used PVP with $M_w = 40,000$ by Sigma-Aldrich Corp., polyethyleneoxide (PEO) with $M_w = 40,000$ by Serva (Germany; Fig. 4.3), tri-block copolymers of ethylene (m)- and propylene-oxide (n) Pluronics by BASF (USA) – F127 with $M_w = 12,600$ ($m = 100$, $n = 60$), P85 with $M_w = 4500$ ($m = 52$, $n = 40$), Pluronic F87 with $M_w = 7700$ ($m = 61$, $n = 40$), and F108 with $M_w = 14,600$ ($m = 133$, $n = 50$).

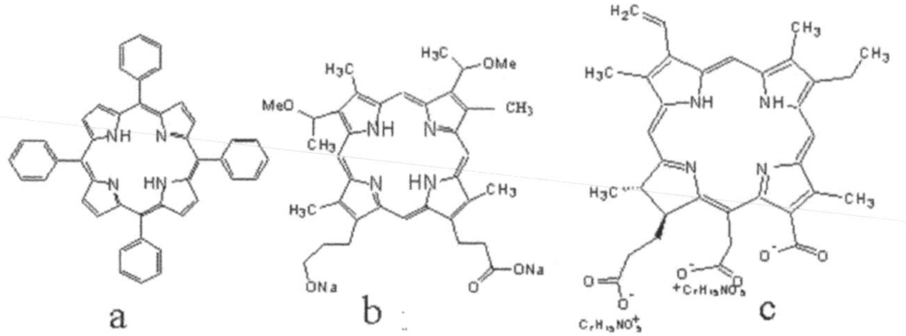

FIGURE 4.1 Structure of (a) Tetraphenylporphyrin, (b) Dimegin, and (c) PHOTODITAZINE.

Water soluble DMG–AP and PD–AP complexes were prepared by the following procedure. PPS (10.0 g) was dissolved in 80 mL of water at a temperature of ~900°C while stirring for 30–40 min. After cooling it to the room temperature, 20 mL of PPS water solution with 5 mg/mL concentration was added to the mixture and stirred to complete the homogenization. This method of complex preparation was applied for all studied polymers.

Solubilization of hydrophobic TPP were prepared by mixing the TPP and Pluronics in a chloroform solution with the porphyrin concentration of 1×10^{-4} M and the polymer concentration of 1.7×10^{-4} M. The porphyrin to Pluronic ratio $q = 0.6$ provided the optimal degree of solubilization. For the AFM imaging, thin films of Pluronics (pure and containing TPP) on silicon substrates were prepared from the chloroform solution by spin coating at a rotation speed of 7000 rpm.

SINGLET OXYGEN MEASUREMENTS

Singlet oxygen photogeneration by DMG was detected directly by measuring time-resolved luminescence at 1270 nm, in D_2O solutions, after excitation of the photosensitizer at 532 nm. Samples in 1 cm fluorescence cuvettes (QA-1000, Hellma Optik, Germany) were excited with 750 ps microjoule pulses generated by a microchip Nd:YAG laser (Pulselas-P-1064-FC, Alphalas GmbH, Germany) operating with 2–10 kHz repetition rate. To filter out first and third harmonics of laser radiation, 50 cm water filter and dichroic mirrors (BK7 series, Eksma Optics, Lithuania) were used. Near-infrared luminescence was measured perpendicularly to the excitation beam in a photon counting mode using a thermoelectric cooled NIR PMT module (model H10330-45, Hamamatsu, Japan) equipped with a 1100 nm cut-off filter and a narrow-band filter centered at 1270 nm (NB series, NDC Infrared Engineering LTD, UK). The data were collected using a computer-mounted PCI-board multichannel scaler (NanoHarp 250, PicoQuant GmbH, Germany). The data collection was synchronized with laser pulses using an ultrafast photodiode (UGP-300-SP, Alphalas GmbH) as a trigger. The data analysis, including first-order luminescence decay fitting by the Levenberg-Marquardt algorithm, was performed using a custom-written software. The quantum yield for singlet oxygen photogeneration by DMG was determined as relative to that of Rose Bengal by measuring initial intensities of the 1270 nm phosphorescence, extrapolated to zero time after laser pulses, as a function of laser pulse energy, which was varied by using a series of optical filters with different transmission at 532 nm.

Porphyrin photosensitizing activity in D_2O was evaluated based on the kinetics of the decrease in oxygen concentration (oxygen consumption). The effective rate constants of histidine photooxidation k_{HIS} were determined in accordance with $k_{HIS} = AC_{O2}/C_{HIS}C_{ph}At$, where C_{ph} is the PPS concentration, C_{HIS} is the substrate concentration, and AC_{O2} is the change in the oxygen concentration for the time At. The error of the effective rate constant measurements was estimated as 10%.

EPR OXIMETRY

Photosensitized oxygen consumption in the presence of histidine was measured by EPR oximetry employing 4-protio-3-carbamoyl-2,2,5,5-tetraperdeuteromethyl-3-pyrroline-1-yloxy (mPCTPO) as a spin probe and using Bruker EMEX-AA spectrometer (Bruker BioSpin, Germany). The spectral parameters of the mPCTPO spin probe were calibrated for dissolved oxygen concentration in a water solution at room temperature.[32] EPR samples in flat quartz cells were irradiated in situ in the resonant cavity with yellow light derived from a Cermax PE300CE-13FM 300 W lamp in air-cooled housing (Perkin Elmer, USA) using a combination of filters (5 cm of aqueous solution of 5 g/L $CuSO_4$, green dichroic and a cut-off <500 nm filter). The sample irradiance, measured by a calibrated photodiode (Hamamatsu, Photonics, Japan), was in the range of 151–174 W/m^2.

Porphyrin photosensitizing activity in H_2O was evaluated based on the kinetics of the decrease in L,D-tryptophan concentration upon irradiation with a phototherapeutic LED apparatus by Polyronic Ltd. (X = 400 nm, power of 210 mW) in the presence of the porphyrin or its mixtures with the polymers at the temperatures of 23 ± 1°C. The effective rate constants of tryptophan photooxidation k_{TRP} were determined in accordance with $k_{TRP} = AC_{TRP}/C_{TRP}C_{ph}At$, where C_{ph} is the PPS concentration, C_{TRP} and AC_{TRP} are the substrate concentration and its change for the time At, respectively. UV-visible absorption spectra of the solutions were obtained using a Cary50 spectrophotometer (Varian, Austria), the error of measurements being 10%.

*H NMR spectra were recorded using a Bruker AVANCE III 500 MHz spectrometer. The samples solutions of dimegin, polymers, and PPS-AP systems (in mass ratio 1:1) in D_2O (Aldrich, 99 atom% D) were placed into standard ampoules (with the outer diameter of 5 mm). ^{1}H NMR spectra were obtained at a temperature of 20.5 0.1°C and the instrument working frequency of 500 MHz. The frequencies of NMR shifts were calibrated using the residual signals of the protons from the deuterated water (4.71 ppm). The ^{1}H NMR data interpretation was performed in accordance with the literature data,[33] ^{13}C NMR (working frequency of 125.8 MHz) data, and 2D spectra correlations 1H–1H COSY, 1H–1H NOESY, 1H–^{13}C HSQC, and 1H–^{13}C HMBC.

Size distributions of Pluronic micelles, porphyrines aggregates, and their mixtures were estimated with dynamic light scattering technique using PhotoCor goniometer (USA) equipped with He-Ne laser (10 mW,

633 nm). Autocorrelation functions of scattered light intensity were evaluated with 288-channel correlation device PhotoCor-SP at scattering angle 90°. Mathematical treatment of autocorrelation functions to obtain size distribution of scattering particles was performed using Dyna L.S. software.

ATOMIC FORCE MICROSCOPY DATA

Atomic force microscopy imaging of the prepared films was performed in the semicontact mode with a Solver P47 AFM instrument (NT-MDT, Russia). We used NSG10 probes (NT-MDT, Russia) with a nominal spring constant of 5.1 N/m, nominal resonant frequency of 150 kHz, and nominal tip radius of 10 nm. Typically, we obtained 6×6 μm images at a scan rate of 1 Hz, with a 512×512 pixels resolution. The images were flattened using the instrument built-in image processing software.

TPP and Pluronics were mixed in a chloroform solution with the porphyrin concentration of 1×10^{-4} M and the polymer concentration of 1.7×10^{-4} M. The porphyrin to Pluronic ratio $q = 0.6$ provided the optimal degree of solubilization, according to the previous studies.[5] For the AFM imaging, thin films of Pluronics (pure and containing TPP) on silicon substrates were prepared from the chloroform solution by spin coating at a rotation speed of 7000 rpm.

Photosensitizing activity of PD–AP complexes in the cell experiments was carried out in Malignant Mouse Fibroblast Cells NIH-3T3-EWS-FLI1. These cells simulate the behavior of Uing human sarcoma (granted by Dr Claude Malvy, Institute Gustave Roussy, Villejuif, France). These cells were made from normal cells with a gene EWS-FLI1 transfection.

Cells were cultivated in DMEM (Sigma) medium, which contained 10% newborn calf serum, 4 mM glutamine, 100 units/mL penicillin, 100 mkg/mL streptomycin, and 2.5 mkg/mL puromycin in CO_2-incubator "NAPCO" (USA) (37єC, 5% CO_2, and 95% humidity). The seeding density of the cells was 5×10^4 cell/mL, and they were reseeded when the density value was reduced to 0.4 billion/mL (usually, it happened in 3 days period). All cells were counted in the Goriachev chamber (microscope "Axiovert 25", "Zeiss", Germany).

For photocytotoxicity measurement experiments, PD and cell suspension (3×10^4 cell/mL) were seeded into 96-hole plane-table (Costar, USA).

Sterile solution of PD with polymer was charged into the plane-table holes after 24 h. Cells were incubated for 3 h and then illuminated with the laser Atkus-0,4 (by the closed corporation of the Scientific-Production Association of Space Instrument Making, Russia; k_{ex} = 660 nm, 65 mWt/cm^2) for 3 min. Later, the plane-table was again placed in the incubator and left undisturbed for 24 h. Each sample was measured in four identical probes.

PD–AP complexes for cell experiments were prepared by the following procedure. PD (10.0 g) was dissolved in 80 mL of water at a temperature of ~900°C, while stirring for 30–40 min. After cooling it to the room temperature, 20 mL of PD water solution with 5 mg/mL concentration was added to the mixture and stirred to complete the homogenization. This method of complex preparation was applied for all studied polymers.

PD–AP complex activity was defined as a ratio of survived cells. Phototoxicity of PD with or without the presence of polymer was determined by using a method based on the ability of mitochondrial ferments to reduce methyltetrazolium blue to formazan that have deep blue color. The number of viable cells was proportional to the quantity of the reduced formazan, which was determined using spectrophotometric analysis $(X_{ex}$ = 550 nm).

PDT therapy of gunshot wounds of rats was carried out in the following method. Animal purulent wounds were washed out with the topical antiseptic and were treated with 0.5% and 1% PD water solution or PD–AP (1% PD) gel drifted on gauze pad. Gauze pad was removed after 24 h of depositing, and wound surface was illuminated for 6 min with the low-energy semiconductor laser Atkus-2 $(X$ = 660 nm, power = 1 W/cm^2, and energy density = 50 J/cm^2). This treatment was carried out for three times. Each time a new gauze pad was used.

STUDY DESIGN BY EARLY STAGES OF THE WOUND HEALING IN RATS

In our experiments on rats, we replicated the standard planar full-thickness skin wound model.[34] The next day after wounding, one of the following six substances, isotonic sodium chloride solution, aqueous photoditazin solution, aqueous PVP solution, aqueous Pluronic F127 solution, photoditazin-PVP complex, and photoditazin Pluronic F127 complex, was injected into the wound bed of animals in the experimental groups. The wounds of the animals in the control group were left untreated. Fifteen minutes after the injection, the wounds were exposed to the laser radiation. The procedure was repeated 3 days after the wounding. Clinical observations were made during the 4 days following the 5 wounding, and then the animals were sacrificed, wound tissue fragments were excised, and sent for the morphological analysis to study the structure of the wound bed and specifics of inflammatory and reparative reactions in the different experimental groups. The histological characteristics of the wounds were evaluated semiquantitatively, and the results of the numerical score of the morphological features were analyzed statistically.

The experiment was performed on laboratory male albino rats with the body mass of 120–140 g. The experiment was approved by the Local Ethical Committee of the I.M. Sechenov First Moscow State Medical University. The animals were housed one per cage under standard animal housing conditions and given laboratory pelleted feed and unrestricted water. The standard planar full-thickness skin wound model was replicated as described earlier by Solovieva and Timashev[35]. A concentric (approximately 10 mm diameter) skin graft, complete with the subcutaneous fat down to the fascia proper, was excised from the preliminarily depilated dorsal skin in the interscapular area. A Teflon collar with a 3 mm wide flange at the bottom was inserted into the skin defect that had formed so that the flange was fit under the skin and the 10 mm high cylindrical portion of the collar projected above it. Standard-size wounds (300 mm^2 in area) were thus obtained in all the animals. The top of the collar was covered with a cellophane film to protect the wound from drying and extraneous contamination. The experimental wounding was performed under combined anesthesia with intramuscular injection of a zoletil solution (Zoletil 100, Virbac S.A., Italy) at a rate of 6.0 mg of zolazepam hypochloride per kilogram of the animal body mass and 0.5 mL/kg of a xylazine hydrochloride solution (Rometar, Spofa, Praha). The wound PDT sessions were conducted with an Atkus-4 semiconductor laser with the wavelength of 661 nm.

All the materials used for the PDT purposes were investigated with respect to their local effect on the wound surface. Depending on the active factor combination to be studied, the animals were arranged in the following seven groups of three animals each: group I – control, untreated; group II – isotonic sodium chloride solution + irradiation; group III – aqueous PD solution + irradiation; group IV – aqueous PVP solution + irradiation; group V – aqueous Pluronic F127 solution + irradiation; group VI – PD–PVP complex + irradiation; group VII – PD–Pluronic F127 complex + irradiation. Experimental groups I through V served as controls for the main experimental groups VI and VII. The control experiments were aimed at revealing the morphological characteristics of the early stage of the wound healing process taking place under the effect of each of the following PDT agents: laser radiation (group II) and its combinations with the aqueous PD solution (group III) and with each of the polymers used as a base/vehicle in the preparation of the PD–PVP (group IV) and PD–Pluronic F127 (group [VII]) complexes.

The treatment of the wounds was started on the next day after the wounding. The procedures included were as follows:

INJECTION OF THE SUBSTANCES UNDER STUDY INTO THE WOUNDS:

With the protective cellophane film remaining in place on top of the Teflon collar, 0.2–0.3 mL of the solution under study was injected with an insulin administration syringe into the soft tissues of the wound bed at 4–5 equally spaced points, so as a to have a total of 1.0 mL of the solution injected. When the injected solution contained PD or PD-amphiphilic polymer complexes, the wound with the collar was covered with a black cloth after the injections in order to exclude illumination with extraneous light.

LASER TREATMENT OF THE WOUNDS:

The procedure was started 15 min after the injections. The wound bed was irradiated with a defocused CW laser beam (with the light spot diameter of around 0.5–0.8 cm) by moving the light guide in scanning circles for 5 min. The radiation power was 11 mW, and the dose came to 1.45 J/cm^2. The procedure was repeated within 3 days after wounding. Thus, the total irradiation dose amounted to 2.9 J/cm^2. In 4 days after the wounding (the next day after the last irradiation session), the animals were sacrificed by an intraperitoneal injection of 2.0 mL of 25% magnesium sulfate solution. The bordering Teflon collars were removed, and the tissues filling the wounds were completely excised down to the superficial fascia.

PROCEDURE OF MORPHOLOGICAL ANALYSIS

Tissue samples intended for the morphological analysis were fixed for 3 days in a 10% neutral formalin solution, passed through a series of alcohols of gradually increasing concentration, and embedded in paraffin. Paraffin sections of 4–5 μm thickness were stained with hematoxylin and eosin and picrofuchsin by the Van Gieson method (to reveal collagen fibers). The microslides thus obtained were examined using an Olympus BX51 optical microscope (Olympus, Japan) equipped with a SDU-252 digital camera (Special Technik, Russia).

Examination of each microslide included evaluation of several characteristics on a 10-point scale, whereon a score of 0 represented the absence of a morphological feature and that of 10 represented the maximum intensity of the feature. We evaluated the intensity of inflammation (edema, exudation, cellular infiltration, and microcirculation injuries), hemorrhagic reaction (erythrocyte concentration in the exudate and tissue), and reparative processes (angiogenesis, fibroblast proliferation, and abundance and maturity of the granulation tissue).

STATISTICAL ANALYSIS OF THE RESULTS

The statistical analysis of the results was performed using the standard program package SPSS 13.0 for Windows. The intensities of morphological features in animals belonging to different groups were compared by means of nonparametric tests for several independent samples (the Kruskal-Wallis [KW] test) and by the way of pairwise comparison of the groups (the Mann–Whitney [MW] test), provided that statistically significant differences were revealed to exist at the stage of group comparison (the KW test). The significance level p of differences was taken at 0.05. In all cases, we used two-way tests.

RESULTS AND DISCUSSION

PORPHYRIN-POLYMER COMPLEXES IN MODEL PHOTOSENSITIZED PROCESSES

PLURONIC F127 EFFECT ON THE EFFICIENCY OF SINGLET OXYGEN PHOTOGENERATION IN THE PRESENCE OF DIMEGIN

As seen from Figure 4.2, the intensity I_A of $^1O_2$ phosphorescence (proportional to $^1O_2$ concentration in the system) in the presence of dimegin (5×10^{-6} M) increases upon introduction of Pluronic into the reaction mixture (illumination with $\lambda = 532$ nm).

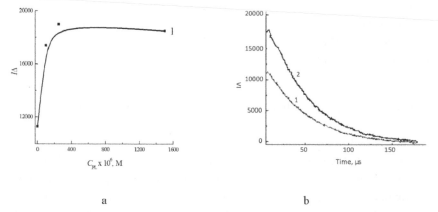

a   b

FIGURE 4.2   (a) Dependence of the Intensity I_A of Singlet Oxygen Phosphorescence ($\lambda = 1270$ nm) on the Pluronic F127 Concentration C_{PL} in D_2O in the Presence of Dimegin. (b) Singlet Oxygen Phosphorescence Kinetics. (1) in the Presence of 5×10^{-6} M of Dimegin; (2) in the Presence of 5×10^{-6} M of Dimegin and 1.5×10^{-3} M of Pluronic.

The growth of the singlet oxygen concentration in the presence of dimegin with Pluronic may be related to the process of dimegin disaggregation. Like all natural porphyrins with a planar spatial structure, dimegin forms tightly bound aggregates in water, which do not noticeably dissociate even at high concentrations of surfactants.[27,28]

For instance, it was shown by Solovieva et al and Zhientaev et al [27,28] that an average size of dimegin aggregates ($\sim 1 \times 10^{-6}$ M) in aqueous buffer solutions (PBS) was 50–100 nm and was only slightly reduced (up to ~50–

70 nm) in the presence of Pluronic F127 ($\sim 8 \times 10^{-5}$ M). Such aggregates of porphyrin molecules have a lower specific (per molecule) photoactivity.[27,35] This is due to the fact that for a "face-to-face" position of porphyrin rings' characteristic for the nonmetallic porphyrins association, an energy transfer can occur between the macrocycles, resulting in the vibration deactivation of the ^{3}PPS* triplet state and the corresponding decrease in the photosensitizing efficiency of porphyrins.[28]

Indeed, it has been shown that introduction of Pluronic increases the quantum yield Φ_Δ of $^1O_2$ photogeneration in the presence of dimegin by 20% ($\Phi_\Delta = 0.55$ in the presence of DMG and $\Phi_\Delta = 0.67$ in the presence DMG with Pluronic F127). Since the intensity of $^1O_2$ phosphorescence does not depend on the Pluronic concentration at the DMG:F127 molar ratio of 1:50 and higher, as shown in Figure 4.2a, the Φ_Δ measurements were performed at the corresponding concentrations of the components. .

The analysis of time dependencies of I_Δ (decay phase of phosphorescence) has shown that introduction of Pluronic did not affect the $^1O_2$ lifetime τ_Δ (Fig. 4.2b). For example, τ_Δ in D$_2$O in the presence of dimegin (curve 1) is 55 μs, in agreement with the data reported elsewhere,[36] while in the presence of dimegin and Pluronic, it is $\sim$50 μs independently of the F127 concentration (curve 2). This fact obviously indicates that Pluronic does not quench singlet oxygen at the concentration tested.

INFLUENCE OF THE NATURE OF APS ON THE PHOTOSENSITIZING ACTIVITY OF DIMEGIN IN AQUEOUS MEDIUM

The increase of the singlet oxygen concentration upon photoexcitation of dimegin in the solution containing Pluronic obviously accounts for the observed growth of the rate of histidine photooxidation in D$_2$O (Fig. 4.3a). Introduction of Pluronic into the dimegin solution in D$_2$O elevates the effective rate constants of histidine photooxidation k_{HIS} by $\sim$2.3 times. The rate constant of histidine photooxidation rises sharply at lower concentrations, while at higher concentrations, k_{HIS} increases only insignificantly, which is in a good agreement with the data in Figure 4.2a, where essential rise of I_Δ, that is, of the singlet oxygen concentration, is observed only at lower concentrations of Pluronic.

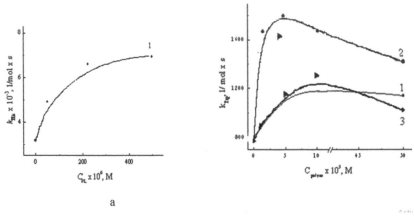

FIGURE 4.3 Dependences of the Effective Rate Constants of DMG-Sensitized Photooxidation of (a) Histidine (k_{HIS}) in D_2O and (b) Tryptophan (k_{TRP}) in Water on the Concentration of (1) F127, (2) PVP, and (3) PEO. The Dimegin Concentration is (a) 2.5 Ч 10^{-6} M and (b) 5 Ч 10^{-6} M.

It is noteworthy that if sodium azide (a physical quencher of singlet oxygen) is added to the dimegin–histidine system in D_2O, then the reaction of the substrate photooxidation almost stops. This fact is obviously an additional evidence (besides the $^1O_2$ phosphorescence measured at 1270 nm) for the singlet oxygen $^1O_2$ formation in the systems under study.[37]

A similar effect of the increase of the rate of dimegin-photosensitized tryptophan photooxidation was observed in water in the presence of Pluronic, PVP, and PEO (Fig. 4.3b). As seen from Figure 4.3b, the highest effect (increase of the rate constant k_{TRP} by ~2.1 times) was found in the presence of PVP. The increase in the efficiency of singlet oxygen generation and in the PPS photocatalytic activity in the presence of the polymers in D_2O and H_2O may be related to the porphyrin disaggregation resulted from the interactions of PPS with AP. PVP appears to affect the PPS photoactivity and, hence, the size of the porphyrin aggregates to the greatest extent. Previously, we observed a high activity of hydrophobic PPS solubilized with Pluronics in the oxidation of tryptophan.[38]

PVP containing both hydrophobic (vinyl) and hydrophilic (pyrrolidone) fragments is known to form hydrogen bonds and complexes with a number of aromatic compounds—amines, dyes bearing a negative charge.[39] Since the studied porphyrin molecules contain hydrophobic porphyrin core and hydrophilic peripheral oxygen-containing anionic groups, it can be expected that the interaction of PPS with PVP proceeds most efficiently. It is also known that hydrophobic interactions determine the solubilization of low-molecular weight substances by Pluronics.[27,28,40] One can assume that the PPS binding to the other APs is also accounted for by hydrophobic interactions. However, the interaction with the vinyl fragments in PVP is apparently more efficient than that with the propyleneoxide or methylene fragments of Pluronic or PEO. Besides, as mentioned before, in the case of PVP, not only binding to the hydrophobic fragments but also binding to the hydrophilic (pyrrolidone) fragments is possible.[39] The value of 1:10 of the porphyrin/polymer molar ratio corresponding to the highest k_{TRP} also points out to a stronger binding in the DMG–PVP systems (Fig. 4.3b). This value is two times greater than the molar ratios (~1:20) corresponding to the highest effect of porphyrin "disaggregation" for the dimegin–Pluronic and dimegin–PEO systems. Thus, the more efficient interaction of the photosensitizer with the polymer in the dimegin–PVP system leads to a more intensive disaggregation and to enhancement of the porphyrin photoactivity. Previously, we saw the same effect of increasing the activity of tetrephenylporphyrins in photooxidation of cholesterol upon binding to fluoropolymers.[41–43]

DETAILS OF PPS INTERACTIONS WITH APS

The interactions between the water-soluble porphyrin and APs' hydrophobic fragments have been detected by ^{1}H NMR spectroscopy. We studied ^{1}H NMR spectra of dimegin, polymers, and dimegin-AP systems in D_2O.[44,45] The chemical shifts of the proton signals are presented in Table 4.1.

TABLE 4.1   Signals of DMG, Polymers, and DMG–AP Systems Protons (ppm)

	DMG	DMG–F127	DMG–PVP	DMG–PEO
–CH-meso	10.43 (1H)	10.43	10.45	10.46
	9.87 (1H[a])	9.99	10.02[a]	9.94[a]
	8.93 (1H[a])	9.16	9.08[a]	9.05[a]
	7.25 (1H)	7.77	7.80[a]	7.48
–CH–CH$_3$	6.31 (2H)	6.25	6.29	6.34
–O–CH$_3$	3.59 (6H)	3.57	3.59	3.63
	3.81 (3H)	3.74	3.78	3.83

−CH₃-*cycle*	3.59 (3H)	1	3.56	3.62
−CH₂−CH₂−	3.28 (6H[a]) 2.65 (4H[a])	> 2.79[a]	About L 2.8–3.4[a]	3.33[a] 2.78[a]
	2.37 (2H)	2.49	2.51	2.44
	2.09 (2H)	2.26	2.25	2.18
−CH−CH₃	2.50 (6H) F-127	2.42	2.42	2.51
EO−CH₂	3.65	3.52	–	–
PO−CH₂−/−CH−	3.43–3.59	3.52	–	–
PO−CH₃	1.11 PEO	0.61	–	–
−CH₂−CH₂−	3.66 PVP	–	–	3.57
−CH₂-*vinyl*	1.67 1.52	–	«1.5[a]	–
−CH-*vinyl*	3.72 3.59	–	«3.5[a]	–
−CH₂ (2) *pyrrolidone*	2.39	–	11.9–2.3[a]	–
	2.27 2.25			
−CH₂ (3) *pyrrolidone*	1.96	–	«1.9[a]	–
−CH₂ (4) *pyrrolidone*	3.24	–	2.9–3.2[a]	–

Notes: [a]Signal broadening.
Abbreviations: DMG, Dimegin; AP, amphiphilic polymer; PVP, polyvinylpyrrolidone; PEO, polyethyleneoxide

It was shown that in all the cases, a downfield shift of the porphyrin cycle meso-protons and propionic acid residues was observed in the¹H NMR spectra of DMG–AP systems, as well as an upfield shift of the proton signals of the polymers' hydrophobic fragments (as compared with the signal positions in the spectra of initial components). These findings indicate disaggregation of initially associated DMG molecules in water and DMG binding to polymer. Note that the ¹H NMR spectrum of initial dimegin contained four signals (singlets) of the meso-protons in the range of ~10.5–7.2 ppm and two signals of the methyl group protons of the porphyrin ring at 3.59 ppm and 3.28 ppm. Three signals at 2.09–2.65 ppm corresponded to the protons of the peripheral methyl groups of the propionic acid residues in the DMG molecule.

In the¹H NMR spectra of Pluronic F127, PEO, and PVP, [30,33] we observed the signals of the protons of the methine (CH–) and methylene (CH₂) groups in the range of 3.43–3.65 ppm. Besides, a signal (triplet) of the methyl group protons of the hydrophobic propyleneoxide block was seen for the Pluronic F127 in the strong field (1.11 ppm), while in the

PVP spectra, we observed the signals of the methylene group protons of the hydrophilic pyrrolidone fragments in the range of 1.96–2.39 ppm and at 3.24 ppm.

As mentioned before, a downfield shift of the signals of the porphyrin cycle meso-protons and of the protons of $-CH_2-CH_2-COON$ groups was found in the ^{1}H NMR spectra of the system DMG–AP, which apparently indicates disaggregation of dimegin.[30] At the same time, in the ^{1}H NMR spectrum of the DMG complex with F127, the greatest variations, namely, broadening and upfield shift up to 0.61 ppm, were observed for the signal of the methyl group protons of the polypropyleneoxide block. Such changes testify the presence of interactions between the porphyrin and the hydrophobic block of the polymer. This conclusion is confirmed by the results obtained earlier, having shown that, in the process of solubilization, DMG localizes in the hydrophobic core of Pluronic micelles.[27,28] A similar upfield shift was observed for the signals of the methylene group protons of PEO in the ^{1}H NMR spectra of the DMG complex with PEO. In the ^{1}H NMR spectrum of PVP, the greatest variations were detected for the signals of the methylene group protons of the vinyl fragment: the signals in the range of 1.52–1.67 ppm degenerated into one signal at 1.5 ppm. This fact pointed out the interaction of PPS mainly with the hydrophobic fragments of the polymer macromolecules. However, besides the mentioned shifts in the spectrum of the DMG–PVP systems, we observed also a shift of the methylene group's signal of the pyrrolidone ring, which obviously indicates the presence of the PPS interactions with the hydrophilic fragments of macromolecules, as well. The obtained data correlated well with the above-noted fact of the highest activity of the PPS–PVP systems in the tryptophan photooxidation compared with that of the PPS complexes with other APs.

The PPS disaggregation and interaction with the polymers were also indicated by the changes in the UV-Vis absorption and fluorescence spectra of dimegin in the presence of AP.[29,36] Indeed, in the presence of AP, we observed mostly bathochromic shifts (by 5–10 nm) and increase in the intensity of bands (by 10–70%) in the UV-Vis absorption and fluorescence spectra of dimegin.[29] It should be noted that the concentrations of the polymers, at which the shifts and intensity increase were observed, depended on the nature of the polymer. For example, PVP affects the intensity of the bands and their positions in the UV-Vis spectra of DMG to the greatest extent (at the porphyrin concentration of 5×10^{-6} M), causing the bands (Soret, I and II) shift by 10 nm and the growth in their intensity by 50–80% already at the polymer concentration of 1×10^{-5} M. In the presence of Pluronic at the concentration of 1×10^{-5} M, we observed only a Soret band shift (5 nm); the intensity grew by 20%–50%. With the further increase of Pluronic concentration, the changes in the UV-Vis spectrum of PPS are similar to those found in the presence of PVP. In the presence of PEO (at the polymer concentration of 1×10^{-5} M), the changes in the UV-Vis spectrum of dimegin were similar to those found in the presence of Pluronic at the same concentration, but increasing the polymer concentration resulted only in a higher intensity growth. These results confirm the above-mentioned data on the stronger DMG binding to PVP compared with PEO and Pluronic F127.

Study of Pluronic–Porphyrins Interaction Using DLS and AFM Technique

Aggregation of porphyrins in water solution is mainly due to hydrophobic and stacking interactions between porphyrins molecules. Therefore, it is more pronounced for highly hydrophobic TPP (Fig. 4.4a) in comparison with DMG containing two anionic groups (Fig. 4.4b).[27] In both cases, aggregates of about 100 nm were observed, and TPP aggregates were characterized by the broader size distribution (Fig. 4.4a) than DMN aggregates (Fig. 4.4b). No light scattering particles were observed in water solutions (from 1×10^{-6} mole/L to 1×10^{-4} mole/L) of even more hydrophilic PD, containing three anionic groups (data not shown). Size of Pluronic micelles was found to be about 7 nm and 11 nm for P85 (Fig. 4.4c) and F127 (Fig. 4.4d) micelles, respectively. Size distribution of the micelles was rather narrow in both cases. These data are consistent with the previously published results obtained by pulse-NMR technique and small-angle neutron scattering.

Addition of Pluronics to the porphyrines induced considerable decrease in the size of particles observed in the solution. Thus, Pluronic P85 caused decrease in the size of TPP aggregates from ~100 nm to 10 nm that was comparable with the size of undisturbed Pluronic micelles. Similarly, addition of Pluronic F127 decreased the size of DMN aggregates from 80 nm to 10 nm. However, in both cases, considerable broadening of size distribution in comparison with empty micelles was observed. Hydrophilic PD did not form aggregates in solution and did not alter size distribution of Pluronic F127 micelles.

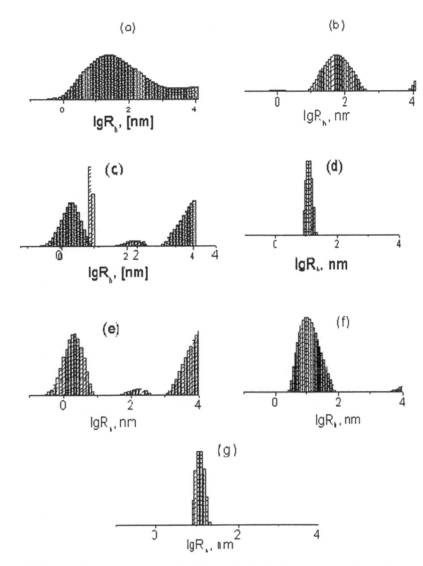

FIGURE 4.4 Size Distribution of (a) TPP and (b) DMN Aggregates (1×10^{-6} mole/L); (c) Pluronic P85 (1.1×10^{-5} mole/L) and (d) Pluronic F127 (7.9×10^{-5} mole/L) Micelles; (e) TPP/Pluronic P85 Mixture, (f) DMN/Pluronic F127 Mixture, and (g) PD/Pluronic F127 Mixture. Buffer PBS, 37°C.

Thus, the question "why PD photocatalytic activity increases in the presence of Pluronics, while no aggregates are detected in its water solution?" arises. We suppose that the observed effect is mainly due to the concentration of photocatalyst and the substrate within micellar volume. Previously, we have shown that L-tryptophan did not solubilize in Pluronic micelles; therefore, the "micellar catalysis" mechanism seems to be very realistic.[28]

So, we found that either water-insoluble porphyrins (*meso*-TPP) and its derivatives, protoporphyrin IX derivatives) or hydrophilic porphyrins dimegin and PD can be efficiently solubilized in the Pluronic micelles, a process that is accompanied by a marked improvement in the catalytic activity of photosensitizers.[27,28] Although only water-soluble porphyrins are generally considered in the PDT drugs development, combination of photosensitizers with Pluronics gives an opportunity to consider very active but water-insoluble porphyrins (Fig. 4.5a, b) as potential PDT drugs. Solubilization of TPP by Pluronics increases its photocatalytic activity by a factor of 50–60.[28] An exceptionally high increase in the photocatalytic activity of TPP is primarily related to its dissolution: in the absence of a Pluronic, TPP produces only dispersion with a vanishingly low photocatalytic activity. Second, solubilization can reduce the sizes of porphyrin aggregates, which are characterized by a lower photocatalytic activity than that of individual molecules. The aggregates are formed in the aqueous medium and stabilized by the hydrophobic and π–π interactions of aromatic rings. The above-mentioned data of dynamic light scattering have shown that, even at 1×10^{-6} mol/L, TPP produces large associates; their dimensions vary from 200 nm to several microns. In contrast, only individual porphyrin molecules or small aggregates are contained within a Pluronic micelle, apparently inside the hydrophobic core.

Finding the characteristic features of PPS–AP interactions is important in order to expand the spectrum of potential PPS–AP systems and to develop novel pharmaceutical formulations. It is reasonable to suppose that porphyrines also influence the structure of Pluronic micelles. We used atomic force spectroscopy imaging to assess the details of the structures grown from Pluronics and Pluronic–TPP solutions on silicon substrates. We attribute the observed alterations in the Pluronic crystals to the formation of complexes between the polymer and TPP, which account for enhanced photocatalytic activity of such systems in the generation of singlet oxygen.[27,28,46,47]

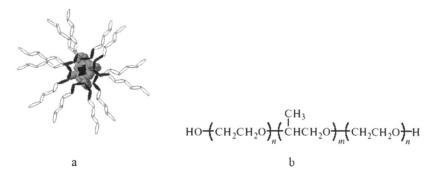

$$HO{-}(CH_2CH_2O)_n(\overset{\displaystyle CH_3}{\underset{\displaystyle |}{CH}}CH_2O)_m(CH_2CH_2O)_n{-}H$$

a   b

FIGURE 4.5   Structures of (a) A Pluronic Polymer, and (b) A Hypothetical Structure of A Pluronic Micelle Containing Solubilized Aggregates of TPP.

The spin-coating technique allows the preparation of ultrathin (5–20 nm) films of Pluronics with controllable thickness and shapes of polymer crystals, depending on the speed of the spinner and the polymer solution concentration. In the conditions used in this study, we observed the formation of similar dendritic structures (Fig. 4.6) for all the three Pluronics, despite their difference in the degree of crystallinity and hydrophobicity (defined by the ratio of EO to PO units). The film thicknesses measured 11–13 nm for F87 and F108 and 9–14 nm for F127.[47]

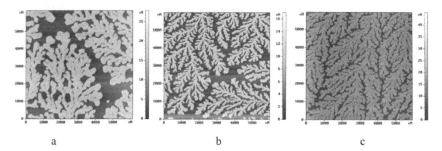

a   b   c

FIGURE 4.6   Typical Appearance of Thin PLURONIC Films Grown from 1×10^{-4} M Chloroform Solutions on Silicon Substrates: (a) Pluronic F87, (b) Pluronic F127, and (c) Pluronic F108.

The observed dendritic patterns of crystallization are characteristic for thin films of PEO[48] and may be attributed to crystallization of PEO units of a Pluronic. It has been shown[49] that in very thin (~15 nm) films, flat-on lamellae of the polymer branch to form dendritic structures. In such dendrit-ic structure, we observed that by crystallization under certain conditions, dewetting of the polymer may occur in thin films leading to the formation of separate "islands".[47,50] Indeed, occasionally, we observed crystallization of Pluronics in the form of "islands" with an elevated rim.

In the presence of TPP, the structure of thin films of Pluronics under-went marked changes (Fig. 4.7, a–c). For all the three Pluronics under study, particular convex structures appeared on top of the dendrites of TPP-containing Pluronics. Such convex structures typically grew to 20–50 nm in height, depending on the Pluronic type, while their diameters measured up to 1000 nm at the base.

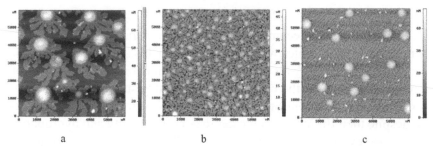

a   b   c

FIGURE 4.7   Effect of TPP on the Growth of Pluronic films. (a) F87, (b) F127, and (c) F108. The Concentration of TPP is 1×10^{-4} M, and the Pluronics' Concentration is 1.7×10^{-4} M.

The biggest "lumps" were usually observed in the case of F87 (Fig. 4.7 and 6a). The convex structures in F127 (Fig. 4.7b) were typically rather small but much more abundant as compared with those in the other two Pluronics. Various analyses of individual convex particles reveal that they consist of multiple flat layers positioned on top of each other (Fig. 4.8, a–b). High resolution topography and phase imaging shows that while the convex particles are apparently formed by the same polymer lamellae as the underlying dendrites, they have a distinct manner of multilayer pack-ing and an invariably rough top surface.

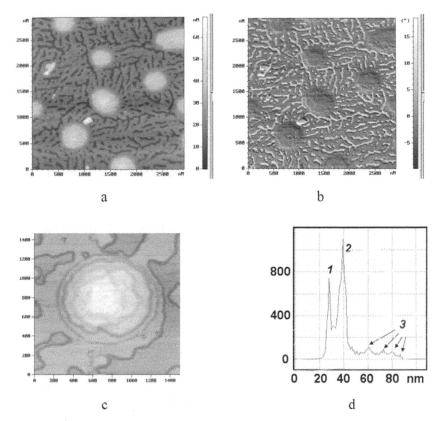

FIGURE 4.8   Convex Particles Formed in the Presence of TPP Consist of Multiple Flat Layers Located on Top of Each Other. (a and b) High Resolution Topography and Phase Images of a F108-TPP Film, (c) Quasi-3D image of an Individual Particle in a F87-TPP Film, (d) Height Histogram Taken Around An Individual Particle. (*1*) Substrate, (*2*) Film (Dendrite) Surface, (*3*) Steps of Multiple Flat Layers Over the Main (Dendrite) Surface.

In Figure 4.8b, a close-up of an individual particle's topography is presented in a coloring regime with gradient lighting (quasi-3D image). At least four flat layers are seen over the basic (dendrite) film level, and the top surface differs from the other layers by increased roughness. Obviously, the polymer packing in the last finishing layer proceeds in a more disordered manner, leaving loose polymer chains on top of the flat crystal and giving rise to the observed rough structure of the top layer. The cross-sections of the convex structures show a stepwise character of the height change, which can be seen even better if a histogram of heights within the area of an individual "lump" is calculated (Fig. 4.8c). The observed steps have a typical height of 10–13 nm, suggesting that they are formed by the same polymer lamellae as those in the dendrites. In some cases, though, we detected thinner (4–5 nm) or thicker (20 nm) steps between the layers in the multilayered convex structures. Interestingly, in the case of relatively small particles in the Pluronic F127, the thickness of the layers was the same as that in the other Pluronics, while the number of layers was diminished to two to three, or even only one flat layer was seen above the surface of the dendrites.

The mechanism of the formation of a thin film for PEO includes adsorption of a monolayer followed by the growth of flat-on lamellae from random nucleation sites.[51] Since Pluronics contain rather long PEO fragments ($\approx$22 nm for F87, $\approx$36 nm for F127, and $\approx$47 nm for F108), their crystallization apparently proceeds by the same mechanism. On hydrophilic surfaces, Pluronics adhere through their PEO units in a pancake manner, in contrast to hydrophobic surfaces which cause them to adopt a brush-like conformation.[52] The silicon wafers we used had a contact angle for water of ~40°, characteristic for silicon oxide, so the first adsorbed polymer layer was formed by extended polymer chains tightly bound to the surface via their hydrophilic PEO ends. The following crystallization of PEO fragments proceeds through the formation of flat-on lamellae, which grow with branching or seaweed-like tip-splitting,[51] resulting in the large dendritic structures. The growth of dendritic structures of a uniform height (10–13 nm) appears to be an almost exclusive way of Pluronics crystallization in a thin film. It should be noted that we observed same dendritic structures with different Pluronic concentrations, different substrates, and also during crystallization from aqueous solutions. Only very rarely we found single convex multilayered particles in the AFM images of pure Pluronics.

The formation of thick multilayered crystals can occur at a certain elevated temperature[53] but is apparently strictly unfavorable in the conditions of Pluronics crystallization from a chloroform solution at room temperature.

However, the presence of a relatively small number of TPP molecules (compared with the number of the polymer units – 275 units of F87, 442 units of F127, and 537 units of F108 per TPP molecule) causes a marked change in the growing pattern of the Pluronic crystals. The three-dimensional multilayered structures form in abundance in the presence of the porphyrin despite the unfavorable conditions for their growth. Apparently, such noticeable alteration of the local crystallization mechanism can be related to certain specific interactions between TPP molecules and macromolecules of a Pluronic. Porphyrins are capable of rather strong complexation with Pluronics, as seen from the studies of their partitioning coefficients (defined as the ratio of porphyrin concentrations in aqueous solution and in micelles).[28] In particular, the partitioning coefficient for TPP in F127 was found to be 3×10^4, a result which indicates a high affinity between the porphyrin and Pluronics. Porphyrin–Pluronic complexes have been recently directly observed in our group by NMR spectroscopy.[44]

The interactions most likely occur between the hydrophobic PPO parts of the amphiphilic Pluronic molecules and small TPP aggregates, which are always present in porphyrin solutions.[54] Both individual and aggregated TPP moieties were found within the hydrophobic layer of a Langmuir-Blodgett film prepared from an amphiphilic copolymer containing TPP pendants.[55] A direct spectroscopic evidence of ZnTPP aggregation in a polymer film prepared by spin coating from a solution has been obtained by O'Brien et al.[56]

We believe that the complexes between TPP and Pluronics create a micelle core upon their transfer from a solid film into the aqueous phase. High efficiency of TPP solubilization and small sizes of TPP aggregates account for the marked improvement of the porphyrin photocatalytic activity seen in the experiment.[28] The AFM imaging data have shown the same porphyrin-driven alterations in the structure of the three studied Pluronics, despite their differences in the molecular weight and degree of hydrophobicity. Thus, one could expect the same mechanism for the subsequent TPP solubilization by F87, F108, and F127. Indeed, we observed similar catalytic activity of TPP–F87, TPP–F108, and TPP–127 systems in the tryptophan photooxidation: the effective rates of oxidation were measured as 2.0×10^3, 1.9×10^3, and 1.9×10^3 L/(mol sec), respectively.

Thus, the collection of the presented data leads to a conclusion that in the presence of APs in aqueous media, a disintegration of the porphyrin molecular associates takes place, due to the hydrophobic interactions between AP and PPS. Such PPS disaggregation results in the enhancement of their photosensitizing activity in the processes of $^1O_2$ generation and oxidation of substrates. This enhancement of the PPS photosensitizing activity in vitro in the presence of APs may be a component of the earlier revealed effect of enhancement of PPS–AP complexes activity in vivo in the PDT in the treatment in rat's purulent and firearm wounds.[25–27,57,58]

PHOTOTOXICAL PROPERTIES OF PORPHYRIN–POLYMER COMPLEXES

We examined the phototoxic activity of photoditazin–polymer complexes in cancer cell culture treatment and by PDT of purulent wounds in laboratory rats. So we analyzed the effects of photoditazin and its complexes with APs, on the early stage of wound healing in a rat model. A skin excision wound model with prevented contraction was developed in male albino rats divided into groups according to the treatment mode. All animals received injections of one of the studied compositions into their wound beds and underwent low-intensity laser irradiation. The clinical monitoring and histological examination of the wounds were performed. It has been found that all the PD formulations have significant effects on the early stage of wound healing. The superposition of the inflammation and/or regeneration is the main difference between groups. The aqueous solution of PD alone induced a significant capillary hemorrhage, while its combinations with APs did not. The best clinical and morphological results were obtained for the PD–Pluronic F127 composition. Compositions of PD and APs, especially Pluronic F127, probably have a great potential for therapy of wounds. Their effects can be attributed to the increased regeneration and suppressed inflammatory changes at the early stages of repair.

Photosensitizing activity of PD–AP complexes was examined preliminarily in the cell experiments.

CELLULAR EXPERIMENTS

The polymers PVP, PEO, and Pluronic F127 were chosen to determine the influence of AP on PD phototoxicity in cell experiments. Figure 4.9 shows dependences of the experimental "survival" results for NIH-3T3-EWS-FLI1 cells after incubation in dark for 3 h with PD ($5Ч10^{-7}$ M) and subsequent illumination (with one exception – dependence [2] Fig. 4.1b) with semiconductor laser (A = 658 nm, 20 mW/cm², exposure time = 3 min) of the PD concentration. (The results were estimated in 3 h after illumination.)

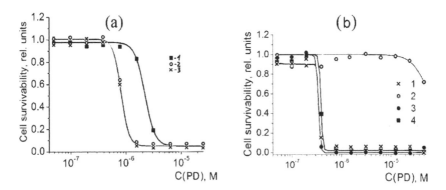

FIGURE 4.9 The NIH-3T3-EWS-FLI1 Cell Survivability Dependence of the PD Concentration After Incubation in Dark With or Without AP in Medium: (a) Without AP (1), with 0.1% PVP (2), and 0.1% PEO (3); (b) with 0.5% Pluronic F127 (1), 0.5% F127 in dark (2), 0.05% Pluronic F127 (3), and 0.1% F127 (4).

In accordance with Figure 4.9, the APs PVP, PEO (Fig. 4.8a, curves 2, 3), and Pluronic F 127 (Fig. 4.8b, curves 1, 3, 4) increase effectively the photoinductive toxicity of PD, and the Pluronic F 127 turn out to be the most active among all studded polymers. The activity of the PD–F127 complex did not depend on Pluronic concentration in solution. It is necessary to note that F127 and other APs do not increase the PD dark activity (Fig. 4.9b).

These results may be connected with the ability of AP to be bound with lipid membranes, as far as the cell membranes are considered as an important target of PDT treatment. The direct bonding of PD and membranes is caused by low lifetime of singlet oxygen ($\sim 10^{-6}$ c), which carries the photodynamic cellular damage.

PDT EXPERIMENTS IN HEALING WOUNDS IN RATS

Curing of soft tissues pyoinflammatory diseases remains a topical problem today in spite of significant advances of clinical medicine and pharmacology because drug treatment on different stages of wound process is not always successful. There are many antibacterial and wound healing agents, but they do not stimulate the processes of regeneration adequately due to the changes of suppurative complication disease-producing factors in modern time. Generally, anti-infective activity of preparations used is not sufficient. Of late years, for the therapy of wound formed after suppurative focus surgical d-bridement, the laser and PDT is applied. PDT is based on the photosensitization (PS) activation of molecular oxygen. Under the action of light, PS gets excited and transfers the molecular oxygen dissolved in the organism to its chemically active singlet state. The singlet oxygen oxidizes cell components and initiates a cascade of processes resulting in the death of the ill cells. Recently, PDT has been used for the treatment of festering wounds and trophic ulcers.[24,57,58] An important advantage of PDT is its ability to affect bacterial cultures that are resistant to antibiotics.

PD–AP COMPLEX BY PDT OF GUNSHOT WOUNDS IN RATS

The PD–AP complex efficiency was demonstrated in local gunshot wound treatment of laboratory rats (Table 4.2, Fig. 4.10). The data in Table 4.2 display the wound healing time values depending on the dosage of PD. The time taken for complete healing of the wound efficiently by PD–Pluronic gel (containing 1% wt PD with Pluronic F127) is higher (up to ~40%) compared with using PD corresponding solution.

TABLE 4.2 Pluronic F127 Effect in the Healing of Rat Burn Wounds

Animal Groups and Type of Treatment	Time of Healing (Days)		Acceleration of Wound Healing Regarding the Control Group (%)
	Granulation	Complete Healing of Wounds	
Control group (the use of antiseptic solutions)	15.2 ± 1.4	29.8 ± 0.8	–
First experimental group (1% PD solution)	11.3 ± 0.7	24.2 ± 0.2	18.9
Second experimental group (1% gel-complex PD – F127)	7.0 ± 0.5	17.2 ± 1.1	42.3

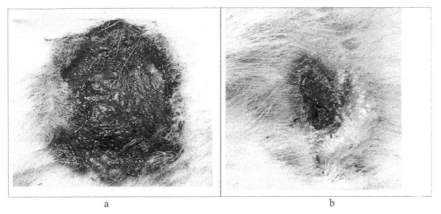

FIGURE 4.10 Rat's Gunshot (a) Wound After 15 Days Conventional Antiseptic Treatment; (b) The Wound After 15 Days of PDT Treatment with 1% gel PD-Pluronic F127.

The observed effects are illustrated with two photos presented at Figure 4.10: (a) wound after 15 days conventional antiseptic treatment, (b) wound after 15 days of PDT treatment with 1% gel PD-Pluronic F127.

EARLY STAGE OF THE WOUND HEALING PROCESS BY PDT USING PD–AP COMPLEXES

The early stage in the healing of wounds has been known to have a principal effect on the progress and results of the further recovery of the skin and determine in particular the probability of pathological scarring. It has recently been demonstrated that it is exactly at the initial stage of reparation that the positive effects of PDT manifest themselves most distinctly.[59–61] We believe that the comparative evaluation of the efficacy of the newly developed photosensitizers with APs on the basis of the results of investigations into morphological changes taking place in wound tissues at the initial stage of the healing process can be a valid groundwork for selecting photosensitizer modification strategies and developing PDT regimens. The objective of this work was an experimental morphological study of the specific features of the early stage of the wound healing process in the case of laser PDT using the photoditazin and its complexes with APs of different nature.

MACROSCOPIC EVALUATION OF THE CLINICAL PRESENTATION OF THE WOUND CONDITION

One day after the surgical procedure, the clinical condition of the wounds in all the animals under study was of the same type: the wounds were moderately moist, their bed was covered with a thin whitish fibrin layer; there was no apparent exudation, and the perifocal reaction in all cases was but weakly evident.

The next day after the first irradiation session, some differences in the condition were observed between the wounds, especially evident in groups III (aqueous PD solution + irradiation) and VI (PD–PVP complex + irradiation). Found in the wound collar cavity of two animals in each of these groups was a hemorrhagic exudate in amounts of 0.5–1.0 mL. The hemorrhage intensity and amount of the exudate were greater in the group III animals. In the rest of the animals of these groups, a small amount of the ordinary rose-tinted exudate was present in the wound collar cavity. In groups I (untreated), II (isotonic sodium chloride solution + irradiation), IV (aqueous PVP solution + irradiation), V (aqueous Pluronic F127 solution + irradiation), and VII (PD–Pluronic F127 complex + irradiation), the clinical presentation of the wounds practically did not change.

Prior to the second irradiation session (day 3 after the wounding, and day 2 after the first irradiation session), the exudative-hemorrhagic reaction in the animals of groups III and VI somewhat intensified: in those wounds where this reaction in the preceding observation was ambiguous, the amount of the exudate increased and its color became more marked. In four animals of these groups, the amount of the exudate was so substantial that it had to be extracted before the wounds could be irradiated. With the exudate removed, a cyanotic wound surface was exposed, with hemorrhage foci at the injection sites (punctures) of the preparations under study; no fibrin was present. The condition of the wounds in the rest of the groups (I, II, IV, V, and VII) was "quiet"; only a slight increase of the fibrin layer, especially at the collar walls, could be observed in some animals. An insignificant amount of a rose-tinted exudate was found in the wound collar cavities of the animals of group VII (PD–Pluronic F127 complex + irradiation). Not numerous petechial hemorrhages could be observed on the wound bed in one animal.

Four days after the wounding and two PDT sessions, the tissue layer excised in the animals of groups III and VI was thicker ($\approx$ 6–8 mm), gelatinous (edematous), and impregnated with a hemorrhagic exudate. In the control animals (groups I, II, IV, and V), the excised wound bed tissue layer was of ordinary density; it was dark pink in color and thinner than that in groups III and VI. The wound bed tissue layer in the animals of group VII (PD–Pluronic F127 complex + irradiation) was thicker, but less dense and edematous than that in groups III and VI.

RESULTS OF HISTOLOGICAL EXAMINATIONS

The histological examinations of the structure of the tissues developing in the wound bed, conducted within four days after the wounding and two PDT sessions, showed that these tissues in all the groups of animals had a similar structure. It consisted of three layers: a fibrinous-leukocytic layer (exudate), a layer of the granulation tissue, and a fatty tissue layer. However, the morphological characteristics of these layers in each of the groups had certain differences, apparently due to photobiological effects. The numerical scores of the intensity of these morphological features are listed in Table 4.3.

Predominant in the wounds in control groups I, II, IV, and V (untreated, isotonic sodium chloride solution + irradiation, aqueous PVP solution + irradiation, and aqueous Pluronic F127 solution + irradiation, respectively) were inflammatory processes: exudation, neutrophil infiltration, edema, blood vessel hyperemia, with stasis and erythrocytic sludge in the lumen, and erythrocyte diapedesis (signs of microcirculation injuries). In group I, the developing granulation tissue did not yet form a continuous layer but was represented by individual foci. In control groups II, IV, and V, the granulation tissue occupied the entire wound surface, but its layer was relatively thin. The tissue itself was markedly immature. It consisted of proliferating spindle-shaped and polygonal fibroblasts, macrophages, lymphocytes, and few vessels. Neoangiogenesis was weakly expressed, and vertical capillaries were practically absent. Collagen fibers were thin and immature (bleak color in Van Gieson staining with picrofuchsin). In the animals of these groups, the reparative regeneration processes that manifest themselves in fibroblast proliferation and neoangiogenesis were much less expressed than the signs of inflammation.

The most significant morphological features of the wound healing process were revealed in group III animals (aqueous PD solution irradiation).

The main distinctive feature was the perceptible accumulation of loose hemorrhagic exudate in the wound. This fact can probably be considered as the evidence that the permeability of vessels has increased in response to the PDT procedure combining the photobiological effects of laser radiation and PD. The structure and functional condition of granulation tissues in group III animals also featured a number of distinctions. The layer of granulations was thicker than that in control groups I, II, IV, and V. The tissue showed a greater maturity: the number of vascular elements therein was increased, which points to intensive neoangiogenesis. Separate vertical capillaries were observed in some areas. The granulation tissue consisted of fibroblasts and thin immature collagen fibers and contained an increased number of lymphocytes and macrophages. This testifies to the activation of cellular immunity reactions under the effect of PDT.

Thus, the wounds in the animals of control group III (aqueous PD solution + irradiation) show more pronounced regeneration processes (more abundant and somewhat more mature granulation tissues) and intensified neoangiogenesis and fibroblast proliferation. At the same time, the signs of inflammation and microcirculation injuries do not differ in intensity from those in the animals of the rest of the control groups. The hemorrhagic character of the exudate and tissue edema point to increased permeability of blood vessel walls.

TABLE 4.3 The 0–10 Point Numerical Scores for the Intensity of Morphological Alterations in the Animals Under Study

Group	Animal No.	Intensity of Morphological Alterations, Score		
		Inflammation	Hemorrhagic Reaction	Reparation
I	1	8	0	2
	2	8	1	2
	3	7	0	3
II	4	7	2	3
	5	7	2	3
	6	6	2	4
III	7	6	5	5
	8	7	6	4
	9	6	5	5

IV	10	6	2	3
	11	6	2	3
	12	7	2	1
V	13	6	2	3
	14	5	2	2
	15	6	2	3
VI	16	6	6	3
	17	3	4	5
	18	3	4	5
VII	19	3	2	5
	20	3	2	5
	21	3	2	6

In group VI (PD–PVP complex + irradiation), the exudate was of hemorrhagic character (Fig. 4.6), but the amount of hemorrhages was perceptibly smaller than in group III (aqueous PD solution + irradiation). At the same time, the granulation tissue occupied a greater volume in the structure of the wound bed. It was more mature and rich in vascular elements. Characteristic of this granulation tissue was also the presence of numerous proliferating fibroblasts and an increased number of macrophages, which points to activation of the reparative processes, and the signs of inflammation (edema, neutrophil infiltration, erythrocyte stasis, and sludge in capillaries) being less marked than in control groups I (untreated), II (isotonic sodium chloride solution + irradiation), and III (aqueous PD solution + irradiation).

The histological study of the wound tissues in group VII animals (PD–Pluronic F127 complex + irradiation) showed that, in contrast to the wounds in the animals of group V (PD–PVP complex + irradiation) and control group III (aqueous PD solution + irradiation), the concentration of erythrocytes in the fibrinous-leukocytic layer was minimal, which was in agreement with the macroscopic examination data. The granulation tissue layer was even thicker than that in the rest of the groups. It is important to note that the granulation tissue formed a continuous layer and extended over the entire wound surface, which was uncharacteristic of the control groups. This tissue was of a sufficiently mature character. Some capillaries assumed vertical orientation. As compared with the control groups, increased number of macrophages was observed. The number of neutro-phils and edema were less than those in control groups I and II and also in group VI.

Thus, the main distinctive feature of the early stage of the wound heal-ing process in the case of PDT using the PD–Pluronic F127 complex is the absence of the hemorrhagic exudate observed to occur in the wounds subject to PDT using the PD–PVP complex. Apparently, the complex of PD and Pluronic F127 reduces the harmful effect of the photosensitizer on the endothelium and basal membrane of capillaries and, as a consequence, prevents erythrocytes from egressing into the surrounding tissue. In addi-tion, inflammatory alterations, microcirculation injuries included, in the wounds in group VII animals were less marked. The signs of regeneration (maturing of the granulation tissue) were more pronounced than those in all the rest of the groups.

The results obtained bear witness to the fact that the photobiological properties of PD in the PD–Pluronic F127 complex remain unimpaired, while its histotoxicity, which in the case of PDT using an aqueous PD solution, manifests itself in the increase of the permeability of capillaries and development of hemorrhagic reactions decreases. Consequently, the use of the PD–Pluronic F127 complex in the PDT of wounds proves more effective than the original PD preparation at the early stage of the wound healing process.

The *statistical analysis of the intensity of morphological features in the groups of animals under study* by the group comparison (KW) proce-dure demonstrated that statistically significant relations existed between the kind of substance injected into the wound bed and the intensity of inflammation ($p = 0.014$), the hemorrhagic reaction ($p = 0.003$), and the reparative reaction ($p = 0.018$). This evaluation is based on the fact that the laser treatment conditions were the same for all the groups of animals under study. The results of the pairwise comparisons of the intensities of morphological features in the groups of animals under study (the relation between the mean ranks of values in the groups being compared) are pre-sented, along with the p values, in Table 4.4.

Injections of sodium chloride into the wound bed tissues, combined with their laser irradiation (group II), cause no statistically significant changes in the intensity of inflammation and hemorrhagic and reparative reactions as compared with group I (untreated).

The use of the aqueous PD solution (group III) for this purpose results in a statistically significant enhancement of the reparative reaction compared with group I (untreated). At the same time, it is accompanied by intensification of the hemorrhagic reaction in comparison with groups I, II, IV, V, and VII.

The morphological effects of injections of aqueous PVP (group IV) and Pluronic F127 (group V) solutions into the wound bed tissues, combined with the laser treatment, are statistically indistinguishable. At the same time, for both these polymers, no statistically significant changes were found in the intensity of the morphological features under study by comparison with group II (isotonic NaCl solution + irradiation). Compared with group I (untreated), injections of the aqueous PVP and Pluronic F127 solutions, combined with laser irradiation, somewhat increased the intensity of the hemorrhagic reaction. Since no statistically significant differences in the intensity of the morphological features existed between groups I and II, it can be assumed that the hemorrhagic reaction owed to the combined effect of laser radiation and injections. Inasmuch as the laser treatment regimen in group II was the same as in groups III through VII, the intensification of the hemorrhagic manifestations can be explained by the alteration of the osmotic balance in the wound tissues occurring against the background of the laser-induced dilation of blood vessels.

The use of the PD–PVP complex (group VI) was observed to cause a statistically significant decrease of inflammation as compared with group I (untreated). At the same time, the intensity of the hemorrhagic reaction increased by comparison with groups I, II, IV, and V. No statistically significant differences in the intensity of inflammation, hemorrhagic reaction, and reparative phenomena were found in comparison with group III (aqueous PD solution).

The injection of the PD–Pluronic F127 complex (group VII) into the wound bed tissue is characterized by a statistically significant decrease of the inflammatory manifestations as compared with groups I, II, III, IV, and V. Statistically significant increase in the reparative activity was observed in comparison with groups I, II, IV, and V. The hemorrhagic reaction was more pronounced than that in group I (untreated), but statistically significantly weakened as compared with groups III (aqueous PD solution + irradiation) and VI (PD–PVP complex + irradiation).

TABLE 4.4 Comparison of the Intensities of Morphological Features by Means of the Mann–Whitney Test

Group	Morphological Features[a]	II	III	IV	V	VI	VII
		NaCl, 0.9%	PD	PVP	Pluronic F127	PD–PVP complex	PD–Pluronic F127 Complex
I (untreated)	In	II < I, $p = 0.099$	III < I, $p = 0.068$	IV < I, $p=0.068$	V < I, $p = 0.043$	VI < I, $p = 0.043$	VII < I, $p = 0.034$
	H	II > I, $p = 0.099$	III > I, $p = 0.043$	IV > I, $p = 0.034$	V > I, $p = 0.034$	VI > I, $p = 0.043$	VII > I, $p = 0.034$
	R	II > I, $p = 0.099$	III > I, $p = 0.043$	IV > I, $p = 0.816$	V > I, $p = 0.456$	VI > I, $p = 0.068$	VII > I, $p = 0.043$
II	In		III < II, $p = 0.456$	IV < II, $p = 0.456$	V < II, $p = 0.099$	VI < II, $p = 0.068$	VII < II, $p = 0.034$
	H		III > II, p 0.034	IV = II, $p = 1.000$	V = II, $p = 1.000$	VI > II, $p = 0.034$	VII = II, $p = 1.000$
	R		III > II, $p = 0.068$	IV < II, $p = 0.197$	V < II, $p = 0.197$	VI > II, $p = 0.239$	VII > II $p = 0.043$
III	In			IV = III, $p = 1.000$	V < III, $p = 0.197$	VI < III, $p = 0.099$	VII < III, $p = 0.034$
	H			IV < III, p 0.034	V < III, $p = 0.034$	VI < III, $p = 0.361$	VII < III, $p = 0.034$
	R			IV < III, $p = 0.043$	V < III, $p = 0.043$	VI < III, $p = 0.796$	VII > III, $p = 0.197$
IV	In				V < IV, $p = 0.197$	VI < IV, $p = 0.099$	VII < IV, $p = 0.034$
	H				V = IV, $p = 1.000$	VI > IV, $p = 0.034$	VII = IV, $p = 1.000$
	R				V > IV, $p = 0.96$	VI > IV, $p = 0.099$	VII > IV, $p = 0.043$
V	In					VI < V, $p = 0.239$	VII < V, $p = 0.034$
	H					VI > V, $p = 0.034$	VII = V, $P = 1.000$
	R					VI > V, $p = 0.099$	VII > V, $p = 0.043$
VI	In						VII < VI, $p = 0.317$
	H						VII < VI, $p = 0.034$
	R						VII > VI, $p = 0.197$

Notes: [a]Morphological features: **In,** inflammation; **H,** hemorrhagic reaction, and **R,** reparative reaction. Statistically significant differences are shown in bold type.

In this part, experiments are described, related with carrying out clinicomorphological studies of the tissue reactions and specific features of the wound healing process in rats on day 4 after the wounding in the case of PDT of wounds with the use of the photosensitizer PD and its combinations with APs. The choice of this time period is nonrandom, for it is exactly during the course of this period that the transition takes place from the inflammatory to the proliferative phase of the wound healing process. The histological examination of the superposition of the inflammation and/or regeneration morphological features developing during this period under the effect of various extrinsic factors allows one to evaluate the activity of the wound healing process and the role of these factors, which either hamper or, conversely, stimulate its course.[62]

The comparative experimental morphological studies on the effect of PDT, with numerical scoring being made for the inflammatory, hemorrhagic, and reparative (regencrative) manifestations, helped us to reveal a number of specific features of the healing process in uncomplicated wounds. Since the laser treatment regimen was the same for all the groups of animals under study, the differences observed were conditioned by the choice of the substances used for photochemical treatment purposes (the compositions injected into the wound bed tissues).

When the isotonic sodium chloride solution was injected into the wound bed and the wound was thereafter exposed to laser radiation, the wound was dominated by inflammatory processes manifest in edema, neutrophil infiltration, and microcirculation injuries (hyperemia, erythrocyte stasis, sludge in the lumen of microvessels, and diapedetic hemorrhages). At the same time, the reparative–regenerative processes (fibroblast proliferation and neoangiogenesis) were but weakly expressed. The absence of the wound healing effect from a low-intensity laser light[61] is apparently explained by the twofold reduction (to 2.9 J/cm^2) of the total irradiation dose by comparison with that provided for by the wound treatment protocol described earlier for a similar model in animals.[61] Based on the fundamental principles of the photodynamic concept of the healing effect of low-intensity optical radiations [61], it can be suggested that to achieve photosensitized free-radical activation of the cells participating in the wound healing process, it is necessary either to increase the wound irradiation dose or to combine wound irradiation with agents catalyzing the photochemical reactions; that is, to use exogenous photosensitizers, for example, PD.

The aqueous solution of PD, combined with laser irradiation, caused perceptible activation of the reparative processes (neoangiogenesis, fibroblast proliferation, collagen fibrillogenesis, granulation tissue growth), as compared with the untreated control. This photosensitizer had no significant effect on the inflammatory processes. At the same time, it should specifically be noted that PD intensified hemorrhagic morphological features, which can be due to the increase in the permeability of blood vessel walls or can be due to the difference in osmotic potential between the PD solution and the tissue fluid. PD has been known to have the ability of accumulating predominantly in neovascular tissues [61], and granulation tissues are essentially this type. The increase of the permeability of blood vessel walls can supposedly be due to the damage caused to the endothelium and basal membrane of the newly formed vessels by the free radicals resulting from PDT. One of the ways to combat the undesirable toxic complications and improve the healing effect is to use photosensitizers in combination with APs.[27,63] It is desirable to create the photosensitizer–polymer complexes using such polymers that do not affect the bioavailability and photocatalytic activity of the chromophore and do not hamper the development of the reparative processes in the wound. In our investigations, we used PVP and Pluronic F127.

When in our experiments we used the amphiphilic polymer PVP or Pluronic F127 wound-bed injections (without PD) in combination with laser irradiation of the wounds, no statistically significant differences were found in the intensity of the inflammatory and hemorrhagic manifestations in the wounds by comparison with control group I. At the same time, a tendency toward weakening of the inflammatory and intensification of the reparative manifestations was observed in the case of Pluronic F127 application as compared with PVP. This fact allows us to state that the above polymers used in combination with laser irradiation do not inhibit the wound reparation process, and at the same time, they even somewhat activate it, as illustrated by the example of Pluronic F127.

When using the PD–Pluronic F127 complex in the PDT of wounds, we observed the most clearly pronounced improvement of the wound healing process at the early stage. The observed improvement involved decrease of inflammation, activation of reparation, and lowering of the intensity of the hemorrhagic reaction manifestations to the minimal values among all the combinations of the therapeutic factors studied (to the level characteristic of sodium chloride injections combined with laser irradiation). The comparative analysis of the intensities of the morphological features of wound healing showed that, when using the PD–PVP complex, the hemorrhagic reaction was manifest approximately to the same extent as in the case of aqueous PD solution. The PDT of wounds with the use of the PD–PVP complex had practically no effect on the intensity of inflammation and reparation.

CONCLUSIONS

We showed that APs, especially block copolymers of EO and PO, can increase photocvatalytic activity of porphyrins photosensitizers – dimegin and PD – an N-methylglucamine salt of chlorine e_6 both in model experiments in water solution and in biological systems. In fact, AP enhances phototoxicity of photoditazin in cell cultures and wound healing by PDT. This effect may be due to various reasons. Moreover, the mechanism of influence of AP on PD photocatalytic activity in water solution and in biological experiments may be quite different. We suppose that AP-induced activation of PD in water solution is mainly due to disaggregation of PPS associates by complex between PPS and AP formation. Probably such mechanism is acting in biological systems too. Besides, we expect that Pluronic and other AP may suppress reparative processes in the cells, for example, via inhibition of membrane ATPases located on cell plasma membranes or causing leakage of intracellular scavengers of singlet oxygen such as glutathione. Anyway, this subject requires further investigation. However, applicability of the observed effect in PDT of infected wounds has been proved herein and offers the challenges for wide investigation of polymer-based formulations in photodynamic therapy.

The observed clinicomorphological details of the early stage of the wound healing process under the PDT applying PD–AP complexes lead us to the conclusion that the polymer Pluronic F127 can weaken (and possibly eliminate) the adverse effect of PD on microvessels and prevent the development of local hemorrhagic reactions that complicate the wound healing process. It is also important to note that laser irradiation in combination with the PD–Pluronic F127 complex is conducive to the weakening of the inflammatory processes and intensification of the reparative ones by comparison with the PD–PVP complex. In this connection, the

PD–Pluronic F127 complex can be used as a photosensitizer for the PDT purposes.

Establishing the mechanisms of the interactions of the components within the PD–Pluronic F127 complex and regulation of those reactions during the light treatment process, as well as the study of the PDT effects dynamics with the use of this complex, will be the subject of our further investigations.

KEYWORDS

- **Photodynamic therapy**
- **porphyrin photosensitizers**
- **singlet oxygen**
- **photogeneration**
- **tryptophan**
- **histidine**
- **photooxidation**
- **photoditazin**
- **cancer cells**
- **purulent wounds**

REFERENCES

1. Dougherty, T. J.; Gomer, C. J.; Henderson, B. W.; Jori, S.; Kessel, D.; Korbelik, M.; et al. Photodynamic therapy. *J. Natl. Cancer Inst.* 1998, 90, 889.
2. Wilson, B. C.; Patterson, M. S. The physics, biophysics and technology of photodynamic therapy. *Phys. Med. Biol.* 2008, 53(9), 61.
3. Krasnovsky, A. A. Jr. *Singlet oxygen and primary mechanisms of photodynamic therapy and photodynamic diseases.* In *Photodynamic Therapy at the Cellular Level*; Uzdensky, A. B., Ed.; Research Signpost: Trivandrum, Kerala, India, 2007; p 17.
4. Chin, W. W.; Heng, P. W.; Thong, P. S.; Bhuvaneswari, R.; Hirt, W.; Kuenzel, S.; Soo, K. C.; Olivo, M. Improved formulation of photosensitizer chlorin e6 polyvinylpyrrolidone for fluorescence diagnostic imaging and photodynamic therapy of human cancer. *Eur. J. Pharm. Biopharm.* 2008, 69(3), 1083.

5. Bellnier, D. A.; Potter, W. R.; Vaughan, L. A.; Sitnik, T. M.; Parsons, J. C.; Greco, W. R.; et al. The validation of a new vascular damage assay for photodynamic therapy agents. *J. Photochem. Photobiol.* 1995, 2(5), 896.

6. Giri, U.; Sharma, S. D.; Abdulla, M.; Athar, M. Evidence that in-situ generated reactive oxygen species act as a potent stage-i tumor promoter in mouse skin. *Bio-chem. Biophys. Res. Commun.* 1995, 209, 698.

7. Lu, Z.-R.; Vaidya, F.Ye.A. Polymer platform for drug delivery and biomedical imaging. *J. Control Release.* 2007, 122(3), 269.

8. Chin, W. W.; Heng, P. W.; Olivo, M. Chlorin e6 - polyvinylpyrrolidone mediated photosensitization is effective against human non-small cell lung carcinoma compared to small cell lung carcinoma xenografts. *BMC Pharmacol.* 2007, 7, 1.

9. Isakau, H. A.; Parkhats, M. V.; Knyukshto, V. N.; Dzhagarov, B. M.; Petrov, E. P.; Petrov, P. T. Toward understanding the high PDT efficacy of chlorin e6-polyvinylpyrrolidone formulations: photophysical and molecular aspects of photosensitizer-polymer interaction in vitro. *J. Photochem. Photobiol., B.* 2008, 92(3), 165.

10. Kojima, C.; Toi, Y.; Harada, A.; Kono, K. Preparation of poly(ethylene glycol)-attached dendrimers encapsulating photosensitizers for application to photodynamic therapy. *Bioconjugate Chem.* 2007, 18(3), 663.

11. Regehly, M.; Greish, K.; Rancan, F.; Maeda, H.; Böhm, F.; Röder, B. Water-soluble polymer conjugates of ZnPP for photodynamic tumor therapy. *Bioconjugate Chem.* 2007, 18(2), 494.

12. Li, B.; Moriyama, E. H.; Li, F.; Jarvi, M. T.; Allen, C.; Wilson, B. C. Diblock copolymer micelles deliver hydrophobic protoporphyrin IX for photodynamic therapy. *J. Photochem. Photobiol.* 2007, 83(6), 1505.

13. Chin, W. W. L.; Heng, P. V. S.; Bhuvaneswari, R.; Lau, W. K. O.; Olivo, M. The potential application of chlorin e6 –polyvinylpyrrolidone formulation in photodynamic therapy. *Photochem. Photobiol. Sci.* 2006, 5, 1031.

14. Rakestraw, S. L.; Ford, W. E.; Tompkins, R. G.; Rodgers, M. A.; Thorpe, W. P.; Yarmush, M. L. Antibody targeted photolysis: In vitro immunological, photophysical and cytotoxic properties of monoclonal antibody-dextran-Sn(IV) chlorine e6 immunoconjugates. *Biotechnol. Prog.* 1992, 8(1), 30.

15. Allison, A. C.; Byars, N. E. An adjuvant formulation that selectively elicits the formation of antibodies of protective isotypes and of cell-mediated immunity. *J. Immunol. Methods.* 1986, 95, 157.

16. Lowe, K. C.; Armstrong, F. H. Oxygen-transport fluid based on perfluorochemicals: effects on liver biochemistry. *Adv. Exp. Med. Biol.* 1990, 277, 267.

17. Carr, M. E.; Rowers, R.; Jones, M. R. Effects of poloxamer 188 on the assembly, structure and dissolution of fibrin clots. *Thromb. Haemostasis.* 1991, 66, 565.

18. Justicz, A. G.; Farnsworth, W. F.; Soberman, M. S.; Tuvlin, M. V.; Bonner, G. D.; Hunter, R. L.; Martino-Saltzman, D. Reduction of myocardial infarct size by poloxamer 188 and mannitol in a canine model. *Am. Heart J.* 1991, 122, 671.

19. Sezgin, Z.; Yuksel, N.; Baykara, T. Investigation of pluronic and PEG-PE micelles as carriers of meso-tetraphenyl porphine for oral administration. *Int. J. Pharm.* 2007, 332(1–2), 161.

20. Bourre, L.; Thibaut, S.; Briffaud, A.; Lajat, Y.; Patrice, T. Formulation and character-ization of poloxamer 407: thermoreversible gel containing polymeric microparticles and hyaluronic acid. *Pharm. Res.* 2002, 45(2), 159.

21. Chowdhary, R. K.; Sharif, I.; Chansarkar, N.; Dolphin, D.; Ratkay, L.; Delaney, S.; Meadows, H. Correlation of photosensitizer delivery to lipoproteins and efficacy in tumor and arthritis mouse models; comparison of lipid-based and Pluronic P123 for-mulations. *J. Pharm. Pharm. Sci.* 2003, 6(2), 198.

22. Patent RF 2314806, 2007.

23. Patent RF 2396994, 2009.

24. Solovieva, A. B.; Tolstih, P. I.; Melik-Nubarov, N. S.; Zhientaev, T. M.; Kuleshov, I. G.; Glagolev, N. N.; et al. Combined laser and photodynamic treatment in extensive purulent wounds. *Laser Phys.* 2010, 20(5), 1068.

25. Patent RF 2457873, 2012.

26. Patent RF 2460555, 2012.

27. Solovieva, A. B.; Melik-Nubarov, N. S.; Zhiyentayev, T. M.; Tolstih, M. I.; Kuleshov, I. I.; Aksenova, N. A.; et al. Development of novel formulations for photodynamic therapy on the basis of amphiphilic polymers and porphyrin photosensitizers. pluronic influence on photocatalytic activity of porphyrins. *Laser Phys.* 2009, 19(4), 817.

28. Zhientaev, T. M.; Melik-Nubarov, N. S.; Litmanovich, E. A.; Aksenova, N. A.; Glagolev, N. N.; Solov'eva, A. B. Effect of the pluronic on catalytic activity of water soluble porphyrins. *J. Polym. Sci. A.* 2009, 51(5), 502.

29. Gorokh, Yu.A; Aksenova, N. A.; Solov'eva, A. B.; Ol'shevskaya, V. A.; Zaitsev, A. V.; Lagutina, M. A.; Luzgina, V. N.; Mironov, A. F.; Kalinin, V. N. Effect of amphiphilic polymers on the photocatalytic activity of watersoluble porphyrin sensitizers. *Russ. J. Phys. Chem.* 2011, 85(5), 871.

30. Solov'eva, A. B.; Aksenova, N. A.; Glagolev, N. N.; Ivanov, A. V.; Volkov, V. I.; Chernyak, A. V. Amphiphilic polymers in photodynamic therapy. *Russ. J. Phys. Chem. B.* 2012, 6(3), 433.

31. Aksenova, N. A.; Zhientaev, T. M.; Brilkina, A. A.; Dubasova, L. V.; Ivanov, A. B.; Timashev, P. S.; et al. Polymers as enhancers of photodynamic activity of chlorin pho-tosensitizers for photodynamic therapy. *Photonics Lasers Med.* 2013, 2(3), 189.

32. Panzella, L.; Szewczyk, G.; D'Ischia, M.; Napolitano, A.; Sarna, T. Zinc-induced structural effects enhance oxygen consumption and superoxide generation in synthetic pheomelanins on UVA/visible light irradiation. *Photochem. Photobiol.* 2010, 86(4), 757.

33. Babushkina, T. A.; Kirillova, G. V.; Ponomarev, G. V.^{1}H NMR spectroscopic study of hydrogen bonds and mobile NH protons in aqueous solutions of porphyrin ions. *Chem. Heterocycl. Compd.* 1998, 34(4), 474.

34. Moan, J. On diffusion length of singlet oxygen in cells and tissues. *J. Photochem. Photobiol. B Biol.* 1990, 6, 343.

35. Solovieva, A. B.; Timashev, S. F. Catalytic system on the base of immobilized porphy-rins and metalloporphyrins. *Russ. Chem. Rev.* 2003, 72(11), 965.

36. Solov'eva, A. B.; Zapadinskii, B. I.; Berlin, A. A. New technology of production of optically transparent polymer compositions containing capsulated nanosized aggre-gates of photoactive compounds. *J. Polym. Sci., Ser. D. Glues Seal. Mater. (in Rus.).* 2011, 4(3), 259.

37. Engl, R.; Kilger, R.; Maier, M.; Scherer, K.; Abels, Ch; Baumler, W. Singlet oxygen generation by 8-methoxypsoralen in deuterium oxide: relaxation rate constants and dependence of the generation efficacy on the oxygen partial pressure. *J. Phys. Chem. B.* 2002, 106, 5776.

38. Vakrat-Haglili, Y.; Weiner, L.; Brumfeld, V.; Brandis, A.; Salomon, Y.; Mcllroy, B.; et al. The microenvironment effect on the generation of reactive oxygen species by Pd-bacteriopheophorbide. *J. Amer. Chem. Soc.* 2005, 127, 6487.

39. Kirsh, Yu. E. *Poly-N-vinylpyrrolidone and Other Poly-N-vinylamides. "Nauka" ("Science", in Rus.)*; Publishing House: Moscow, 1998.

40. Bugrin, V. S.; Kozlov, M.Yu; Baskin, I. I.; Melik-Nubarov, N. S. Intermolecular interactions determine the solubilization in Pluronic micelles. *Russ. J. Polym. Sci., Ser. A.* 2007, 49, 463.

41. Solovieva, A. B.; Lukashova, E. A.; Vorobiev, A. V.; Timashev, S. F. Polymer sulfofluoride films as carriers for metalloporphyrin catalysts. *React. Polym.* 1991, 16(1), 9.

42. Belyaev, V. E.; Solovieva, A. B.; Glagolev, N. N.; Vstovsky, G. V.; Timashev, S. F.; Krivandin, A. B.; et al. Factors determining the photocatalytic activity of immobilized porphyrins in organic and water media. Current papers "Laser use in oncology III". *Proc. SPIE.* 2005, 5973, 148.

43. Glagolev, N. N.; Belyaev, V. E.; Rzheznikov, V. M.; Solov'eva, A. B.; Golubovskaya, L. E.; Kiryukhin, Yu.I; Luzgina, V. N. Effect of the nature of polymer matrix on the process of cholesterol photooxidation in the presence of immobilized tetraphenylporphyrins. *Russ. J. Phys. Chem. A.* 2009, 83(13), 168.

44. Aksenova, N. A ; Oles, T.; Sarna, T.; Glagolev, N. N.; Chernjak, A. V.; Volkov, V. I.; Kotova, S. L.; Melik-Nubarov, N. S.; Solovieva, A. B. Development of novel formulations for photodynamic therapy on the basis of amphiphilic polymers and porphyrin photosensitizers. Porphyrin-polymer complexes. in model photosensitized processes. *Laser Phys.* 2012, 22(10), 1642.

45. Aksenova, N. A.; Alexanyan, K. V.; Glagolev, N. N.; Solov'eva, A. B.; Rogovina, S. Z. The influence of the nature of the polymers on the processes of photo-oxidation catalyzed by porphyrins. *Pharm. Chem. J.* 2012, 40(7), 3.

46. Solovieva, A.; Vstovsky, G.; Kotova, S.; Glagolev, N.; Zav'yalov, B.; Belyaev, V.; Erina, N.; Timashev, P. The effect of porphyrin supramolecular structure on singlet oxygen photogeneration. *Micron.* 2005, 36(6), 508.

47. Kotova, S. L.; Timofeeva, V. A.; Belkova, G. V.; Aksenova, N. A.; Solovieva, A. B. Porphyrin effect on the surface morphology of amphiphilic polymers as observed by atomic force microscopy. *Micron.* 2012, 43, 445.

48. Hobb, J. K. Insights into polymer crystallization from in-situ atomic force microscopy. *Lect. Notes Phys.* 2007, 714, 373.

49. Chan, C.-M.; Li, L. Direct observation of the growth of lamellae and spherulites by AFM. *Adv. Polym. Sci.* 2005, 188, 1.

50. Sommer, J.-U.; Reiter, G. A. Generic model for growth and morphogenesis of polymer crystals in two dimensions. *Lect. Notes Phys.* 2003, 606, 153.

51. Mirsaidov, U.; Timashev, S. F.; Polyakov, Yu.S; Misurkin, P. I.; Musaev, I.; Polyakov, S. V. Analytical method for parametrizing the random profile components of nanosurfaces imaged by atomic force microscopy. *Analyst.* 2011, 136, 570.

52. Nejadnik, M. R.; Olsson, A. L. J.; Sharma, P. K.; van der Mei, H. C.; Norde, W.; Busscher, H. J. Adsorption of pluronic F-127 on surfaces with different hydrophobicities. *Langmuir*. 2009, 25(11), 6245.

53. Zhang, G.; Jin, L.; Ma, Zh; Zhai, X.; Yang, M.; Zheng, P.; Wang, W.; Wegner, G. Dendritic-to-faceted crystal pattern transition of ultrathin poly(ethyleneoxide) films. *J. Chem. Phys.* 2008, 129, 224708.

54. Ogi, T.; Ito, S. Migration and transfer of excitation energy in homogeneously dispersed porphyrin monolayers prepared from amphiphilic copolymers. *Thin Solid Films*. 2006, 500, 289.

55. Hunter, C. A.; Sanders, J. K. M. The Nature of π-π Interactions. *J. Am. Chem. Soc.* 1990, 112, 5525.

56. O'Brien, J. A.; Rallabandi, S.; Tripathy, U.; Paige, M. F.; Steer, R. P. Efficient S_2 state production in ZnTPP–PMMA thin films by triplet–triplet annihilation: Evidence of solute aggregation in photon upconversion systems. *Chem. Phys. Lett.* 2009, 475, 220.

57. Koraboev, Y. M.; Tolstih, P. I.; Geinits, A. V.; Urinov, A. I.; Uldasheva, N. E. *Photodynamical Therapy of Purulent Wounds and Trophic Ulcers*; Andzian: Minsdrav Uzbekistan, 2005; p 127.

58. Tolstih, P. I.; Derbenev, V. A.; Kyleshov, I. U.; Azimshoev, A. M.; Eleseenko, V. I.; Solovieva, A. B. Theoretical and practical aspects of laser photochemistry treatment of purulent wounds. *Biother. J. (in Rus.)*. 2008, 7(4), 20.

59. Meirelles, G. C.; Santos, J. N.; Chagas, P. O.; Moura, A. P.; Pinheiro, A. L. A comparative study of the effects of laser photobiomodulation on the healing of third-degree burns: a histological study in rats. *Photomed. Laser Surg.* 2008, 26, 159.

60. Garcia, V. G.; de Lima, M. A.; Okamoto, T.; Milanezi, L. A.; Junior, E. C.; Fernandez, L. A.; de Almeida, J. M.; Theodoro, L. H. Effect of photodynamic therapy on the healing of cutaneous third-degree-burn: histological study in rats. *Lasers Med. Sci.* 2010, 25, 221.

61. Rudenko, T. G.; Shekhter, A. B.; Guller, A. E.; Aksenova, N. A.; Glagolev, N. N.; Ivanov, A. V.; Abayants, R. K.; Kotova, S. L.; Solovieva, A. B. Specific features of the early stage of the wound healing process occurring against the background of photodynamic therapy using fotoditazin photosensitizer-amphiphilic polymer complexes. *Photochem. Photobiol.* 2012, 90, 1413–1422.

62. Shekhter, A. B. *Inflammation and regeneration.* In *Inflammation (in Rus.)*; Serov, V. V., Ed.; Meditsina: Moscow, 1995; p 200.

63. Fabris, C.; Valduga, G.; Miotto, G.; Borsetto, L.; Jori, G.; Garbisa, S.; Reddi, E. Photosensitization with zinc (II) phthalocyanine as a switch in the decision between apoptosis and necrosis. *Cancer Res.* 2001, 61(20), 7495.

CHAPTER 5

STRUCTURE AND PROPERTIES OF RUBBER POWDER AND ITS MATERIALS

E. V. PRUT, O. P. KUZNETSOVA, and D. V. SOLOMATIN

Semenov Institute of Chemical Physics of Russian Academy of Sciences, Moscow, Russia

Email: evprut@chph.ras.ru

CONTENTS

ABSTRACT

The treatments, technologies, and materials of rubber are presented in this review. The review gives a comparative analysis of process characteristics, rheological and mechanical properties, structural transformations, curing behavior, and diffusion processes taking place in unfilled and filled rubbers during their treatment by the method of high-temperature shear deformation (HTSD). Sulfur-cured rubbers are considered, along with details of degradation mechanisms of the rubber network by mechanical and thermal exposure. Properties of blends and composites of devulcanized and virgin rubbers are presented. Future directions and further development of the technology HTSD are discussed.

INTRODUCTION

New properties of macromolecules with important practical results have been discovered in polymer chemistry and physics. Basic advances reveal key relationships in the development of new materials that exhibit previously inconceivable combinations of properties. A principal difference between fundamental science and technologies is intent.

Macromolecular science covers a fascinating field of research, focused on the creation, understanding, and tailoring of materials formed out of very high molecular weight molecules. Such compounds are needed for a broad variety of important applications. Due to their high molar masses, macromolecules show particular properties not observed in any other class of materials.

The history of synthetic polymers is incredibly short. Considering the uneven rate of polymer consumption around the world, polymer production has the potential to increase tenfold by the mid-21st century. Polymers are the fastest growing structural materials. Nowadays, polymers are extensively used in different fields.

Many new multicomponent polymeric materials have been developed during the past two decades. Polymer compositions grow to take on a very significant role in the major application areas for polymers. This is connected with the significant expansion of the polymer applications with comparatively limited assortment of monomers. The combination of polymers using different methods provides one with a means of varying their properties. These combinations of polymers allow one to obtain polymer blends. Mixing of different polymers is an effective and economic way for production of new materials with desired properties. Performance of such multicomponent polymeric materials depends on their compatibility and morphology developed during the processing. The large numbers of scientific papers, industrial patents, scientific meetings, and exhibitions devoted to this class of materials are a sufficient witness to their strategic importance.[1-4]

Blends based on thermoplastics and rubbers occupy a prominent position among multicomponent polymeric materials. The mechanical properties of these materials can be changed in a wide range by varying the ratio of blend components. When the content of rubber is small (5–20%), blends behave as high-impact thermoplastics. If the rubber content varies from ~50% to 80%, thermoplastic elastomers (TPEs) can be prepared. Both classes of materials find wide application that requires detailed consideration of different factors, such as the influence of the chemical and molecular structures of the elastomer, the molecular-weight distribution of the properties of the finished products.

Toughened plastics are usually multiphase polymers; the dispersed phase consists of rubbery or thermoplastic domains, and the continuous phase is a thermoplastic polymer matrix. The basic reason for toughening a plastic via this route is to improve its toughness, or crack resistance, without significantly decreasing other important properties such as the modulus and creep resistance. Such toughened plastics that are widely used in many diverse industries form the basis for engineering plastics, structural adhesives, and matrices for fiber-composite material.[5]

The TPEs constitute a commercially relevant and fundamentally interesting class of polymeric materials. The emergence of TPEs provided a new horizon to the field of polymer science and technology. They combine the properties of cross-linked elastomers, such as impact resistance and low temperature flexibility, with the characteristics of thermoplastic materials, for example, the ease of processing. In general, TPEs are phase separated systems consisting of a hard phase, providing physical cross-links, and a soft phase, contributing to the elastomeric properties. In many cases, the phases are chemically linked by block or graft copolymerization. In other cases, a fine dispersion of the hard polymer within a matrix of the elastomer by blending also results in TPEs like behavior. Consequently, TPEs exhibit mechanical properties that are, in many ways, comparable to those of a vulcanized rubber, with the exception that the network and hence the properties of the TPEs are thermally reversible. This feature makes TPEs ideally suited for high-throughput thermoplastic processes, such as melt extrusion and injection molding. Mainly three classes of commercial TPEs can be distinguished: polystyrene-elastomer block copolymers, multiblock copolymers, and polymer-elastomer blends. Thus, TPEs combine the mechanical properties of rubbers at ordinary temperatures with processability above their melting temperatures inherent of linear thermoplastic polymers.

The properties of TPEs can be substantially improved by the method of dynamic vulcanization. The feature of this method is that rubber component is chemically cross-linked during the process of mixing. This gives rise to the heterogeneous structure in which the cross-linked elastomer particles of the order of 1–3 μm in size are dispersed in the continuous matrix of the thermoplastic component. TPEs prepared by dynamic vulcanization are identified as thermoplastic vulcanizates (TPVs). The characteristics of elastomeric phase can have significant impact on the mechanical properties of TPVs. At room temperature, the TPVs exhibit the mechanical properties typical of elastomers, and at temperatures above the melting temperature of thermoplastics, their rheological behavior is inherent to thermoplastic polymers. The processing behavior of TPVs is essentially influenced by the rheological properties of their melt. In order to reduce the viscosity and to improve the processability of TPVs, some modifiers (e.g., paraffinic oil) are added to the material.[6]

One of the various problems of the 21st century is the problem of waste disposal management.[7–9] A lot of waste rubber is produced every year in the world. The main sources of waste rubber products are discarded waste tires, rubber pipes, rubber belts, rubber shoes, edge scraps, and waste products that are produced in the rubber processes and others. Since polymeric materials do not decompose easily, disposal of waste polymers is a serious environmental problem. The three-dimensional cross-linked structure of waste rubber makes it infusible, insoluble, and difficult to recycling. Typical methods have been developed to treat the waste rubber: combustion, landfilling, biodegradation, and recycling. Among them, recycling is the most attractive method. Recycling is a major issue for most plastics processors and waste disposal authorities in the new century. However, the technology for recycling of rubbers is complex and costly.

The three-dimensional network of the thermoset polymer system must be broken down either through the cleavage of cross-links or through the carbon–carbon linkage of the chain backbone. As was mentioned by Tobolsky,[10] the C–C bond dissociation energy upon peroxide vulcanization is D_{C-C} = 390 kJ/mol, and in the case of vulcanization with sulfur-containing agents, the dissociation energy of monosulfide, disulfide, and polysulfide bonds is D_{C-S-C} = 210–251 kJ/mol, $D_{C-S-S-C}$ = 172 kJ/mol, and $D_{C-S-S-S}$ = 113 kJ/mol, respectively.

At present, the major effort in waste rubber recycling is to reuse it as a finely crumb rubber, produced by mechanical grinding. A lot of effort is put into the improvement of the quality of recycled rubber, reflecting, on the one hand, the high interest in rubber recycling, and on the other hand, indicating the difficulties of the recycling process. The processes used for grinding of rubber are based on cutting, shearing, or impact, depending on the equipment (knife, shredder, granulator, extruder, disk grinder, or impact mill) and the grinding conditions (ambient, wet, or cryogenic grinding). The choice of the process is based on the requirements for the final product, such as particle size and particle size distribution, morphology of the particles, and purity of the rubber powder.[7,9]

Waste rubber has been recycled and reused for a long time, through the application of various techniques involving size reduction by grinding, at either ambient or cryogenic temperature. Each of these grinding techniques produces crumb rubber of different particle size, particle-size distribution, and shape. Crumb rubber is added as a partial replacement of virgin rubber in various blends and composite. However, the amount of crumb rubber used in these compositions is limited, due to some loss in its physical properties. The search for better technologies that will allow larger quantities of used waste rubber to be incorporated into new products continues, and several new approaches have been successful.

One of the promising methods developed in the past two decades is the elastic deformation grinding technique (EDG). This technique is based on the degradation of a material in a complex strained state by the action of uniform compression pressure and shear forces under elevated temperatures. EDG makes it possible to obtain fine powders of comminuted rubbers, thus allowing the valuable properties of elastomer materials to be realized to a considerable extent.[11]

EFFECT OF JOINT ACTION OF HIGH PRESSURE AND SHEAR DEFORMATION ON PHYSICAL AND CHEMICAL PROPERTIES OF SOLID

The pioneer works by R. Bridgman[12] initiated the systematic studies on the high-pressure effect on physical and chemical properties of solid. This effect, being rather strong, ensured a sufficient improvement of the properties of materials treated in these conditions. In this respect, it became necessary to develop basically new technological processes.

Today, the advances in polymeric materials can be achieved through the development of totally new methods of mechanical treatment that will result in a sharp rise of productivity and quality of articles with a simultaneous reduction of metal consumption and specific power inputs of equipment. A successful implementation of this field is impossible without applying to the processing technology the latest achievements of polymer physics and chemistry, rheology, and high-pressure physics.

A highly effective method of controlling polymer properties under processing and improvement of articles' characteristics is an action of mechanical fields on the material causing the change in parameters of the micro- and macrostressed state of the system.[13]

In 1970, at the Semenov Institute of Chemical Physics of Academy of Sciences in Moscow, Russia, the scientists under the direction of the academician, Enikolopyan, studied the behavior of solids under high pressure and shear deformation.[14,15]

Possible procedures for physical modification polymer materials involve elastic-deformation grinding and plastic deformation at high pressure using Bridgman anvils.

Experiments under shear deformation and high pressure conditions were carried out on an anvil-type apparatus, which is based on the fact that the compression of a solid specimen between two plates did not lead to the flow out of the substance if a sufficiently thin specimen was taken. A version of this apparatus is shown in Figure 5.1. The device consists of a pair of metal anvils, having the form of a truncated cone and made of a special alloy. The bottom anvil is supported by a ball bearing, permitting rotation of this anvil relative to the top anvil. The sample is placed between polished working surfaces of the anvil. Shear deformation is created by rotation of the bottom anvil. Its magnitude is characterized by the rotation angle.

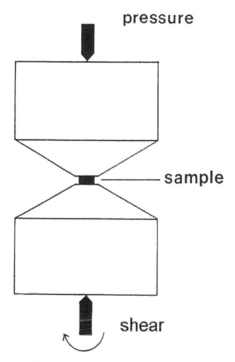

FIGURE 5.1 Bridgman Anvils.

The degradation of isotactic polypropylene (IPP) subjected to the joint action at high pressure and shear deformation on the Bridgman anvils was studied. It was shown that weight-average molecular weight (M_w) and number-average molecular weight (M_n) values decrease with shear deformation. The M_n value drops to higher extent with pressure compared with the value of M_w At 1 GPa, the M_w/M_n ratio is virtually independent of the shear angle, whereas at pressures of 2 GPa and 5 GPa, molecular-weight distribution becomes wider and, eventually, takes a bimodal character. The melting temperature was found to decrease linearly with decrease in M_n value.[16,17]

X-ray diffraction measurements of IPP/high-density polyethylene (HDPE) blend showed that intensive plastic deformations involved in the solid-state mixing and elastic deformation treatment lead to a breakage of the initial crystalline structure until reaching a grain size of 2–4 nm in the component representing the dispersed phase.[18–21]

The effect of intensive plastic deformation on the structure of thermodynamically incompatible polymer blends was studied on a mixture containing IPP and up to 10 wt% HDPE.[22] The mixing was carried out in the following regimes: (i) in the melt at 473 K in the absence of any significant shear deformation; (ii) in the course of elastic deformation of components in the rotor disperser under considerable shear deformation, whereby the mixing occurs during the crystallization of components at 409–411 K (Fig. 5.2); (iii) during the joint action at high pressure and shear deformation on the Bridgman anvils.

This unit is a rather simple apparatus consisting of a transport screw and a grinding head. The screw is 32 mm in diameter. The ratio of the length to diameter L/D is equal to 11. The extruder has two working zones with autonomic temperature control. The gradual compression of material (the degree of compression was as high as 3.5) and its heating are realized in the transport zone. The grinding head presents a cam rotor rotating inside a channeled cylinder, that is, the grinding of material is performed in a narrow annular gap under shear deformations, which were constant over the entire surface of the annular gap. The shear deformation level was determined by the product of the average shear rate and the residence time in the grinding zone. In this case, the stationary temperature regime in the head and extruder zones was maintained. Thus, in this unit, the pressure is produced by screw, and the necessary shear deformations are formed between the walls of grinding chamber and the rotating rotor.

The intensive shear deformations involved both in the solid-phase blending of the IPP/HDPE system and in the elastic deformation blending in the rotor disperser lead to the formation of blends with a homogeneous structure of the amorphous regions. The critical content at which this phenomenon is manifested depends on the magnitude of shear deformation. The formation of homogeneous amorphous regions in the blend leads to an increase in the elongation at break.[21]

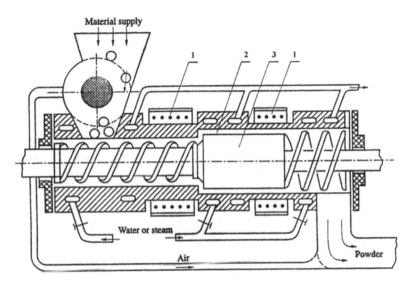

FIGURE 5.2 Schematic Representation of Rotor Disperser: (1) Heating Elements, (2) Concentric Transport Gap of the Dispersion Chamber, (3) Grinding Rotor.

ELASTIC-DEFORMATION GRINDING

The researchers of the Semenov Institute of Chemical Physics of Academy of Science of SSSR observed a formation of fine powder in a high-shear Banbury mixer when a low-density polyethylene (LDPE) was subjected to high pressure and shear accompanied by cooling (instead of heating). This discovery resulted in the development of a new grinding process that was initially termed elastic deformation grinding (EDG). This technique is based on the degradation of a material in a complex strained state by the action of uniform compression pressure and shear forces under elevated temperatures.[23,24]

Over the years, EDG was also termed elastic-strain powdering (ESP). When modified single- or twin-screw extruders were used, EDG was also referred to as extrusion grinding (EG), solid-state shear extrusion (SSSE), and solid-state shear pulverization (SSSP or S³P). In the more recent past, we used the term intensive stress action compression and shear (ISAC&S) or high-temperature shear deformation (HTSD).[7]

The extensive studies have provided a basic understanding of the grinding of polymers under high shear and compression. So, it has been shown that in the range of the melting of LDPE is observed drop in elongation at break, that is, a transition from ductile to brittle fracture. Temperature range depends essentially on the molecular-weight distribution and thermal history of the sample. It was found that on EDG in the melting temperature range, a highly dispersed LDPE powder is formed (Fig. 5.3).[25] The EDG method was characterized by rather low energy consumption, which was two to three times lower than that of conventional grinding methods.[26]

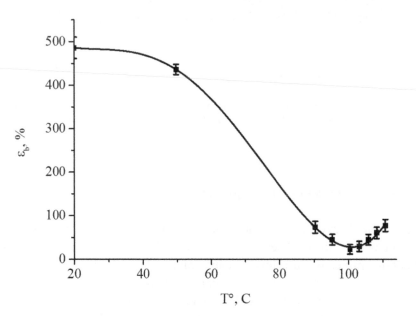

FIGURE 5.3   Elongation at Break ε_b of Low-Density Polyethylene as a Function of the Temperature.

Solid-state extrusion was successfully employed for rubber powder production.[27] In such materials, elastic deformation dominates over plastic deformation. The defects shaped as helical cracks arise at some critical value of accumulated rubberlike strain. Under extrusion of cross-linked elastomers, the extrudate is not destroyed at the initial stage of the extrusion draw ratio $\lambda < \lambda_{max}$ (Fig. 5.4a).[27] At increased strain, helical defects become more distinct, after which secondary destruction occurs in the form of small helical cracks branching from the main helices (Fig. 5.4b). Fracture in this case has multiple characters, and the extrudate is a loose agglomerate of small particles. Therefore, at complex loading, when deformation is less than its critical value corresponding to a particular temperature–time regime, the deformation is reversible, and there the critical value, accumulation of macroscopic defects takes place (Fig. 5.4c). This process is accompanied by a partial release of elastic energy. The remaining energy makes the material strained, which forms the basis for a further sharp increase of surface in the course of disintegration. Multiple fractures occur when there are no other ways to release the elastic energy except for the formation of a new surface. Surface area should increase as the number of elementary acts of fracture increases, that is, at higher deformation, lower values of critical deformation, and increased contribution of the rubberlike component to the total deformation.[28]

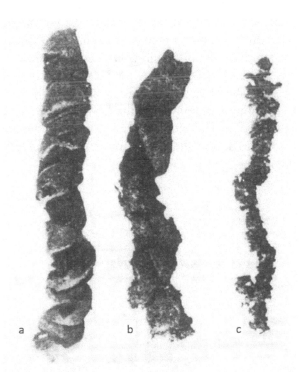

FIGURE 5.4 Characteristic Structures of Extrudate as a Function of Extrusion Draw Ratio λ.

Pavlovskii and coworkers[29] observed some peculiarities during grinding of used tires by the EDG method, where fracture occurred at elevated temperature under complex deformation of compression and shear. They used a modified co-rotating twin-screw extruder consisting of five zones capable of maintaining independent temperature of each zone during processing (Fig. 5.5).

Twin screw extruders can be used for the grinding of rubber. A grinding extruder is a twin-screw extruder equipped with two independent co-rotating screws with a set of grinding elements. Starting with a material having a granulometry of some millimeters, the final powder can have an average particle size of a few hundred micrometers.[7]

In the first zone, the material is subject to compression, and in the remaining zones, high shear deformations are realized with cam elements. The grinding of material occurs in gaps between the rotating faceted cam elements as well as in gaps between side and end faces of neighbor grinding elements located at different screws. As is shown, the best design of shear grinding elements is based on the combination of direct and reverse cams. Such design gives the optimum combination of shear deformations and the counter pressure effect. In this case, the material is densified as it passes through extruder. The screw rotation velocity, the rate of material input, and the grinding temperature influence the degree of grinding and the energy consumption. Therefore, in the second unit, the pressure is created by two screws, and the shear deformation itself is realized between the shear elements.

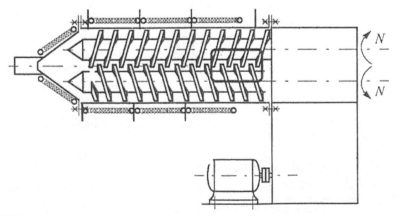

FIGURE 5.5   Schematic Representation of Twin-Screw Extruder.

Ground rubber tire (GRT) is a complex composite containing various elastomers, carbon black, zinc oxide, stearic acid, processing oils, and other curatives. For producing rubber powders by the HTSD method, the characterized vulcanizates of ethylene-propylene-diene rubbers (EPDM) with different cross-link density and paraffin oil contents were also used. This allowed one to study and analyze physicochemical processes occurring during the grinding of EPDM and to propose a mechanism for grinding. Moreover, the effect of the properties of EPDM vulcanizates on the characteristics of the resulting rubber powders was estimated.[30]

The grinding of cross-linked oil-expended EPDM was carried out by HTSD method in a rotor disperser.[31] The operation of this disperser is based on repeated deformations of grinding material in gaps between the grinding elements (cams) and the housing walls. It was found that the compacted material is fed into a grinding head and is subjected to further grinding.

During grinding, two competitive processes take place: the fracture and formation of particles smaller in size compared with pristine particles (grinding) and aggregation of particles (Fig. 5.6).

Chaikun and colleagues used both single- and twin-screw extruders and obtained powder of various sizes as a function of temperature and clearance between screws and the barrel. Analysis of rubber powder showed that finer fractions were agglomerated irrespective of the type of extruder used.[32]

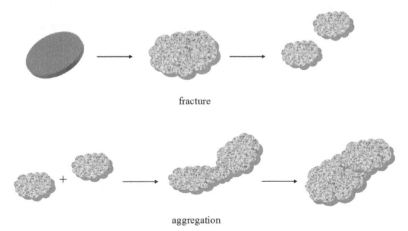

fracture

aggregation

FIGURE 5.6 Model of Grinding Process.

The analysis of the rubber crumb structure during EDG showed that in front of the entrance to the rotor disperser head, the compacting of rubber takes place; the smaller its initial particle size, the higher the degree of compaction. The density of the material before grinding the elements determines the particle size distribution of the obtained powder to a great extent. The coarse initial rubber fraction yielded a looser material before the entrance to the rotor disperser head. This material was comminuted better. A reduction in the size of the powder particles resulted in a material with better mechanical characteristics. Upon grinding, at least two competing processes occur: the formation of particles with a size less than the initial size (disintegration) and agglomeration of particles. Powder particles were asymmetric in shape and had a fairly developed surface area from 750 cm^2/g to 1200 cm^2/g. Two characteristic regions with a distinct boundary were observed at the surface of particles: a region of plastic failure with a highly developed surface and the region of brittle fracture with a smooth surface.

Since during EDG two processes take place simultaneously, namely, fracture and agglomeration, it was important to determine which of these processes is dominant. Pavlovskii and coworkers studied the mechanism of grinding cross-linked elastomers by EDG using twin-screw extruders.[33] Two mechanisms of cross-linked rubber using EDG have been suggested. The first mechanism leads to the formation of small particles, is related to viscoelastic properties of rubber, and is insensitive to a change in deformation rate. The authors suggested particles and the extruder barrel. The second mechanism is sensitive to a deformation rate and follows established physics of fracture by forming microcracks at the stressed sites, accumulation of elastic energy, and its release with a creation of new surface (i.e., powder).

The grinding of cross-linked rubbers occurs under the joint mechanical and thermal action. The thermal and mechanical degradation of polymer materials is inevitable under such conditions and results in the degradation products. During EDG, the action of mechanical factors initiates various chemical processes, such as oxidation, degradation, secondary structuring, and so on. Intensity of mechanochemical processes during EDG was studied by analysis of sol- and gel-fractions. It was found that low-molecular-weight compounds of sol-fraction were concentrated on the surface of particles obtained by the EDG. It was found that the surface of crumb particles contains up to 80% sol-fraction. This conclusion was confirmed by the results of determination of the amount of double chemical bonds on the surface of particles.[34] The concentration of double bonds on the particle surface increases from 2.45×10^{-6} mol/g for the virgin rubber crumb up to 8.13×10^{-6} mol/g for the powder. Removal of this sol-fraction from surface of particles significantly improved mechanical properties of the cross-linked rubber/GRT composites.

It is concluded from a chemical analysis of the powders that grinding leads to the formation of oxidized oligomer products with molecular weight from 300 to 2000 and causes secondary structuring, which is more intensive in large particles.[33] The presence of oligomer products of degradation in the material strongly affects the mechanism of grinding because of changes in the character of friction. The results of grinding can also depend on the structural inhomogeneity of the sample. It is energetically preferable to fracture through the boundaries of structural inhomogeneity. This can lead to fast formation of particles with sizes close to that of inhomogeneity. In particular, this is the reason for the increased content of sulfur after the grinding of carbon black-filled vulcanizates.[32]

These results were also observed for grinding of cross-linked oil-expended EPDM rubber by the HTSD.[30] In this study, EPDM was cured by a sulfur-based system up to different degrees of cross-linking. The cross-link density in rubber powder was higher compared with that for cross-linked rubbers (vulcanizates).

Chaikun and coworkers[32] investigated the influence of sulfur as a vulcanizing agent during EDG on properties of recycled vulcanizates. They found that presence of sulfur–sulfur bonds resulted in the formation of finer rubber. The authors obtained rubber with higher tensile strength than that of the used tire rubber that was ground without the vulcanizing agents.

The results of the work[35] suggest that mechanochemical processes occurring during the preparation of rubber powders from EPDM vulcanizates are accompanied by the formation of structurally and dynamically inhomogeneous networks, with the local mobility of polymer chains in such networks being substantially different in different areas, that is, the preparation of rubber powders produces a network with an essentially inhomogeneous degree of cross-linking.

Polyakov and coworkers[36] showed properties of recycled vulcanizates obtained by EDG technique and by conventional milling. They stated that during EDG, rubber is subjected to multiple shear deformations that allow production of much finer powder than that produced by using roll mills.

The authors indicated that agglomeration of fine powder during EDG was a function of the particle size of resultant tire rubber powder (the smaller the particle size, the more agglomerates were observed). Physical properties of EDG-made rubber were higher for a powder containing finer particle-size powder.

Trofimova et al[34] analyzed used tire rubber crumb ground by two methods: so-called "ozone-knife" and EDG, and investigated the effect of grinding method on the mechanical properties of resultant materials. It is shown that the surface activity of the "ozone-knife" particles is markedly lower as compared with that of the EDG particles.

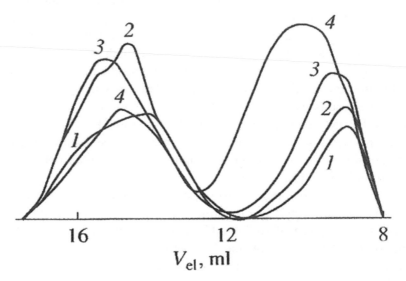

$$V_{el}, \text{ ml}$$

FIGURE 5.7   Grinding Process Curves for the Sol-Fraction of (*1*) the Initial Crumb Rubber and the Crumb Passed Six Times through the Rotor Disperser and Thermally Treated for (*2*) 5, (*3*) 15, and (*4*) 25 min. Extraction with Tetrahydrofuran.

The influence of the shear deformation value in EDG and of the heat treatment time at 150C on the structure of GRT was analyzed by Trofimova et al.[37] The value of shear deformations was varied by changing the number of repeated GRT passes through a rotor disperser. After each pass through the rotor disperser, GRT was divided into two portions, of which one was subjected to thermal treatment at 150°C for 5–30 min and the other was ground further. It is found that the surface of GRT passed four times through the rotor disperser becomes more developed. The proportion of sol-fraction components with a large molecular mass increases with an increase in shear deformation and heat treatment time (Fig. 5.7).

Pavlovskii and coworkers[38] studied the properties of mixtures of tread rubber with up to 60% of used rubber obtained by EDG with a rotor disperser. Ground tread rubber (consisting of three different rubbers) was converted into powder at 140°C with bimodal particle size distribution from 0.315 μm to 0.8 μm. Intensity of mechanochemical processes during EDG was studied by analysis of sol- and gel-fractions. The authors found that gel-fraction was decreased for finer particle-size powder, while sol-fraction localized on the entire surface of particles; large particles with less dense networks were capable of absorbing more of the sol-fraction and had more of toluene extract.

An increase in mechanical properties of materials based on mixtures of tread rubber with used tire rubber was explained by an interaction of dispersed phase and a matrix. Recycled rubber obtained by EDG had the larger surface area, by a factor of two, as compared with that made by conventional grinding. The authors reported that tear strength and flexural modulus of mixtures containing EDG-made used rubber powder were higher than those of virgin tread rubber compound, while a small decrease in plasticity was observed. When sol-fraction was removed from used rubber prior to mixing with virgin rubber, both flexural modulus and elongation at break were increased.

Their work confirmed results obtained by Pavlovskii's earlier work[29,33] and studies by other researchers that EDG process is not simply mechanical grinding, since several chemical processes such as oxidation take place. It was concluded that the EDG process is a mechanochemical grinding process by which an application of mechanical energy causes several chemical changes to occur.

In the work by Prut et al,[30] rubber powder was prepared by the HTSD method via fourfold passing of EPDM through a rotor disperser. While studying the process of EPDM vulcanizate grinding, it was found that, on the particle surface, there were two characteristic zones with the well-defined interface: the zone of plastic fracture with the highly developed surface and the brittle fracture zone with the smooth surface. It was shown that in the course of vulcanizate grinding, rubber powder particles characteristic of the gradient of the cross-link density directed from the periphery regions of particle to the particle center with a higher cross-link density may be formed.

A mechanism of grinding under elastic deformation conditions was first proposed to describe the extrusion in the solid state.[39] Depending on the temperature, rate, and amplitude of deformation, there are three stages of imperfection formation: clouding of extrudate, appearance of helical cracks and small defects, and total fracture of the material. The transition between these stages depends on the character of plastic flow.

A flow of polymer is accompanied by concurrent processes: accumulation of reversible elastic and irreversible viscous flow. The material can accumulate reversible elastic deformation only to a certain limit, and after reaching this limit, a transition takes place to another type of defect, or, as was formulated by the authors, the flow becomes unstable. Intensity of this transition depends on the viscoelastic properties of the material. For example, a rheological explosive is possible.[40]

It is clear that critical deformation depends on the character of mechanical loading. Reaching this deformation is a necessary but not a sufficient condition for multiple fractures. For example, deformation of PP using a Bridgman anvil (pressure 2 GPa and rotation angle 1000°) does not cause fracture, although the conditions in this case are tougher as compared with grinding using rotor disperser.[41]

It follows that a condition no less important is the creation of an optimal temperature regime.

At increased temperature, melting and crystallization start playing an important role. The formation of defects near the phase transition depends on the thermal history of the sample.[42]

The physical principle underlying EDG method consists in the fact that the energy, accumulated in the sample after the application of pressure, is consumed on forming new surface under the influence of shear deformation.

This process is not a gradual grinding of the sample in which particle sizes gradually decrease. It takes place in confined space and a limited time interval. This indicates the spontaneous nature of degradation and allows us to suggest a branching mechanism in the evolution of degradation centers. Under the appropriate conditions of interaction with the mechanical field, the structural elements of the substance can intensively absorb supplied energy. Then, the grinding can proceed in explosive fashion for both polymers and low-molecular-weight substances. The sample undergoes grinding to small, rapidly disintegrating particles.

HIGH-PRESSURE AND HIGH-TEMPERATURE SINTERING

High-pressure and high-temperature sintering (HPHTS) is a novel recycling technique that makes it possible to recycle vulcanized rubber powders made from waste rubber through only the application of heat and pressure. The design of new materials using secondary raw materials is a promising line of research in polymer chemistry. Both economic and environmental problems are solved in this case, which is especially important for the technical rubber goods industry in which wastes amount to millions of tons.[43-45]

The influence of the structure of GRT particles and their surface structure can be analyzed using 100% GRT material prepared by sintering.[46] An inspection of microphotographs has shown that there is no substantial difference in the surface structure of the crumb rubber particles examined. For both compositions, the surface of GRT particles exhibited both mirror and uneven areas. However, there is a small difference in the amount of the sol-fraction. Consider the effect of this factor on the properties of a sintered sheet. In this case, a sintered sheet can be obtained by sintering without the addition of a supplementary vulcanizing agent. Figure 5.8 shows the scheme of sintering. Step 1 corresponds to the approach of particles to one another; step 2 corresponds to the intermixing and repacking of particles and to the degradation of aggregates of particles; step 3 refers to the mutual displacement (shift) of particles, which is accompanied by their partial degradation, changes in shape and internal porosity, and disappearance of boundaries between particles; and step 4 is the emergence of bonds between particles and the formation of a monolithic plate.

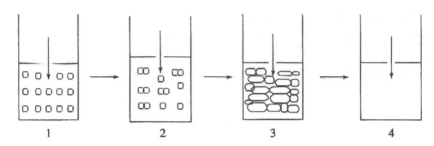

FIGURE 5.8   Scheme of Sintering. (Refer Text for Details.)

When GRT undergoes pressing, particles first approach each other and the compaction of crumb rubber takes place, which is accompanied primarily by structural deformations, that is, by the mutual displacement and repacking of particles and the degradation of weak aggregates (associates) of particles. Basically, only elastic strains develop in this case. As the load increases, structural deformations slow down and the mutual displacement (shear) of particles, which is accompanied by their partial degradation and changes in shape and internal porosity, begins to prevail. The boundary between the particles disappears. This leads to an increase in the number of contacts and entanglements between structural elements; as a result, a monolithic plate should be formed.

It is obvious that the sintering of rubber particles is a rather complex process. When GRT is pressed in the absence of additional vulcanizing agents, the strength σ_b of the final sintered sheet is pressure and temperature-dependent and varies very slightly with time (Table 5.1). There are two limiting models used to describe the sintering process, the Hertz model for elastic solids[47] and the Frenkel model[48] for the coalescence of two types of viscous spherical particles. According to the Hertz model, the compaction time should not affect the mechanical characteristics. According to the Frenkel model, the neck radius between particles χ is related to the compaction time τ by $\chi \sim \tau^{1/2}$. However, as shown in Table 5.1, these models are not quite appropriate for describing the compaction of ground rubber particles.

TABLE 5.1   Influence of Ground Rubber Tire Sintering Parameters (Without Additional Vulcanizing Agents) on the Strength $_b$ of the Final Sintered Sheets[43,49]

T (°C)	Time (min)	σ_b (MPa) at Various Pressures (MPa)		
		10	20	30
160	3	0.6	1.3	1.6
	10	0.6	1.0	1.5
	20	0.6	1.0	1.7
200	3	0.8	1.3	2.0
	10	1.0	1.7	2.1
	20	1.0	1.6	2.4
220	3	1.1	1.6	2.0
	10	1.0	1.7	2.2
	20	1.1	2.0	2.7

The plates produced are often porous and underpressed and have a large number of imperfections. These defects are removed by several means: increasing the compression of ground rubber, multiple refresh pressings in the course of sintering, varying the process temperature, and introducing modifier additives to improve the final sintered sheet.[49]

The complexity of sintering of ground rubber particles is aggravated by the fact that molding in the absence of any chemical additives can lead to degradation processes that will prevail over structuring processes. This degradation will increase the amount of low-molecular-weight fractions and decrease the number of cross-links. As the degradation of ground rubber proceeds, the formation of contact between particles is facilitated by the sol-fraction flow. However, significant degradation of ground rubber is undesirable because of the strong deterioration of the mechanical properties of the final sintered sheet. The properties of sintered sheet prepared with or without the addition of vulcanizing agents are given in Table 5.2. The initial GRT contains a certain residual amount of vulcanizing agents. From Table 5.2, it is seen that the strength σ_b is independent of the size of GRT particles when no vulcanizing agents are additionally introduced. However, the Young's modulus E and the stress at 100% extension σ_{100} are somewhat greater for the molding compound prepared from particles size $d < 0.4$, whereas the extension at break ε_b is smaller. Measurement of the equilibrium swelling ratio showed that Q_∞ is lower for sintered sheets made from smaller particles; that is, the cross-link density of these materials is higher.

TABLE 5.2   Mechanical Properties of Sintered Sheets

Particle Size (d, mm)	E (MPa)	σ_b (MPa)	σ_{100} (MPa)	ε_b (%)	Q_∞ (%)
< 0.4	3.0/9.0	2.9/5.1	2.1/3.9	150/150	1.9/1.8
0.4 < d < 0.7	2.3/9.1	2.9/6.5	1.8/3.3	180/200	2.4/2.0

Notes: E, Young's modulus; σ_b, tensile strength; $_{100}$ stress at elongation of 100%; ε_b, elongation at break; Q_∞, equilibrium swelling ratio. The values given in the numerator and denominator refer to the presence or absence of additional vulcanizing agents, respectively.

TABLE 5.3   Mechanical Properties of Sintered Sheets (Without Additional Vulcanizing Agents) Based on EPDM[29]

Rubber Powder (RP)[a]	Cross-link Density of Rubber Powder ($\nu \times 10^5$ mol/cm³)	Cross-link Density of Sintered Sheets ($\nu \times 10^5$ mol/cm³)	E (MPa)	σ_b (MPa)	ε_b (%)
EPDM-4044					
RP-1	5.2	2.2	2.1	0.9	110
RP-2	5.5	2.5	2.5	0.8	80
RP-3	6.1	2.6	2.7	0.6	60
RP-4	6.2	3.0	3.0	0.6	40
EPDM-4334					
RP-1	4.2	1.2	1.6	0.8	160
RP-2	4.4	1.6	1.7	0.7	130
RP-3	5.1	1.8	1.9	0.6	90
RP-4	5.4	2.0	1.9	0.5	90
EPDM-4535					
RP-1	2.9	0.4	0.9	0.6	260
RP-2	3.1	0.6	0.9	0.6	190
RP-3	3.6	0.8	1.0	0.55	170
RP-4	4.1	1.1	1.0	0.5	160

Notes: E, Young's modulus; σ_b, tensile strength; EPDM, ethylene-propylene-diene rubbers.
[a]Type of RP differs content vulcanizing agent (sulfur) in the original vulcanizates: 1 wt part (1), 2 wt. part (2), 3 wt. part (3), 4 wt. part (4) per 100 wt. part EPDM.

The addition of vulcanizing agents to GRT favors an increase in the cross-link density for larger particles and almost does not affect the same for smaller particles. The introduction of additional vulcanizing agents leads to an increase in the number of the cross-link density inside the particles and between the particles, as well as to the enhancement of mechanical properties. The size of the effect in this case depends on the surface area and the amount of the low-molecular-weight fraction of raw rubbers at the surface.

However, in practice, the effect of cross-link density of rubber powder and plasticizer has not been studied. The effect of these factors is presented in Table 5.3. It is seen that the Young's modulus varies slightly, and the tensile strength and the elongation at break decreased with increasing of cross-link density of rubber powder. The presence of paraffin oil leads to a decrease in modulus, tensile strength and higher elongation at break. In this case, the density of cross-links is about 2–3 times lower than that for oil-extended material than in the absence of oil. Therefore, reducing the cross-link density and increasing the sol-fraction improved microrheology processes that lead to an increase in elongation at break.

TABLE 5.4 Mechanical Properties of Sintered Sheets Based on EPDM[29] Without Additional Vulcanizing Agents

Rubber Powder[a] (RP)	E (MPa)	σ_b (MPa)	ε_b (%)
EPDM-4044			
RP-1	3.0	1.3	100
RP-2	3.1	0.8	70
RP-3	3.1	0.7	50
RP-4	3.5	0.6	40
EPDM-4334			
RP-1	1.8	0.95	140
RP-2	1.9	0.8	100
RP-3	1.9	0.7	80
RP-4	1.9	0.6	70
EPDM-4535			
RP-1	0.9	0.85	230

TABLE 5.4   *(Continued)*

Rubber Powder[a] (RP)	E (MPa)	σ_b (MPa)	ε_b (%)
RP-2	0.9	0.7	220
RP-3	1.0	0.7	200
RP-4	1.1	0.5	170

Notes: E, Young's modulus; σ_b tensile strength; ε_b elongation at break; EPDM, ethylene-propylene-diene rubbers. [a]Type of RP differs content vulcanizing agent (sulfur) in the original vulcanizates: 1 wt part (1), 2 wt part (2), 3 wt part (3), 4 wt part (4) per 100 wt part EPDM.

It was shown that the sintering of the rubber powder in the absence of any chemical additives leads to the predominance of degradation processes. This is reflected in the increase in the proportion of low-molecular-weight fractions and decrease in the cross-link density. The degradation of rubber powder facilitated the formation of the contact between the particles due to microrheological processes caused by the flow of the sol-fraction. However, significant degradation of the rubber powder is undesirable because of the sharp deterioration in the mechanical properties of sintered sheets.

Additional introduction of the vulcanizing system does not change the nature of the dependence of the mechanical properties as the function of the oil content in the rubber and cross-link density of rubber powder. Thus, there is a slight increase in the absolute values of the Young's modulus and tensile strength (Table 5.4).

Thus, mechanical properties of sintered sheets significantly depend on plasticizer content and the cross-link density of the rubber powder. At the same time, the mechanical characteristics of the sintered sheets were about 50% of the initial values of the vulcanizates

It should be noted that in the literature, there are practically no data on the effect of grinding on the properties of the sintered sheets from the rubber powder. Shape and size of particles significantly affect their packaging at sintering, as well as the number and size of pores in the materials is obtained. The particle size is mainly influenced by the speed and temperature of sintering. Proper distribution of particle size is important for high packing density.

Most often assemblies are connected in agglomerates and aggregates. Agglomerates are formed due to the weak mutual attraction between the particles of the powder and can be destroyed with little external influence. In aggregates, particles are bonded more strongly, for example, as a result of mutual attraction in the production process of the powder. Aggregates, during compaction of powders, behave as large particles with high surface area. In the step of sintering, the compacts manufactured from the aggregate powder, hard aggregates as well as larger particles, give rise to a coarse grain structure in the final product.

A problem associated with the aggregation of particles occurs when sintering. Thus, for compacting the aggregated powder by sintering to achieve certain material properties, high temperature is required, and the larger units should be in the powder particles. Due to aggregation and agglomeration of the particles, it is difficult to obtain compact materials. Large mechanical forces or an increase in temperature (sintering) are required to overcome the forces of agglomeration.

The main problem with the compaction of rubber powder in the manufacture of bulk samples is the uneven distribution of the density by the volume of powder products.

Thus, in this section, the experimental results on the preparation of sintered sheets show the feasibility and promise of this technology for the production of various rubber-technical products.

APPLICATIONS OF RUBBER POWDER IN THERMOPLASTICS

Considerable efforts have been made in finding new applications for GRT. Fine GRT particles may be used as fillers and property modifiers in thermoplastic, elastomer, and thermoset blends. Addition of rubber to plastics also improves some of the key properties of the plastics, particularly the impact resistance. Although the use of GRT as fillers in polymer composites is a potentially attractive approach, it is fraught with a number of difficulties. Thermoplastics such as polyethylene and polypropylene are not only cheap, but also are available in a wide range of melt index and microstructure, which can be used for mixing with recycled rubber. Karger-Kocsis et al[9] recently published a comprehensive review regarding the difficulties to produce high-quality GRT-filled compounds. The mechanical properties of such composites depend on the content of GRT, polymer matrix type, and adhesion between the GRT and the polymer matrix, as well as the particle size and their dispersion and interaction between the GRT and the matrix. However, the incorporation of GRT particles into a number of polymer matrices significantly deteriorates the mechanical properties of the composites due to very weak interfacial adhesion between the GRT particles and the matrix-forming polymer. Similar poor behavior of GRT/thermoplastic composites is often reported.[7,9]

Since most polymers and elastomers are thermodynamically immiscible with each other, their blends undergo phase separation, with poor adhesion between the matrix and dispersed phase. Therefore, mechanical properties of multicomponent polymer systems depend on their two-phase structure and blending conditions. By varying mixing conditions, especially, the intensity of shear strains, it is possible to control the level of heterogeneity of blends at different structural scales and, thus, to prepare materials that substantially differ in characteristics from their individual components.

By mixing thermoplastic polymers with GRT, new blends referred to as thermoplastic rubbers were prepared.[50] By their composition, thermoplastic rubbers are similar to TPEs in which vulcanized rubber particles are distributed in the polymer matrix.[51] However, by their structure and properties, thermoplastic rubbers are particulate-filled polymer composites.[52] The dissimilarity between composites containing rubber particles and mineral filler consists in a difference in the rigidity of disperse phase and matrix. As was shown by Bazhenov et al and Goncharuk et al,[53,54] it is possible to prepare composites containing up to 90 parts by volume of rubber particles. In this case, a thermoplastic polymer serves as a matrix.

However, the influence of blending conditions on the mechanical properties of thermoplastic rubbers containing different amounts of a rubber component remains an open question.

In the work by Trofimova et al,[55] the effect of mixing conditions on the mechanical properties of thermoplastic rubbers based on IPP and GRT prepared from the tread rubber by the method of HTSD was studied. Melt blending of IPP and GRT was carried out using a Brabender internal mixer at 190°C for 10 min (rotor speed of 100 rpm) and the method of HTSD in a rotor disperser (temperature 190°).

The mechanical properties of the blend were shown to be independent of mixing conditions (Figs. 5.9 and 5.10). Depending on the amount of crumb rubber, three regions that differ in the mechanism of deformation of thermoplastic rubbers are distinguished: below 0.1, 0.1–0.75, and above 0.75 parts by volume. According to Bazhenov et al,[53] the successive change of deformation, a mechanism from plastic macro-heterogeneous deformation to brittle fracture and then to macro-homogeneous deformation, takes place when the GRT content in the blend increases.

Thus, the content of GRT is an important factor that influences on the structure and properties of composites. So, when the content of GRT in composite is lower than 10–20%, the properties of composite are satisfactory, while the content is higher than 20%, the properties of composite are not satisfactory.

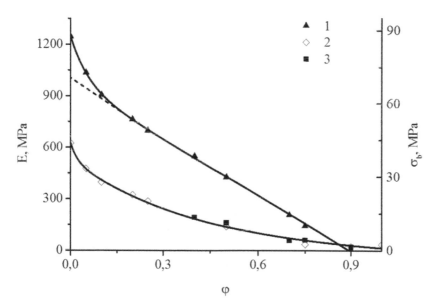

FIGURE 5.9 (1) Elastic Modulus E and (2, 3) and Tensile Strength σ_b as Functions of the GRT Loading φ for the Materials Blended (1, 2) in a Brabender Mixer and (3) by the Method of High-Temperature Shear Deformation.

It is very interesting to explore the possibility of modifying the surface of GRT particles so that new reactive groups are introduced for enhanced miscibility with the thermoplastic phase.

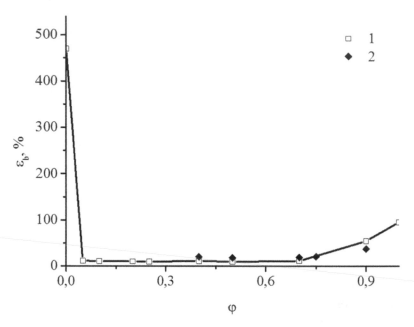

FIGURE 5.10   Elongation at Break ε_b as a Function of the Ground Rubber tire Loading φ for the Materials Mixed (1) in the Brabender Mixer and (2) by the Method of High-Temperature Shear Deformation.

Prut et al[56] have studied mixing of GRT with functional polymers to improve the compatibility and extension at break of GRT/polymer composite. Four different polymers have been used: LDPE, ethylene-vinyl acetate copolymers with content of vinyl acetate 9–14% (EVA-1) and 24–30% (EVA-2), and ethylene-vinyl acetate-maleic anhydride terpolymer OREVAC® T 9318 (Orevac). Composites with different ratio components, viz., GRT/polymer = 20/80, 30/70, 50/50, 60/40, 70/30, 80/20, were prepared by melt mixing the components in a Plastograph EC at 160°C, 100 rpm for 10 min.

The addition of the GRT significantly lowered the mechanical properties. The detailed mechanical properties of all the composites are summarized in Table 5.5. It is observed that incorporation of GRT in polymer matrix decreases the tensile strength, Young's modulus, and elongation at break of all the composites. However, the magnitudes of the tensile strength, Young's modulus, and elongation at break are significantly different depending on the polymer matrix. So, the LDPE/GRT composites have very poor mechanical properties, especially elongation at break. The elongation at break ε_b is seen to drop by about 80% even at 20 wt% filler loading. The ε_b continues to drop until at 30 wt% filler. The Young's modulus E and tensile strength $_b$ are seen to decrease steadily with increasing GRT content. At 30 wt% GRT, the tensile strength σ_b has decreased by about 46% and the Young's modulus E by about 21%.

A potentially even more attractive route is the use of matrices that increase interaction (adhesion) between the matrix and the GRT particles. Data in Table 5.5 indicate that elongation at break decreases with increase in the GTR content up to 80 wt%. Changes in tensile strength and Young's modulus followed the same trend. All the composites have higher elongation at break compared with the composites without functional polymers at 70 wt% GTR. This can be explained by the assumption that the functional polymers increase interaction (adhesion) between the matrix and the GRT particles. Table 5.5 indicates that the highest elongation at break was obtained with the recipe containing OREVAC. The elongation at break of the GRT/OREVAC composites containing 70 wt% GRT is more than two times of that of GRT/LDPE. The enhancement in elongation may suggest the formation of an interfacial region. The possible chemical interaction between OREVAC and GRT may lead to improved compatibility and also to dispersion of the GRT in the matrix, thereby improving the product's elongation at break.

TABLE 5.5 Effect of Functional Polymers on the Mechanical Properties (Young's Modulus [E], Tensile Strength [σ_b], Elongation at Break [ε_b]) and MFI of GRT/Polymer Blends

Polymer Matrix	Composition GRT/Polymer, wt part	Method of Molding	E (MPa)	σ_b (MPa)	ε_b (%)	MFI
LDPE	0/100	Compression molding	140	12.3	380	1.4
	20/80		150	7.8	80	2.1
	30/70		110	6.6	70	1.8
	100/0	Compression molding	5.8	1.1	50	–
EVA-1	0/100	Compression molding	55	16.4	780	3.8
	30/70		28	4.7	155	5.4
	50/50		25	4.0	115	3.1
	70/30		11	2.3	75	0.5

EVA-2	0/100	Compression molding Injection molding	7.4	6.3	790	Flow without load
	30/70		6.5	2.4	230	Flow
	50/50		5.8	2.4	170	13.6
	70/30		6.4	2.5	135	1
	80/20		4.3	1.6	90	0.1
	50/50		8.4	2.7	160	12.6
	70/30		6.6	2.7	140	1
OREVAC	0/100	Compression molding	30.5	20.6	880	6–8
	60/40		17	5.1	180	1.4
	70/30	Injection molding	11	3.8	130	0.3
	80/20		7.5	3.0	120	No flow
	60/40		14.5	5.0	170	1.4
	70/30		13.6	4.5	140	0.3
	80/20		8	4.1	155	No flow

Notes: MFI, melt flow index; GRT, Ground rubber tire.

Samples of GRT/EVA-2 and GRT/OREVAC in the composition of 60/40, 70/30, and 80/20 were prepared by injection molding and compression molding (Table 5.5). As can be seen from Table 5.5, the method of injection molding allows obtaining the samples with the higher elongation at break in comparison with compression molding.

Consequently, different mechanical behavior between composites can be attributed to the difference of bonding between the rubber powders and polymer matrix. The higher percent vinyl acetate has a good compatibility with other materials.

In order to maximize the effectiveness, it was desired for the greatest percentage of GRT possible to be used in the composites.

As can be seen from Table 5.5, the GRT content does not cause significant changes in viscosity of GRT/LDPE composites within the experimental error. The melt flow index (MFI) of GRT/LDPE composites can be explained in terms of the extremely low adhesion in these systems. This may be attributed to the high cross-link density of a tire, which does not allow for any interfacial interpenetration, resulting in a sharp interface. The MFI of the other composites decreases with increasing filler content (Table 5.5). Decreasing of MFI is ascribed to the formation of an interfacial region.

The composites, in which the GRT content was as high as 70 wt%, were characteristic of a good processability.

Prut and coworkers[57] have studied rheological properties of GRT filled IPP of different molecular-weight characteristics. IPP of different molecular weights were used as thermoplastic polymer matrices (Table 5.6).

GRT formed at the output of the rotor disperser was fractionated with a sieve set. GRT with a particle size of d = 0.1−0.4 mm was incorporated into the polypropylene by melt blending.

TABLE 5.6 Molecular Weight Values, Polydispersity Index, and MFI for IPP

Polymer	M (g/mol)	M_w (g/mol)	M_n (g/mol)	IP = M_w/M_n	MFI, 190°C, 2.16 kg
IPP-1	360,000	150,000	40,000	3.8	10.0
IPP-2	535,000	230,000	63,000	3.7	3.0
IPP-3	520,000	280,000	95,000	2.9	0.3

Notes: **IP,** polydispersity index; MFI, melt flow index, M_z z-average molecular weight; M_w, weight-average molecular weight; M_n number-average molecular weight.

The dynamic viscoelastic properties, such as storage and loss moduli, complex viscosity, and loss tangent of IPP/GRT composites, were determined in function of GRT loading and compared with those of the parent IPP. The complex viscosity log η^* of the neat IPPs and IPP/GRT composites decreased with increasing frequency. All the composites and polymers exhibited a pseudoplastic or shear thinning. The higher the GRT loading in the composites, greater is the deviation from the Newtonian behavior.

The storage and loss moduli of virgin IPPs and IPP/GRT composites were shown to increase as the frequency increased. The value of critical crossover frequency ω_c, for which G'' is equal to G', increased with decreasing IPP molecular weight (from 1.6 rad/s for PP-3 to 63.1 rad/s for IPP-1).

A significant improvement in the melt flowability of IPP/GRT composites was found to depend on the IPP molecular weight and the content of GRT. The higher was the GRT loading in the composites, the greater was the deviation from the Newtonian flow behavior.

The addition of a relatively small amount of GRT (5–10 wt%) reduced G', G", and η* of the corresponding two-phase system compared with the matrix (Figs. 5.11–5.13). This decrease depended on the IPP molecular weight. The values of log G', log G", and η* of IPP–2/GRT and IPP–3/GRT composites went through a minimum at a loading of about 10 wt% GRT. The addition of GRT to IPP-1 did not change log G', log G", and η* for all frequencies. The flowability of IPP/GRT composites depended on the IPP molecular weight and GRT loading.

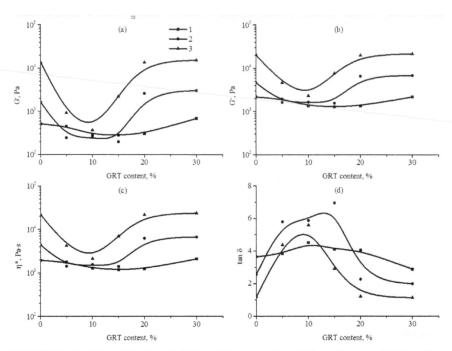

FIGURE 5.11   Plots of (a) log G', (b) log G", (c) log η*, and (d) tan δ versus GRT Loading for (1) IPP-1, (2) IPP-2, (3) IPP-3. Frequency ω = 1 rad/s; GRT, Ground Rubber Tire; IPP, Isotactic Polypropylene.

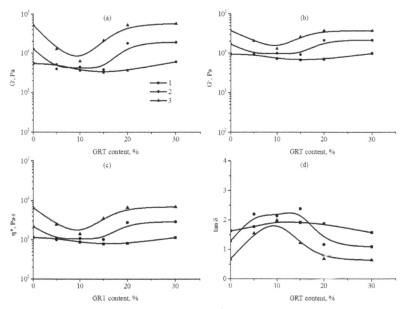

FIGURE 5.12 Plots of (a) log G′, (b) log G″, (c) log η*, and (d) tan δ versus GRT Loading for (1) IPP-1, (2) IPP-2, (3) IPP-3. Frequency ω = 10 rad/s; GRT, Ground Rubber Tire; IPP, Isotactic Polypropylene.

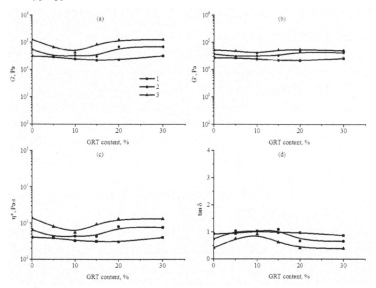

FIGURE 5.13 Plots of (a) log G′, (b) log G″, (c) log η*, and (d) tan δ Versus GRT Loading for (1) IPP-1, (2) IPP-2, (3) IPP-3. Frequency ω = 100 rad/s; GRT, Ground Rubber Tire; IPP, Isotactic Polypropylene.

The dynamic rheological behavior of composites depended strongly on the degree of agglomeration of the powdery filler. The appearance of a minimum in the viscosity curve was reasoned by supposing the formation of additional free volume in the interphase adjacent to filler particles.

A monotonic decrease of the loss tangent tan δ was observed with increasing ω for all composites. The values of tan δ of both the IPP–2/GRT and the IPP–3/GRT composites went through a maximum at GRT loading of about 10 wt%. The addition of GTR increased the range of solid-like behavior in comparison to that of the neat IPP. The shift of the critical crossover frequency for IPP–3/GTR composites indicated a transition from a more viscous to a more elastic behavior.

There were significant differences in the low-frequency data between the composites. The IPP/GRT composites with high molecular weight of IPP exhibited solid-like behavior. In contrast, the low-frequency response for the IPP/GRT composites with low molecular weight revealed liquid-like behavior. The values of the storage modulus G′, the loss modulus G″, complex viscosity η^*, and the loss tangent tan δ of the composites were controlled by the IPP characteristics.

The rheological properties of the composites showed complex behavior, which is not readily predictable. On the other hand, the rheological properties are highly sensitive to the structure of the composites, in our case to the GRT dispersion in the IPP matrix. This study suggests that it is possible to obtain composites with good rheological properties at low cost.

The influence of dynamic vulcanization on the mechanical properties of ternary (IPP–rubber–crumb rubber) and binary (rubber–crumb rubber) blends was studied.[58] Two types of ethylene–propylene–diene elastomer were used as the rubber component, the oil-free elastomer Buna EP G 6470 (EPDM-6470) and the oil-extended elastomer Buna EP G 3569 (EPDM-3569). The blends were vulcanized in the presence of a sulfur accelerating system.

Five types of crumb rubber that differed in the particle size d were used:

SAARGUMMI crumb rubber (Germany) prepared via cryogenic treatment from commercial rubber goods, having a particle size of $d < 0.1$ mm (GRT-1).

SCANRUB crumb rubber (Denmark) prepared from spent car tires by ultrasonic treatment; $d < 0.4$ mm (GRT-2).

SCANRUB crumb rubber (Denmark) prepared from spent car tires by ultrasonic treatment; 0.4 d < 0.7 mm (GRT-3).

SCANRUB crumb rubber (Denmark) passed at 150–155°C for 5 min through a rotary disperser designed on the basis of a single-screw extruder at the Institute of Chemical Physics; 0.4 < d < 0.63 mm (GRT-4).

Crumb rubber (Russia) passed four times at 150–155°C for 5 min through the rotary disperser;[6] 0.4 < d < 0.63 mm (GRT-5).

Table 5.7 presents the characteristics of binary blends of EPDM-6470 and EPDM-3569 with GRT.

Vulcanization of the blends leads to the formation of a network structure in the elastomer phase and to enhancement of interfacial interaction between the components, a situation that can have an effect on the mechanical properties of EPDM blends with crumb rubber.[58]

TABLE 5.7   Cross-link Density v and Mechanical Characteristics of EPDM-6470/GRT and EPDM-3569/GRT Blends (GRT Content of the Blends Is 25 wt%)

GRT Type	E (MPa)	σ_{100} (MPa)	σ_{300} (MPa)	σ_b (MPa)	ε_b (%)	$v \times 10^5$ mol/mL
EPDM-6470/GRT						
None	4.0/3.3	1.2/1.1	1.8/2.1	7.3/3.7	1130/505	22.6
GRT-1	7.1/12.3	1.75/2.8	3.0/4.7	14.7/5.1	160/330	11.4
GRT-2	4.7/6.9	1.4/2.7	1.9/4.5	4.0/5.15	850/350	3.6
GRT-3	6.25/4.8	1.3/1.5	1.9/1.9	2.6/2.1	610/360	8.3
GRT-4	4.7/4.8	1.2/2.0	1.7/–	3.4/3.1	950/270	5.4
GRT-5	4.6/7.0	1.3/1.9	2.3/2.9	2.9/3.0	510/350	4.5
EPDM-3569/GRT						
None	0.6/0.8	0.2/0.4	0.2/0.6	0.7/1.8	>1800/710	7.7
GRT-1	1.4/2.5	0.3/0.9	0.4/1.9	2.2/4.0	1500/750	8.6
GRT-2	0.8/1.5	0.2/0.6	0.2/0.9	0.5/1.7	1730/760	6.3
GRT-3	1.0/1.0	0.25/0.4	0.25/0.5	0.5/1.1	1230/950	8.8
GRT-4	1.0/1.5	0.2/0.45	0.15/0.5	0.3/1.2	>1700/990	8
GRT-5	0.7/1.6	0.1/0.6	0.1/1.0	0.3/1.5	>1800/630	7.1

Notes: [a]The numerator and denominator refer to the characteristics of the material in the absence and in the presence of vulcanizing agents, respectively. E, Young's modulus, σ_b, tensile strength; ε_b, elongation at break.

Figure 5.14 presents tensile stress–strain curves for neat EPDM-6470 and its blends with GRT. When 25 wt% GRT is introduced into EPDM-6470, the Young's modulus E of the blends increases relative to the initial elastomer, and the breaking strength σ_b and elongation at break ε_b decrease. The exception is the uncured blend of EPDM-6470 with GRT-1 having a particle size of $d < 0.1$ mm (Fig. 5.14, curve 2; Table 5.7). For this blend, the parameters E and σ_b have values 1.8–2.0 times above those for the virgin elastomer. However, the elongation at break ε_b is practically the same. It may be assumed that this behavior is due to more homogeneous GRT-1 particles in the elastomer phase. The σ–ε plots for EPDM-6470 blends with other types of GRT have identical patterns. However, the mechanical parameters differ. For example, for the blend with GRT-3, the Young's modulus E is higher and the ultimate strength σ_b and elongation at break ε_b are lower than those for the blends with GRT-4 ($0.4 < d < 0.63$ mm) and GRT-2 ($d < 0.4$ mm). The lowest value of ε_b is observed for the EPDM-6470 blend with GRT-5 ($0.4 < d < 0.7$ mm). It is difficult to analyze these results, because GRT-3 particles ($0.4 < d < 0.7$ mm) have a less developed surface than GRT-4 ($0.4 < d < 0.63$ mm) and a larger size than GRT-2 particles ($d < 0.4$ mm). Furthermore, GRT-4 and GRT-2 particles differ not only in size but also in the structure of the surface layer owing to different preparation procedures.

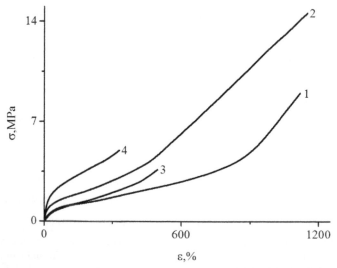

FIGURE 5.14   Tensile Stress–Strain Curves for (1, 3) EPDM-6470 and (2, 4) its Blends with GRT-1: (1, 2) the Uncured and (3, 4) the Cured Materials. The EPDM/GRT ratio is 75/25. GRT, Ground Rubber Tire; EPDM, Ethylene-Propylene-Diene Rubbers.

The vulcanization of these blends varies in the form of σ–ε curves and the mechanical parameters (Fig. 5.14, curves 3, 4); the plots lack the portion that corresponds to crystallization during stretching. After vulcanization, the Young's modulus E becomes higher and the elongation at break becomes smaller relative to the uncured blends. The value of ε_b depends on neither the particle size of GRT nor its preparation procedure in this case (Table 5.7). At the same time, the values of E and ε_b are higher for blends not only with GRT-1 ($d < 0.1$ mm) but also with GRT-2 ($d < 0.4$ mm) and GRT-5 ($0.4 < d < 0.63$ mm). Consequently, the parameters E and σ_b are dependent on the particle size of crumb rubber (Table 5.7).

When oil-extended EPDM-3569 is used in blends, the appearance of –ε curves and the mechanical parameters change as well (Fig. 5.15). The curve portion corresponding to rubber crystallization is less pronounced than that for EPDM-6470 (Fig. 5.15, curve 1): σ for EPDM-3569 increases with an increase in ε to a lesser extent than for EPDM-6470, a difference that is most likely due to the presence of oil. As a result, EPDM-3569 has lower values of E and σ_b and a higher value of ε_b than EPDM-6470 (Table 5.7). This effect is characteristic of both uncured and cured blends. From the data presented in Table 5.7, it is seen that, in the presence of crumb rubber, the Young's modulus increases and the elongation at break ε_b drops relative to those parameters of the uncured initial elastomer EPDM-3569. The maximal growth in E and σ_b characterizes the blend with GRT-1 ($d < 0.1$ mm), as in the case of the EPDM-6470 blend. It is noteworthy that the particle size of GRT and its preparation procedure have a weak effect on the other blends, vulcanized or unvulcanized. The lack of the effect is due to a more homogeneous structure of distribution of crumb rubber in oil-extended EPDM-3569.

The σ–ε tensile stress–strain curves of the uncured EPDM-3569/GRT blends exhibit a yield point (Fig. 5.15). A similar effect is observed for GRT blends with oil-extended Dutral TER 4535 EPDM (Italy).[59] According to Nielsen,[60] the yield point appears in elastomer compositions with hard fillers. Its appearance is associated with either the formation of microcracks or the debonding of the polymer from the filler upon the failure of the adhesive bond between them and a dramatic decrease in the elastic modulus of the composition. It is likely that, owing to the presence of oil, interfacial interaction between EPDM and GRT is very weak in the given case and drawing causes the formation of vacuoles.

Thus, the replacement of 25% elastomer with GRT leads to an increase in the Young's modulus of E of blends, regardless of the type of GRT and the nature of EPDM. The ultimate strength σ_b and the elongation at break ε_b drop in the presence of GRT for both cured and uncured EPDM-6470 blends. However, for the systems with oil-extended elastomer EPDM-3569, the ultimate strength σ_b insignificantly increases and a change in the elongation at break ε_b depends on vulcanization of the blend.

Ternary IPP/EPDM/GRT blends were obtained at the two component ratios 50/37.5/12.5 and 30/52.5/17.5. The EPDM/GRT ratio was three in both cases. The PP/rubber (EPDM blend with GRT) ratios of 50/50 and 30/70 are typical of TPEs.[6]

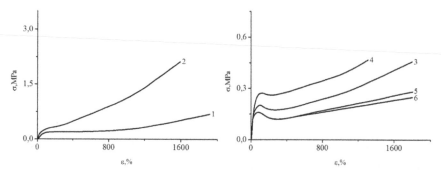

FIGURE 5.15 Tensile Stress–Strain Diagrams for Uncured EPDM-3569 and its Blends with GRT: (*1*) EPDM, (*2*) EPDM/GRT-1, (*3*) EPDM/GRT-2, (*4*) EPDM/GRT-3, (*5*) EPDM/GRT-4, and (*6*) EPDM/GRT-5. The EPDM/GRT ratio is 75/25. GRT, Ground Rubber Tire; EPDM, Ethylene-Propylene-Diene Rubbers.

From the results (Tables 5.8 and 5.9), it follows that the presence of GRT in a blend, regardless of its type, increases the Young's modulus E relative to the GRT-free blend and decreases the breaking strength σ_b and elongation at break ε_b, with ε_b decreasing by a factor of ~10.

From the data presented in Table 5.7, it is seen that the highest values of E, σ_b, and ε_b were obtained for EPDM-6470 with GRT-1 ($d < 0.1$ mm). In this case, vulcanization has almost no effect on ε_b (Fig. 5.16). When other types of GRT are used in the uncured PP/EPDM-6470/GRT blends at a component ratio of 50/37.5/12.5, the elongation at break ε_b is less than 100% (Table 5.7) and the specimens experience quasi-brittle fracture. The vulcanization of these blends increases ε_b to 100% or more. The stress–

strain plots exhibit a small growth in the plastic region, an increase in ε without a noticeable rise in σ (Fig. 5.15).

TABLE 5.8 Mechanical Characteristics and MFI of 50/37.5/12.5 IPP/EPDM-6470/GRT and IPP/EPDM-3569/GRT Blends

GRT Type	E (MPa)	σ_{100} (MPa)	σ_{300} (MPa)	σ_b (MPa)	ε_b (%)	MFI g/10 min 2.16 kg	10.6 kg
IPP/EPDM-6470/GRT							
None	205/250	8.6/11.6	11.4/17.9	15.6/32.0	580/620	0.50/–	7.15/–[a]
GRT-1	400/430	11.1/11.3	13.9/15.3	16.7/20.1	450/460	0.40/–	6.00/1.05
GRT-2	350/330	–/10.7	–	10.0/11.8	60/185	0.40/–	7.00/0.15
GRT-3	320/300	–/10.3	–	9.4/10.2	35/100	0.55/–	8.70/0.20
GRT-4	380/310	–/10.0	–	10.0/10.0	45/100	0.95/–	11.25/1.00
GRT-5	300/350	10.7/–	–	10.9/9.8	120/60	0.40/0.85	–
IPP/EPDM-3569/GRT							
None	190/210	7.7/8.8	10.1/12.6	12.2/14.8	450/390	2.00/0.30	59.0/47.8
GRT-1	270/290	9.5/9.5	12.0/–	13.6/10.4	400/180	1.20/1.40	35.0/40.65
GRT-2	250/240	–	–	8.9/8.8	40/50	1.40/1.35	43.5/42.6
GRT-3	220/430	–	–	7.8/11.7	30/10	2.65/2.15	63.5/43.9
GRT-4	215/240	–	–	8.4/8.5	35/40	2.30/1.90	46.05/50.4
GRT-5	200/210	–	–	7.6/8.7	30/40	4.30/2.40	–

Notes: GRT, ground rubber tire; E, Young's modulus; σ_b, tensile strength; σ_{100}, stress at elongation of 100%; ε_b, elongation at break; MFI, melt flow index; IPP, isotactic polypropylene; EPDM, ethylene-propylene-diene rubbers.

An unusual picture is observed for the MFI of vulcanized blends. The vulcanized IPP/EPDM-6470 = 50/50 blend in the absence of GRT does not flow even under a load of 10.6 kg. When GRT is added, the blend begins to flow (Table 5.7). A similar result is obtained for the uncured 30/70 IPP/EPDM-6470 blend, which does not flow under a load of 2.16 kg in the absence of GRT and begins to flow when GRT is added (Table 5.7). Thus, partial replacement of rubber with GRT in the IPP/EPDM blend leads to a decrease in the viscosity of the system, a change that is most likely due to the character of surface of crumb rubber particles.

TABLE 5.9 Mechanical Characteristics and MFI of 30/52.5/17.5 IPP/EPDM-6470/GRT and IPP/EPDM-3569/GRT Blends

GRT Type	E (MPa)	σ_{100} (MPa)	σ_{300} (MPa)	σ_b (MPa)	ε_b (%)	MFI g/10 min	
						2.16 kg	10.6 kg
IPP/EPDM-6470/GRT							
None	26/61	3.6/6.6	6.0/14.5	8.8/19.0	660/380	–[a]/–[a]	2.20/–[a]
GRT-1	86/139	5.7/7.1	8.6/10.5	10.3/16.2	485/570	0.10/–[a]	2.20/–[a]
GRT-2	63/101	5.8/7.6	–	6.8/10.0	210/200	0.20/–[a]	3.40/–[a]
GRT-3	44/71	4.4/6.2	–	4.65/8.0	120/190	0.20/–[a]	3.12/–[a]
GRT-4	50/64	4.8/5.4	–	5.0/8.2	145/290	0.25/–[a]	3.84/–[a]
GRT-5	34/65	4.35/5.7	5.7/8.4	5.8/8.4	260/315	0.30/–[a]	3.76/0.10
IPP/EPDM-3569/GRT							
None	30/58	2.6/4.0	3.8/6.1	4.6/7.0	520/400	0.20/–[a]	29.0/1.36
GRT-1	69/96	4.2/5.2	5.6/6.7	6.0/7.9	355/430	0.40/0.02	18.0/5.13
GRT-2	60/73	4.1/4.6	–	4.2/4.6	130/90	0.60/–[a]	28.8/2.50
GRT-3	46/62	–	–	2.9/3.6	70/55	0.60/–[a]	28.9/1.20
GRT-4	43/58	3.3/4.2	–	3.1/4.0	95/75	0.56/–[a]	32.0/1.12
GRT-5	41/52	3.2/4.3	–	3.3/4.7	125/170	1.00/0.07	36.5/–[a]

Notes: GRT, ground rubber tire; E, Young's modulus; σ_b, tensile strength; $_{100}$ stress at elongation of 100%; ε_b, elongation at break; MFI, melt flow index; IPP, isotactic polypropylene; EPDM, ethylene-propylene-diene rubbers.

The MFI of the blends increases with an increase in load. At a IPP/EPDM-6470/GRT = 50/37.5/12.5, the vulcanized blends flow only under a load of 10.6 kg (Table 5.8).

In the blends of oil-extended EPDM-3569 with GRT, the MFI dramatically increases (Tables 5.8 and 5.9). The oil in the rubber acts as a plasticizer, and the MFI grows. For the uncured and cured IPP/EPDM-3569/GRT = 50/37.5/12.5 blends, the MFI is improved to 35–65 and 40–50 g/10 min (at a load of 10.6 kg), respectively. As the IPP content decreases, the MFI decreases to 18–32 g/10 min for the vulcanized blends or to 1–5 g/10 min for the uncured blends (Tables 5.8 and 5.9). Upon vulcanization, the presence of oil in the rubber has no effect on the MFI, unlike the case of oil–free rubber.

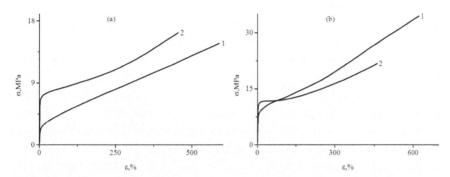

FIGURE 5.16 Tensile Stress–Strain Curves for (a) Uncured and (b) Cured IPP/EPDM-6470/GRT Blends Having a Component Ratio of 50/37.5/12.5: (*1*) IPP/EPDM and (*2*) IPP/EPDM/GRT-1. GRT, Ground Rubber Tire; IPP, Isotactic Polypropylene; EPDM, Ethylene–Propylene–Dienc Rubbers.

Thus, the use of oil-extended rubber and introduction of GRT into the blends lead to the situation that the rheological properties of the blends, and hence the processability of these materials are improved. It is likely that this effect is due to a change in the interfacial tension and sliding on the interface.[61]

Thus, vulcanization improves the mechanical characteristics. This tendency is observed regardless of the type of EPDM rubber or GRT. The highest values of the mechanical characteristics are observed for the IPP/EPDM-3569/GRT blends in which the particle size of GRT is $d < 0.1$ mm.

Consequently, the best mechanical and rheological characteristics were obtained for the ternary blends using GRT with a particle size of $d < 0.1$ mm.

The addition of GRT powder promotes the flowability of uncured IPP/EPDM = 50/50 and these TPVs (Table 5.10). The MFI values of TPVs-containing GRT are higher than those of TPVs without GRT. One can see that blends with IPP/EPDM = 30/70 do not flow at low and medium loads, but an increase of a load to 10 kg leads to a weak flow.[62]

As can be seen from Table 5.10, rubber particle size is an important factor affecting the MFI of TPVs. The reduction of the rubber particle size increases MFI of uncured blends and dynamically cured TPVs. So, the

TPVs processability is strongly influenced by a partial substitution of the EPDM rubber by the GRT powders. GRT contains mineral fillers such as carbon black and other additives, which could vary rheological properties of TPVs. Hence, use rubber powder of cured virgin elastomer without any additives will be good. For these reasons, Prut et al used the powder of EPDM-4044 vulcanizates without mineral fillers.

Figure 5.17 shows plots of IPP/EPDM-4044 complex viscosity versus rubber powder content. One can see that a partial replacement of virgin EPDM-4044 by rubber powder increases η^* value of uncured IPP/EPDM-4044/RP blends. But growth of rubber powder content in TPVs steadily decreases their complex viscosity. So, the addition of 25wt% rubber powder results in one order reduction in η^* value.

Therefore, the partial replacement of traditional rubbers used in TPV by GRT will make it possible to improve the processability of TPVs and reduce the environmental pollution.

TABLE 5.10 MFI Values of Uncured and Dynamically Cured TPV

IPP/EPDM/ GRT	Particle Size of GRT d, (mm)	MFI (dg/min)					
		2.16 kg		5.00 kg		10.00 kg	
		Uncured	Cured	Uncured	Cured	Uncured	Cured
50/50/0	–	–	–	–	–	–	–
50/37.5/12.5	without fraction	0.05	0.4	0.4	1.6	2.4	8.4
50/37.5/12.5	$d < 0.1$	0.3	0.4	0.4	1.5	2.4	8.4
50/37.5/12.5	$0.1 < d < 0.4$	0.1	0.2	0.2	1.0	1.8	6.0
50/37.5/12.5	$0.4 < d < 0.63$	–	0.05	0.1	0.6	1.3	5.1
30/70/0	–	–	–	–	–	–	–
30/52.5/17.5	without fraction	–	–	–	0.1	0.1	0.6
30/52.5/17.5	$d < 0.1$	–	–	–	0.1	0.1	1.2
30/52.5/17.5	$0.1 < d < 0.4$	–	–	–	–	0.1	0.4
30/52.5/17.5	$0.4 < d < 0.63$	–	–	–	–	–	0.05

Notes: IPP, isotactic polypropylene; EPDM, ethylene-propylene-diene rubbers; GRT, ground rubber tire; MFI, melt flow index; TPV, thermoplastic vulcanizates.

Therefore, a partial replacement of a virgin elastomer by rubber powders can essentially change a rheology of TPVs.

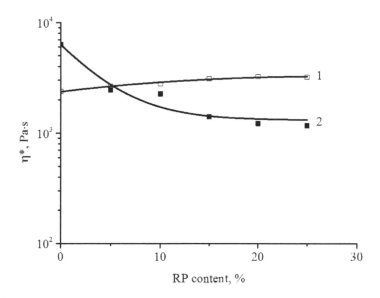

FIGURE 5.17 Plots of Complex Viscosity log η^* Versus Rubber Powder Content for (1) Uncured and (2) Dynamically Cured by Sulfur System IPP/EPDM-4044 = 50/50. IPP, Isotactic Polypropylene; EPDM, Ethylene-Propylene-Diene Rubbers.

CONCLUSION

This review provides an overview of the science and technology of grinding of waste plastics. The review describes the types of rubbers that are suitable for recycling, the mechanism of degradation of various rubbers, and characterization of the grinding products.

Today, the process HTSD is developed as a new size-reduction method for producing rubber powder. The use of rubber powder has increased rapidly over recent years due to their unique combination of processability and performance. Powders made with the HTSD technology would enhance the manufacturing of parts due to their unique morphology and surface characteristics. Because HTSD deals with the behavior of solids, sometimes referred to as particulates or powders, it is important to study the relationship between powder properties, such as particle size, shape, and bulk density and their motions during processing.

In addition to the powder-production capability, it has been demonstrated that the process allows for blending incompatible polymers and incorporating and homogenizing additives. With HTSD, these processes occur in the solid state, resulting in powders with properties superior to those of melt-processed materials. Efficient mixing in the solid state will permit the development of new polymer blends with stable morphologies while eliminating phase inversion associated with conventional melt-mixing techniques.

Progress in rubber separation and sorting is a major factor in the future development of both mechanical and feedstock recycling. As methods of rubber sorting are improved, the purity and homogeneity of the rubber wastes are increased, which favor the application of more advanced grinding processes, specific for each type of rubber. Therefore, more intensive work on rubber separation should be carried out in the next few years.

Increasing the value of the products derived from feedstock recycling of rubber wastes will also help the process economy.

As a final conclusion, it can be stated that, although feedstock recycling of rubber wastes has been studied for more than 20 years, due to the need for suitable alternatives for dealing with the increasing generation of these types of wastes and the number of technological challenges still present, this topic will continue to be the subject of much research activity for the beginning of the 21st century.

ACKNOWLEDGMENTS

The authors gratefully acknowledge the Russian Science Foundation (Contract No 14−13−00803) for the financial support.

KEYWORDS

- **Rubber powder**
- **ground rubber tire**
- **elastic deformation grinding**
- **high-temperature shear deformation**
- **blend**
- **composite**

REFERENCES

1. In *Polymer Blends*; Paul, D. R., Bucknall, C. B., Eds.; John Wiley & Sons, Inc: New York, 2000; Vol. 1 and 2.
2. In *Polymer Blends*; Paul, D. R., Newton, S., Eds.; Academic Press: New York, 1978.
3. Utracki, L. A. *Polymer Alloys and Blends*; Hanser: Munich, 1990.
4. Harrats, C.; Thomas, S.; Groeninckx, G. In *Micro- and Nanostructured Multiphase Polymer Blend Systems: Phase Morphology and Interfaces*; Harrats, C., Thomas, S., Groeninckx, G., Eds.; CRC Press: Boca Raton, 2006; p 456.
5. Bucknall, C. B. *Toughened Plastics*; Applied Science Publ: London, 1977.
6. In *Key Engineering Materials*, Chapter 17; Kajzar, F., Pearce, E. M., Turovskij, N. A., Mukbaniani, O. V., Eds.; Apple Academic Press: Toronto, NJ, 2014; Vol. II, p 337.
7. Isayev, A. I.; Sayata Ghose. *Ultrasonic devulcanization of used tires and waste rubbers*. In *Rubber Recycling*; De, S. K., Isayev, A. I., Khait, K., Eds.; CRC Press: Boca Raton, 2005; pp 311–384.
8. Adhikari, B.; De, D.; Maiti, S. Reclamation and recycling of waste rubber. *Prog. Polym. Sci.* 2000, 25, 909.
9. Karger-Kocsis, J.; Meszaros, L.; Barany, T. Ground tyre rubber (GTR) in thermoplastics, thermosets, and rubbers. *J. Mater. Sci.* 2013, 48, 1.
10. Tobolsky, A. V. *Properties and Structure of Polymers*; John Wiley & Sons, Inc: New York, 1961.
11. Prut, E. V. Plastic flow instability and multiple fracture (grinding) of polymers: a review. *Polym. Sci.* 1994, 36, 493.
12. Bridgman, P. W. *Studies in Large Plastic Flow and Fracture*; McGraw-Hill: New York, 1952.
13. Fridman, M. L.; Prut, E. V. Physical and physico-chemical effects on the rheological properties of the polymer during processing. *Usp. Khim. (in Rus.).* 1986, 53, 309.
14. Zharov, A. A.; Kazakevich, A. G.; Enikolopyan, N. S. The mechanism of the polymerization of acrylamide in conditions of high pressure and shear deformation. *Rep. Acad. Sci. USSR. (in Rus.).* 1976, 230, 354.
15. Zhorin, V. A.; Kissin, Yu.V; Luizo, Yu.V; Fridman, N. N.; Enikolopyan, N. S. Structural changes in polyolefins at high pressures in combination with the shear strain. *Vysokomol. Soedin., Ser. A. (in Rus.).* 1976, 18, 2677.
16. Kompaniets, L. V.; Dubnikova, I. L.; Kuptsov, S. A.; Zharov, A. A.; Prut, E. V. Effect of high pressure on mechanical degradation of isotactic polypropylene under plastic flow. *Polym. Sci., Ser. A.* 2001, 43, 332.
17. Kompaniets, L. V.; Kuptsov, S. A.; Erina, N. A.; Dubnikova, I. L.; Zharov, A. A.; Prut, E. V. Effect of joint action of high pressure and shear deformation on mechanical degradation of isotactic polypropylene. *Polym. Degrad. Stab.* 2004, 84, 61.
18. Kuptsov, S. A.; Erina, N. A.; Minina, O. D.; Zhorin, V. A.; Prut, E. V.; Antipov, E. M. Influence of large plastic deformation on the structure of the plastic phase in the bicomponent mixtures of polypropylene-high density polyethylene. *Vysokomol. Soedin., Ser. B. (in Rus.).* 1991, 33, 529.

19. Kuptsov, S. A.; Zhorin, V. A.; Erina, N. A.; Minina, O. D.; Prut, E. V.; Antipov, E. M. Phase diagrams for polypropylene-high density polyethylene after high pressure-induced plastic flow. *Polym. Sci., Ser. B*. 1993, 35, 414.

20. Kuptsov, S. A.; Erina, N. A.; Minina, O. D.; Antipov, E. M.; Prut, E. V. An X-ray scattering study of polyolefins subjected to elastic strain powdering. *Polym. Sci., Ser. A*. 1993, 35, 307.

21. Kuptsov, S. A.; Erina, N. A.; Zhorin, V. A.; Antipov, E. M.; Prut, E. V. The effect of plastic flow at high pressure on the structure of polyolefins. *Polym. Sci., Ser. A*. 1995, 37, 1037.

22. Kompaniets, L. V.; Krasotkina, I. A.; Erina, N. A.; Zhorin, V. A.; Nikol'skii, V. G.; Prut, E. V. Effect of intensive plastic deformation on the relaxation transitions in polyolefins and related blends. *Polym. Sci., Ser. A*. 1996, 38, 491.

23. Enikolopyan, N. S.; Akopyan, E. L.; Nikol'skii, V. G. Some problems of strength and fracture of polymer materials. *Macromol. Chem. Suppl*. 1984, 6, 316.

24. Enikolopyan, N. S. Physico-chemical aspects of plastic flow. *Macromol. Chem. Suppl*. 1984, 8, 109.

25. Akopyan, E. L.; Nikol'skii, V. G.; Enikolopyan, N. S. Elastic deformation grinding of thermoplastics. *Rep. Akad. Sci. USSR. (in Rus.)*. 1986, 291, 133.

26. Enikolopyan, N. S. Some aspects of chemistry and physics of plastic flow. *Pure Appl. Chem*. 1985, 57, 1707.

27. In *High-Pressure Chemistry and Physics of Polymers*, Chapter 9; Kovarskii, A. L., Ed.; CRC Press: Boca Raton, 1993.

28. Pershin, S. A.; Kryuchkov, A. N.; Knunyants, M. I.; Dorfman, I.Ya; Khoutimskii, M. N.; Prut, E. V.; Matkarimov, S.Kh. Influence of the structure of the vulcanized elastomer to flow instability and the nature of their fracture. *Vysokomol. Soedin., Ser. B. (in Rus.)*. 1986, 31, 100.

29. Pavlovskii, L. L.; Zeleneskii, S. N.; Polyakov, O. G.; Chaikun, A. M.; Prut, E. V.; Matkarimov, S.Ch; Enikolopyan, N. S. Some trends during grinding of used tires by elastic deformation grinding. *Proizvod. Ispol'z. Elastomerov. (in Rus.)*. 1990, 4, 23.

30. Prut, E. V.; Solomatin, D. V.; Kuznetsova, O. P.; Tkachenko, L. A.; Zvereva, U. G.; Rochev, V.Ya; Bekeshev, V. G. Processes of grinding of ethylene-propylene-diene rubbers and properties of rubber powder. *eXPRESS Polym. Lett*. in press.

31. Prut, E. V.; Khalilov, D. A.; Kusnetsova, O. P.; Tkachenko, L. V.; Solomatin, D. V. Grinding of EPDM vulcanizates: high-temperature sintering of rubber powder. *J. Elastom. Plast*. 2015, 47, 52–68.

32. Chaikun, A. M.; Polyakov, O. G.; Prut, E. V.; Razgon, D. R. Extrusion-grinding of crumb rubber from tires. *International Conference "Rubber–94*. 1994, 3, 681. (in Rus.).

33. Pavlovskii, L. L.; Dorfman, I.Ya; Kumpanenko, E. N.; Prut, E. V. About grinding mechanism of vulcanized elastomers. *Vysokomol. Soedin., Ser. B. (in Rus.)*. 1991, 32, 784.

34. Trofimova, G. M.; Novikov, D. D.; Kompaniets, L. V.; Medintseva, T. I.; Yan, Yu.B; Prut, E. V. Effect of the method of tire grinding on the rubber crumb structure. *Polym. Sci., Ser. A*. 2000, 42, 825.

35. Kuznetsova, O. P.; Khalilov, D. A.; Aliev, I. I.; Yastrebov, B. V.; Vasserman, A. M.; Prut, E. V. Structural and dynamic microheterogeneity of rubber powder. *Russ. J. Phys. Chem.* 2009, 3, 1004.

36. Polyakov, O. G.; Chaikun, A. M.; Razgon, D. R.; Struzhkova, N. G.; Prut, E. V.; Kryuchkov, A. N. Strength of used vulcanizates made by extrusion-grinding of tread tire rubber. *Proizvod. Ispol'z. Elastomerov. (in Rus.).* 1992, 12, 17.

37. Trofimova, G. M.; Novikov, D. D.; Kompaniets, L. V.; Shashkova, V. T.; Medintseva, T. I.; Chaikun, A. M.; Prut, E. V. Modification of crumb rubber. *Polym. Sci., Ser. A.* 2003, 45, 537.

38. Pavlovskii, L. L.; Kuznetsova, O. P.; Kumpanenko, E. N.; Prut, E. V. Investigation of properties of tread rubber with used vulcanizates. *Proizvod. Ispol'z. Elastomerov. (in Rus.).* 1992, 8, 18.

39. Dorfman, I.Ya; Kryuchkov, A. N.; Prut, E. V.; Enikolopyan, N. S. The phenomenon of loss of stability of plastic flow in the solid-state extrusion of polymers. *Rep. Akad. Sci. USSR. (in Rus.).* 1984, 278, 141.

40. Knunyants, M. I.; Kryuchkov, A. N.; Dorfman, I.Ya; Prut, E. V.; Enikolopyan, N. S. Features of rheological explosive during solid-state extrusion. *Rep. Akad. Sci. USSR. (in Rus.).* 1988, 299, 147.

41. Erina, N. A.; Potapov, V. V.; Kompaniets, L. V.; Knunyants, M. I.; Prut, E. V.; Enikolopyan, N. S. Mechanism of elastic-deformation grinding of isotactic polypropylene. *Vysokomol. Soedin., Ser. A. (in Rus.).* 1990, 32, 766.

42. Knunyants, M. I.; Dorfman, I.Ya; Kryuchkov, A. N.; Prut, E. V.; Enikolopyan, N. S, Multiple fracture of low density polyethylene in the extrusion in the phase transition. *Rep. Akad. Sci. USSR. (in Rus.).* 1987, 293, 1409.

43. Accetta, A.; Vergnaud, J. M. Upgrading of scrap rubber powder by vulcanization without new rubber. *Rubber Chem. Technol.* 1981, 54, 302.

44. Accetta, A.; Vergnaud, J. M. Rubber recycling. Upgrading of scrap rubber powder by vulcanization. II. *Rubber Chem. Technol.* 1982, 55, 961.

45. Tripathy, A. R.; Morin, J. E.; Williams, D. E.; Eyles, S. J.; Farris, R. J. A novel approach to improving the mechanical properties in recycled vulcanized natural rubber and its mechanism. *Macromolecules.* 2002, 35, 4616.

46. Kuznetsova, O. P.; Zhorina, L. A.; Prut, ÉV. Blends based on ground tire rubber. *Polym. Sci., Ser. A.* 2004, 46, 151.

47. Timoshenko, S. P.; Goodier, J. N. *Theory of Elasticity*; McGraw-Hill: New York, 1970.

48. Frenkel', Ya.I. *Kinetic Theory of Liquids*; Nauka: Leningrad, 1975. (in Rus.).

49. Polyakov, O. G.; Chaikun, A. M. *Revulcanizates of Crumb Rubber*; TsNIITÉNeftekhim: Moscow, 1993. (in Rus.).

50. Knunyants, M. I.; Chepcl', L. M.; Kryuchkov, A. N.; Zelenetskii, A. N.; Prut, E. V.; Enikolopyan, N. S. An influence of processing conditions on properties of composites based on polyethylene and vulcanized elastomers. *Mekh. Kompoz. Mater. (in Rus.).* 1988, 5, 927.

51. Prut, E. V.; Zelenetskii, A. N. Chemical modification and blending of polymers in an extruder reactor. *Russ. Chem. Rev.* 2001, 70, 65.

52. Serenko, O. A.; Avinkin, V. S.; Bazhenov, S. L. The effect of strain hardening of a thermoplastic matrix on the properties of a composite containing an elastic filler. *Polym. Sci., Ser. A.* 2002, 44, 286.

53. Bazhenov, S. L.; Goncharuk, G. P.; Knunyants, M. I.; Avinkin, V. S.; Serenko, O. A. The effect of rubber particle on the fracture mechanism of a filled HDPE. *Polym. Sci., Ser. A.* 2002, 44, 393.

54. Goncharuk, G. P.; Bazhenov, S. L.; Obolonkova, E. S.; Serenko, O. A. Effect of concentration of rubber particles on the fracture mechanism of filled polypropylene. *Polym. Sci., Ser. A.* 2003, 45, 584.

55. Trofimova, G. M.; Kompaniets, L. V.; Novikov, D. D.; Prut, E. V. Strain properties of blends based on isotactic polypropylene and crumb rubber. *Polym. Sci., Ser. B.* 2005, 47, 159.

56. Prut, E. V.; Zhorina, L. A.; Novikov, D. D.; Kompanietz, L. V.; Gorenberg, A. Ya. Effect of polymer matrix on mechanical properties of ground rubber tire/polymer composites. *J. Appl. Polym. Sci.* 2012, 42(7), 825–830.

57. Prut, E.; Kuznetsova, O.; Karger-Kocsis, J.; Solomatin, D. Rheological properties of ground rubber tire filled isotactic polypropylenes of different molecular weight characteristics. *Reinf. Plast. Compos.* 2012, 31, 1758.

58. Dementienko, O. V.; Kuznetsova, O. P.; Tikhonov, A. P.; Prut, E. V. The effect of dynamic vulcanization on the properties of polymer–elastomer blends containing crumb rubber. *Polym. Sci., Ser. A.* 2007, 49, 1218.

59. Kuznetsova, O. P.; Chepel', L. M.; Zhorina, L. A.; Kompaniets, L. V.; Prut, E. V. Static and dynamic vulcanizates containing ground tire rubber. *Prog. Rubb., Plast. Recycl. Technol.* 2004, 20, 85.

60. Prut, E.; Kuznetsova, O.; Chepel', L.; Zhorina, L. Thermoplastic elastomers based on ground rubber tire. The eighth international conference on thermoplastic elastomers. Conf. Proc. TPE 2005. Paper 13, Rapra, 2005.

61. Nielsen, L. E. *Mechanical Properties of Polymers and Composites*; Marcel Dekker: New York, 1974.

62. Prut, E.; Medintseva, T.; Solomatin, D.; Kuznetsova, O. Rheological behavior of thermoplastic vulcanizates. *Macromol. Symp.* 2012, 59, 321.

CHAPTER 6

BIODEGRADABLE COMPOSITIONS OF POLYLACTIDE WITH ETHYL CELLULOSE AND CHITOSAN PLASTICIZED BY LOW-MOLECULAR POLY(ETHYLENE GLYCOL)

S. Z. ROGOVINA*, K. V. ALEKSANYAN, A. V. GRACHEV, A. YA. GORENBERG, A. A. BERLIN, and E. V. PRUT

Semenov Institute of Chemical Physics of Russian Academy of Sciences, Moscow, Russia, Email: s.rogovina@mail.ru

CONTENTS

ABSTRACT

Compositions of polylactide (PLA) with polysaccharides ethyl cellulose and chitosan are obtained at different initial ratios of components under conditions of shear deformation in a Brabender mixer. It has been shown that the addition of a given amount of low-molecular poly(ethylene glycol) (PEG) leads to an increase in the elongation of rigid polysaccharide–PLA compositions. The influence of molecular weight and amount of PEG on the thermal behavior of PLA is investigated by differential scanning calorimetry method. The biodegradability of films prepared from the blends under investigation is estimated by weight loss after holding in soil and tests on the fungus resistance, and it is shown that the compositions have good biodegradability. The changes in the film morphology after holding in soil revealed by the SEM method additionally confirm that compositions are subjected to biodegradation.

INTRODUCTION

The aim of this study is to synthesize biodegradable polymer materials for the utilization of synthetic polymer wastes that can be degraded under the environmental conditions (sunlight, humidity, microorganisms, etc.) with formation of water and carbon dioxide[1-3], as degradation of polymers is one of the important problems of modern polymer physics and chemistry.

Among different methods of synthesis of biodegradable polymer composite materials, the most efficient and economically profitable is the production of polymer materials via mixing synthetic polymers with natural biodegradable ones or mixing natural polymers of different classes. Such polymer compositions in which the properties of each component are useful can be successfully applied in different areas, especially in the production of packaging materials, films for foodstuffs, and articles for short-term application.

There are a lot of works on the production and investigation of biodegradable compositions based on synthetic polymers and natural polysaccharides (see, for example, reviews of Arvanitoyannis and Rinaudo [2006 and 2008] [1,3,4]). Under conditions of shear deformation in the absence of solvents, the biodegradable compositions of low-density polyethylene with different polysaccharides (starch, cellulose, chitin, ethyl cellulose, and chitosan) were obtained, and their structure, morphology, and biodegradability were studied. It was shown that the biodegradability depends on the blend composition and the nature of polysaccharide.[5-7]

In recent years, polylactide (PLA) – a product of lactic acid polymerization – attracts a great interest. PLA is isotactic polymer with optical activity of polymers and has L- and D-isomers with quite high degree of crystallinity (30–80%) depending on the production method. PLA density comprises 1.27 g/cm^3; densities of amorphous and crystalline regions are equal to 1.248 and 1.290 g/cm^3, respectively. The glass transition point of PLA lies in the region of 57–60°C, while the crystallinity degree and melting point depend on the isomer composition. The statistic polymers containing different isomers are less crystalline and soften before melting at lower temperatures.

In comparison with other polyesters based on plant raw material, thanks to thermal and mechanical properties of the complex, PLA is most promising for the production of plastics and fibers with the given characteristics, especially taking into account serious technological and economic problems at their production, application, and utilization. Due to its high mechanical characteristics, oil and UV radiation stability, the capability of retaining the shape, etc., PLA is in competition with traditional synthetic polymers. However, the possibilities of its application are restricted by low values of elongation at break (<10%) and impact strength (~5 kJ/m^2).

The mixing of PLA with synthetic and natural polymers allows one to impart new properties to the materials based on it. This explains the wide range of works dedicated to investigation of blends of PLA with polymers of different classes. For example, the compositions of PLA with polybutylene carbonate,[8] ethylene copolymers,[9-11] and thermoplastic polyurethanes[12] were studied. The blends based on PLA and natural rubber are characteristic of a significant increase in the elongation at break compared with PLA.[13]

As is known, pure PLA is biodegradable only under specific conditions (elevated temperature, humidity, etc.),[14,15] unlike its blends with different polymers. The PLA-based compositions with improved biodegradability and high mechanical characteristics were produced with the following polymers: aliphatic polyesters,[16,17] poly([R,S]-3-hydroxybutyrate),[18] poly(3-hydroxybutyrate-co-4-hydroxybutyrate),[19] etc., aliphatic–aromatic copolyesters,[20-22] hyaluronic acid,[23] etc.

A lot of works is dedicated to the study of the compositions of PLA with chitosan – a biodegradable polymer formed by deacetylation of a natural polysaccharide chitin[24–28]. Chitosan is used as a packaging polymer or coating owing to its good film-forming capacity. Among the numerous application areas of the compositions based on chitosan, the medicine is of great importance. The possibility of using chitosan–PLA compositions as a nerve conduit was shown by Xie et al, [28] and this composition was used as a scaffold material in the study by Li et al.[25] The compositions with montmorillonite additives, which can be used for the production of biodegradable devices, which should not be removed from organism, were described by Nanda et al.[26]

One of the most widely used plasticizer for PLA is poly(ethylene glycol) (PEG).[29–33] The PLA–PEG compositions were studied in detail.[30,34–36] However, there are no literature data on the influence of different low-molecular PEG on PLA properties.

The aim of this work was the production of biodegradable binary compositions of two polysaccharides – ethyl cellulose and chitosan – with PLA, as well as their ternary compositions containing low-molecular PEG, and investigation of the mechanical characteristics, thermal properties, morphology, and biodegradation capacity. The films prepared from compositions of PLA and chitosan may have increased water resistance in comparison with the films prepared from pure chitosan that may improve their performance as packaging materials. The compositions based on thermoplastic ethyl cellulose and PLA can be used for molding and casting of the articles for different purposes: spectacle frames, tool handles, toothbrushes, etc.

EXPERIMENTAL

MATERIALS

Chitosan (deacetylation degree = 0.87, and molecular weight = 4.4 10^5 [Bioprogress, Russia]), ethyl cellulose (ethoxy group content = 46.6% and dynamic viscosity = 57), PLA (density = 1.24 g/cm^3, T_m = 150–175°C [Hycail® HM 1011, Hycail, the Netherlands]), and PEG of different molecular weights (600, 1000, and 3000) were used.

BLEND PRODUCTION

Ethyl cellulose and chitosan were blended with PLA in a Brabender mixer (Brabender GmbH & Co. KG, Dusseldorf, Germany) at 160°C for 10 min. The ternary compositions with PEG were obtained in the same apparatus under similar conditions.

METHODS OF INVESTIGATION

PRESSING

For the mechanical tests, the SEM examination, and tests on biodegradability, the films 0.18–0.25 mm thick were pressed at 160°C and 10 MPa for 10 min followed by cooling under the same pressure at the rate of 15 K/min. For mechanical tests, the obtained samples were shaped as dumbbell with the size of gauge of 35×5 mm^2.

MECHANICAL TESTS

The mechanical tests were performed on an Instron-1122 tensile test machine in tension mode at the rate of the upper traverse of 50 mm/min at room temperature. Based on the stress (σ)–strain (ε) diagrams, the initial elastic modulus (E), ultimate tensile strength (σ_b), and elongation at break (ε_b) were calculated. The results were averaged for 7–10 samples.

THERMOGRAVIMETRIC ANALYSIS AND DIFFERENTIAL SCANNING CALORIMETRY

The thermal stability of individual polymers and their compositions were analyzed on a STA 449 F3 Jupiter synchronic thermal analyzer (NETSCH, Germany) in platinum open pans in the temperature range of 30–560°C with the rate of temperature change of 10 K/min in air. The sample weight was of 5–8 mg, and the rate of gas consumption was 10 mL/min.

INVESTIGATION OF COMPOSITION BIODEGRADABILITY

INVESTIGATION OF BIODEGRADABILITY BY WEIGHT LOSS AFTER HOLDING IN SOIL

The biodegradation of the polymer compositions was studied by holding the samples in soil. The films were placed into a container with wet soil consisting of biohumus, wood ash, finely grinded clay gravel, etc., at pH = 7.5. The films were placed into soil on different levels that minimizes the probability of partial loss of films. The containers were kept in a thermostat at 30°C for several months. The rate of biodegradation was controlled by the weight losses of samples measured at intervals.

TESTS ON FUNGUS RESISTANCE

The tests on fungus resistance were performed according to the procedure based on the exposition of the materials infected by fungus spores under the optimum conditions for their growth in aqueous solutions of mineral salts followed by the estimation of fungus resistance by the degree of their growth (State Standard 9.049–91). The samples were shaped as plates 50 × 50 mm in size. The concentration of different fungus spores in suspension was 1–2 billion/cm^3. The test time was 28 days; the fungus resistance in terms of the intensity of fungus growth on the samples was evaluated with a six-number scale.

SCANNING ELECTRON MICROSCOPY

The surface morphology of films prepared from initial binary ethyl cellulose–PLA and chitosan–PLA compositions and from their ternary blends with PEG, as well as the film surface morphology after holding in soil, were analyzed using a JSM-7001F JEOL scanning electron microscope at accelerating voltages of 1 kV, without any treatment of surface.

RESULTS AND DISCUSSION

PRODUCTION OF BLENDS AND MECHANICAL TESTS

To endow the biodegradable materials obtained from PLA with new properties and to extend their application areas, the biodegradable polymer compositions – ethyl cellulose–PLA and chitosan–PLA – were produced. The binary compositions of ethyl cellulose and chitosan with PLA containing 30 wt.% and 70 wt.% of polysaccharide (in the case of compositions based on chitosan only 70 wt.%) were obtained by a solid-phase blending of components under conditions of shear deformation in a Brabender mixer.

The results of mechanical tests of films obtained by pressing of compositions at 160°C are given in Tables 6.1 and 6.2. As is seen from Table 6.1, the introduction of 30 wt.% ethyl cellulose practically does not change the elastic modulus (E), whereas the addition of 70 wt.% ethyl cellulose leads to a significant drop of E. In the case of chitosan–PLA composition, E slightly increases, since chitosan is more rigid polymer than ethyl cellulose (Table 6.2). The introduction of 70 wt.% ethyl cellulose leads to a decrease in ultimate tensile strength σ_b (about six times in comparison with pure PLA); the effect observed is probably explained by a poor compatibility of these polymers. For compositions containing 30 wt.% ethyl cellulose or chitosan, this parameter decreased insignificantly compared with PLA. At the same time, as is seen from the data of Tables 6.1 and 6.2, the addition of ethyl cellulose and chitosan to PLA leads to a substantial decrease of elongation at break ε_b. Thus, the change of mechanical characteristics of the polysaccharide blends with PLA depends both on the blend composition and on the polysaccharide nature.

TABLE 6.1 Influence of Composition of Blends Based on Ethyl Cellulose and PLA on Mechanical Characteristics

Composition	Blend Composition (wt.%)	E (MPa)	σ_b (MPa)	ε_b (%)
PLA	–	2625 ± 65	52 ± 1.0	4.7 ± 0.05
Ethyl cellulose–PLA	70:30	1650 ± 96	8.7 ± 0.6	0.9 ± 0.09
Ethyl cellulose–PLA	30:70	2620 ± 90	32.7 ± 1.0	1.9 ± 0.15
Ethyl cellulose–PLA–PEG	30:60:10	1500 ± 74	12.9 ± 0.4	1.8 ± 0.13
	20:60:20	252 ± 17	4.9 ± 0.2	20.1 ± 1.60

With the aim to increase the elasticity of the films obtained from ethyl cellulose–PLA and chitosan–PLA compositions, a plasticizer – PEG (M = 600) – was added as a third component into the compositions.

The introduction of 10 wt.% PEG leads to a slight change in the elongation at break of blends, while the addition of 20 wt.% PEG results in a significant increase of ε_b, especially for compositions containing chitosan. So, the elongation value for chitosan–PLA–PEG (20:60:20 wt.%) composition increases up to 57.5% compared with the binary chitosan–PLA composition with the elongation at break of 2.5% (Table 6.2). However, further increase in PEG content up to 27 wt.% leads to a decrease of ε_b to 29.5% that can be connected with the phase separation of the components.

TABLE 6.2 Influence of Composition of Blends Based on Chitosan and PLA on Mechanical Characteristics

Composition	Blend Composition (wt.%)	E (MPa)	σ_b (MPa)	ε_b (%)
PLA	–	2625 ± 65	52 ± 1.0	4.7 ± 0.05
Chitosan–PLA	30:70	3110 ± 90	45.5 ± 2.2	2.5 ± 0.21
Chitosan–PLA–PEG	30:60:10	1370 ± 61	17.6 ± 0.3	3.2 ± 0.23
	20:60:20	106 ± 11	8.9 ± 0.1	57.5 ± 2.00
	21:52:27	181 ± 28	3.6 ± 0.05	29.5 ± 2.20

ANALYSIS OF BLENDS BY DIFFERENTIAL SCANNING CALORIMETRY AND THERMOGRAVIMETRIC ANALYSIS METHODS

The thermal behavior of PLA in the presence of PEG of different molecular weight was investigated by differential scanning calorimetry (DSC) method.

Figure 6.1 shows the DSC thermograms of PLA and its compositions with PEG_{600} at different plasticizer contents. Compared with T_g of pure PLA (69.3°C), T_g of blends diminishes with increase of PEG_{600} content up to 44.1°C in the presence of 10 wt.% PEG. The obtained dependence can be explained by a gain in the segmental mobility of PLA with the number of plasticizer molecules leading to the enhancement of PLA molecular mobility. The cold crystallization of both PLA and its blend containing 5 wt.% PEG_{600} was not observed, but with the increase in PEG content, the cold crystallization appears. Moreover, the cold-crystallization temperature T_{cc} of PLA decreases as the PEG content increases, that is, connected with a growing number of the crystallization centers. T_m of $PLA–PEG_{600}$ blends weakly depending on their composition. The analogous results were obtained for blends of PLA with PEG_{1000} and PEG_{3000}.

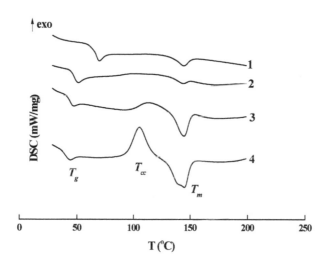

FIGURE 6.1 DSC Curves of PLA (1) and $PLA–PEG_{600}$ Compositions at Different Component Ratio: 95:5 (2), 93:7 (3), and 90:10 wt.% (4) Obtained in Air.

The determined T_g values, as well as crystallization and melting temperatures (T_{cc} and T_m) of initial PLA and PLA plasticized by PEGs of different molecular weights (600, 1000, and 3000), are presented in Table 6.3. As is seen from Table 6.3, for PEGs of different molecular weights, T_g and T_{cc} of PLA–PEG blends decrease with increase of the PEG content. The melting temperature weakly depends on the blend composition, but the heat of melting increases with the increase of content of all PEGs differing in molecular weight that may be explained by increase of the composition crystallinity.

As was found, molecular weight of PEG influences the thermal properties of PLA–PEG blends. Thus, the lower the PEG molecular weight, the stronger the decrease of T_g of PLA–PEG compositions. This is especially evident when PLA–PEG$_{600}$ blends are compared with PLA–PEG$_{1000}$ and PLA–PEG$_{3000}$ blends (Table 6.3).

TABLE 6.3   Characteristic Temperatures of PLA and PLA–PEG compositions

Composition	Blend Composition (wt.%)	T_g (°C)	T_{cc} (°C)	T_m (°C)	H_m (J/g)
PLA	–	69.3	–	144.4	4.5
PLA–PEG$_{600}$	95:5	50.7	–	144.2	1.8
	93:7	46.8	113.5	145.0	12.7
	90:10	44.1	105.4	145.5	18.6
PLA–PEG$_{1000}$	95:5	52.1	–	143.9	4.1
	93:7	49.3	121.5	144.8	6.4
	90:10	46.0	105.9	145.9	21.6
PLA–PEG$_{3000}$	95:5	52.1	–	143.6	4.7
	93:7	49.1	120.5	144.9	11.6
	90:10	44.8	110.3	143.1	19.8

Figure 6.2 shows the thermograms of PLA compositions containing 7 wt.% of PEG of different molecular weight. As can be seen from Figure 6.2, the cold crystallization heats depend on PEG molecular weight and increase with their decrease. This fact can be explained by increase of number of PEG molecules, which are the crystallization centers for PLA, with a decrease of PEG molecular weight. On the other hand, a possibility of interaction of PEG terminal hydroxyl groups with PLA in a Brabender mixer under conditions of shear deformation cannot be excluded. In this case, the lower the PEG molecular weight, the greater is the amount of hydroxyl groups capable of reacting with PLA and the higher the heat effect of reaction. The possibility of some chemical reactions in polymers under these conditions was demonstrated in our previous works[37,38]. However, this problem calls for further investigation.

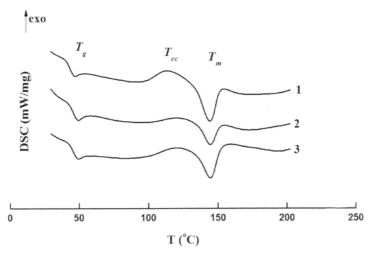

FIGURE 6.2   DSC Curves of (1) PLA–PEG$_{600}$, (2) PLA PEG$_{1000}$, and (3) PLA–PEG$_{3000}$ (93:7 wt.%) Compositions Obtained in Air.

When considering DSC curves of the chitosan–PLA and ethyl cellulose–PLA compositions, the absence of cold crystallization in the curves of the binary compositions of PLA with polysaccharides was found, that is, the PLA crystallization does not occur in the presence of polysaccharides (Fig. 6.3a,b curve 1). However, the addition of PEG leads to the appearance of a peak of cold crystallization for chitosan–PLA–PEG$_{600}$ and ethyl cellulose–PLA–PEG$_{600}$ compositions (Fig. 6.3a,b curve 2). At the same time, some decrease in T_g for the binary compositions in comparison with the initial PLA is observed (Fig. 6.3a,b).

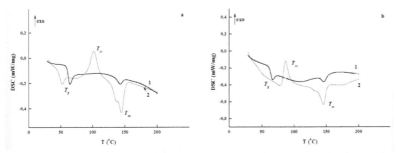

FIGURE 6.3   DSC Curves of Binary and Ternary Compositions Obtained in air: (a) Ethyl Cellulose–PLA (30:70 wt.%; curve 1) and Ethyl Cellulose–PLA–PEG$_{600}$ (30:60:10 wt.%; curve 2); (b) Chitosan–PLA (30:70 wt.%; curve 1) and Chitosan–PLA–PEG$_{600}$ (30:60:10 wt.%; Curve 2).

The compositions obtained were also investigated by thermogravimetric analysis (TGA) method. The TG curves of individual chitosan, PLA, PEG_{600}, and their binary and ternary compositions are shown in Figure 6.4. As can be seen from the presented data, the temperatures of the onset of degradation of binary chitosan–PLA and ternary chitosan–PLA–PEG_{600} (30:60:10 wt.%) compositions are slightly lower compared with that for pure chitosan.

Thus, the thermal stability of the investigated compositions depends on their composition and the polymer nature. The introduction of PEG into the compositions of polysaccharide with PLA leads to a decrease in the material thermal stability.

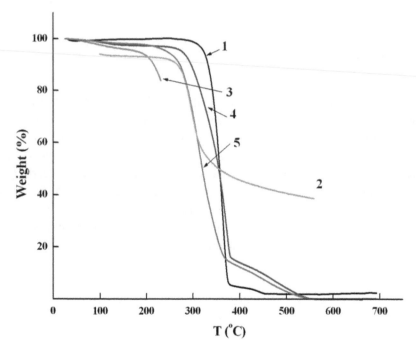

FIGURE 6.4 TG curves of (1) PLA, (2) Chitosan, (3) PEG_{600}, (4) chitosan–PLA (30: 0 wt.%), and (5) Chitosan–PLA–PEG_{600} (30:60:10 wt.%) Compositions.

INVESTIGATION OF BIODEGRADABILITY

The biodegradability of obtained compositions was studied by three independent methods.

The tests on biodegradability were performed on the films prepared from the binary ethyl cellulose–PLA and chitosan–PLA (30:70 wt.%) and ternary ethyl cellulose–PLA–PEG$_{600}$ and chitosan–PLA–PEG$_{600}$ (20:60:20 wt.%) compositions.

INVESTIGATION OF BIODEGRADABILITY BY WEIGHT LOSS AFTER HOLDING IN SOIL

The changes occurred in the samples after their holding in soil at 30°C for several months were estimated through weight losses. Figure 6.5 shows the weight loss curves for the samples held in soil. As can be seen from Figure 6.5, the weight of a binary film containing ethyl cellulose remains almost unchanged throughout the entire experiment, while the binary films containing chitosan and ternary films with PEG are subjected to biodegradation with the weight loss up to 20 wt.%. The appearance of microcracks and spots on the film surface can be clearly seen by the naked eye, especially noticeable changes were observed for the samples containing chitosan; moreover, the samples became significantly more fragile after holding in soil for 12 months (see insert in Figure 6.5).

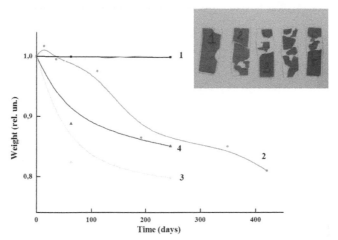

FIGURE 6.5   Weight Loss Curves of Films from (1) Ethyl Cellulose–PLA (30:70 wt.%), (2) Chitosan–PLA (30:70 wt.%), (3) Ethyl Cellulose–PLA–PEG$_{600}$ (20:60:20 wt.%), and (4) Chitosan–PLA–PEG$_{600}$ (20:60:20 wt.%) Blends After Holding in Soil for ~12 Months. Insert: Photographs of Films from Chitosan–PLA Blend (30:70 wt.%) After Holding in Soil.

TESTS ON FUNGUS RESISTANCE

The tests on fungus resistance were performed with the films obtained from the investigated binary and ternary compositions (Fig. 6.6). On examination of films from the compositions containing ethyl cellulose, the materials were not subjected to deep degradation by fungi and the intensity of mold fungus growth was measured according to 2 in the six-number scale. In this case, deep sprouting of the fungus mycelium hyphae inside the film thickness were not observed at magnification ×40 (Fig. 6.6b,d).

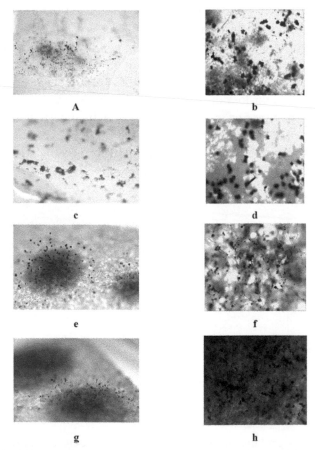

FIGURE 6.6 Micrographs of Surface of Films from (a, b) Ethyl Cellulose–PLA (30:70 wt.%), (c, d) Ethyl Cellulose–PLA–PEG$_{600}$ (20:60:20 wt.%), (e, f) Chitosan–PLA (30:70 wt.%), and (g, h) Chitosan–PLA–PEG$_{600}$ (20:60:20 wt.%) Infected with Fungus Spores for 28 days at Different Magnifications: 15× (a, c, e, g), 40× (b, d, f, h).

At the same time, the films from chitosan–PLA (30:70 wt.%) composition were subjected to effective degradation by fungi. At the greater magnification, it is clearly seen that the fungus mycelium hyphae sprouted deeply into the film structure (Fig. 6.6f). The intensity of fungus growth was estimated by the maximum value 5. The most efficient degradation by the mold fungi was noted for the films from ternary compositions of chitosan: the deep sprouting of the fungus mycelium hyphae into the film structure was observed (Fig. 6.6h). The fungus growth intensity of these films was also measured by 5. However, it should be noted that although in all samples, fungus growth is clearly seen by the naked eye, only the compositions containing chitosan showed the maximum biodegradability.

RESULTS OF SEM EXAMINATION OF SAMPLES AFTER HOLDING IN SOIL

The surfaces of the films obtained from compositions of PLA with ethyl cellulose and chitosan after holding in soil for several months were investigated by the scanning electron microscopy (SEM; Figs. 6.7 and 6.8). The cross-sections of films cannot be investigated, since the samples after holding in soil were highly fragile.

As can be seen from micrographs of chitosan–PLA film surface after holding in soil, in contrast to the initial compositions, at a low (×50) magnification, a network of microcracks that apparently leads to further cracking and fragmentation of the material was observed. At the same magnification, for chitosan–PLA–PEG$_{600}$ compositions, the microcracks are deeper and wider that testifies the more intensive biodegradation of the ternary compositions in the presence of PEG$_{600}$ in comparison with the binary ones (Fig. 6.7c,d). At the medium (×500) and high (×2000) magnifications (Fig. 6.7e,f), the PLA structure as multilayer shell and sponge is clearly seen. At the same magnifications, in the ternary compositions with PEG$_{600}$, it can be seen that the film structure is more uniform (Fig. 6.7g,h). The observed cavities in the PLA matrix suggest the presence of biodegradation.

Different results were obtained for films from the ethyl cellulose–PLA compositions (Fig. 6.8). At medium (×300) magnification, the structure of the PLA matrix containing individual fibrous elements of ethyl cellulose is observed (Fig. 6.8e). The addition of PEG$_{600}$ into composition leads to the appearance of microcracks similar to those observed for chitosan–PLA binary blends, that is, the biodegradability of the compositions based on ethyl cellulose in comparison with chitosan compositions is pronounced in the presence of PEG_{600} (Fig. 6.8f). At high (×1000) magnification, a cavity, which is apparently formed as a result of PLA biodegradation, is clearly seen (Fig. 6.8g,h). Hence, the process of biodegradation occurs not only as a result of polysaccharide degradation, but also is connected with the biodegradation of PLA.

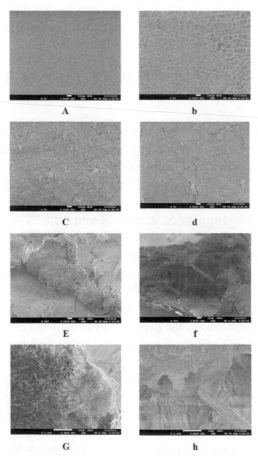

FIGURE 6.7 Electron Micrographs of Surface of Films Obtained from (a, c, e, g) Chitosan–PLA (30:70 wt.%) and (b, d, f, h) Chitosan–PLA–PEG (20:60:20 wt.%) Blends Before (a, b) and After (c, d, e, f, g, h) Holding in Soil at Different Magnifications: ×50 (a–d), ×500 (e, f), ×2000 (g, h).

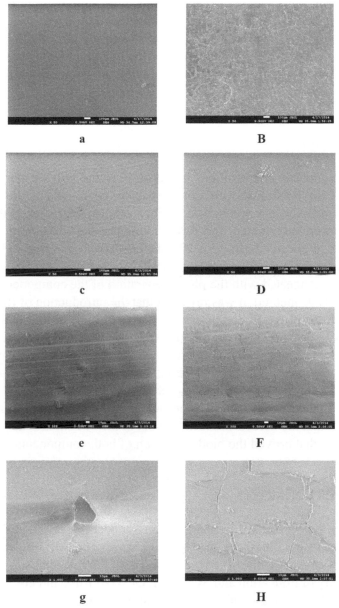

FIGURE 6.8   Electron Micrographs of Surface of Films Obtained from (a, c, e, g) Ethyl Cellulose–PLA (30:70 wt.%) and (b, d, f, h) Ethyl Cellulose–PLA–PEG (20:60:20 wt.%) Blends Before (a, b) and After (c, d, e, f, g, h) Holding in Soil at Different Magnifications: ×50 (a–d), ×300 (e, f), ×1000 (g, h).

Thus, the SEM data show the differences in morphology of the films containing chitosan or ethyl cellulose after their holding in soil, so it can be concluded that the mechanism and, consequently, the rate of biodegradation are connected with the nature of the polysaccharides used. These data confirm the results obtained on studying the composition resistance to fungus action.

CONCLUSION

The biodegradable compositions based on PLA and polysaccharides chitosan and ethyl cellulose were obtained under conditions of shear deformation in a Brabender mixer. The mechanical characteristics of the compositions were determined, and it was found that the addition of 20 wt.% of PEG_{600} plasticizer leads to a 20-fold increase in ε_b values. However, further increase in the PEG content up to 27 wt.% leads to a decrease of ε_b that can be connected with the phase separation of the components.

Using DSC method, it was revealed that the introduction of PEG leads to an increase of the PLA macromolecule mobility resulting in a change of T_g and T_{cc} depending on the amount and molecular weight of PEG used. The study of biodegradability by tests on fungus resistance showed that the compositions containing chitosan have improved biodegradability compared with the blends based on ethyl cellulose. Using SEM method for investigation of the sample morphology after their holding in soil, it was found that the defects are formed both in the PLA matrix and in polysaccharides that proved the biodegradation of both components.

ACKNOWLEDGMENTS

This work was supported by the Russian Foundation for Basic Research, project no. 13-03-12070-ofi_m. This work was produced using equipment of MIPT Center of Collective Use (CCU) with the financial support from the Ministry of Education and SCIENCE OF THE Russian Federation. We are grateful to Yu. I. Deryabina (Bach Institute of Biochemistry) for helping to do microbiological experiments.

KEYWORDS

- **Biodegradability**
- **polylactide (PLA)**
- **blends**
- **ethyl cellulose**
- **chitosan**
- **poly(ethylene glycol)**

REFERENCES

1. Arvanitoyannis, I. S. Totally and partially biodegradable polymer blends based on natural and synthetic macromolecules: preparation, physical properties, and potential as food packaging materials. *J. Macromol. Sci., Part C: Polym. Rev.* 1999, 39(2), 205.
2. Nyambo, C.; Mohanty, A. K.; Misra, M. Polylactide-based renewable green composites from agricultural residues and their hybrids. *Biomacromolecules.* 2010, 11(6), 1654.
3. Rinaudo, M. Main properties and current application of some polysaccharides as biomaterials. *Polym. Int.* 2008, 57(3), 397.
4. Rinaudo, M. Chitin and chitosan: properties and applications. *Prog. Polym. Sci.* 2006, 31(7), 603.
5. Rogovina, S. Z.; Aleksanyan, K. V.; Novikov, D. D.; Prut, E. V.; Rebrov, A. V. Synthesis and investigation of polyethylene blends with natural polysaccharides and their derivatives. *Polym. Sci., Ser. A.* 2009, 51(5), 554.
6. Rogovina, S.; Aleksanyan, K.; Prut, E.; Gorenberg, A. Biodegradable blends of cellulose with synthetic polymers and some other polysaccharides. *Eur. Polym. J.* 2013, 49(1), 194.
7. Rogovina, S. Z.; Alexanyan, Ch.V; Prut, E. V. Biodegradable blends based on chitin and chitosan: production, structure, and properties. *J. Appl. Polym. Sci.* 2011, 121(3), 1850.
8. Wang, X.; Zhuang, Y.; Dong, L. Study of biodegradable polylactide/poly(butylene carbonate) blend. *J. Appl. Polym. Sci.* 2013, 127(1), 471.
9. Anderson, K. S.; Lim, S. H.; Hillmyer, M. A. Toughening of polylactide by melt blending with linear low-density polyethylene. *J. Appl. Polym. Sci.* 2003, 89(14), 3757.
10. Feng, Y.; Hu, Y.; Yin, J.; Zhao, G.; Jiang, W. High-impact poly(lactic acid)/poly(ethylene octene) blends prepared by reactive blending. *Polym. Eng. Sci.* 2013, 53(2), 389.
11. Oyama, H. I. Super-tough poly(lactic acid) materials: reactive blending with ethylene copolymer. *Polymer.* 2009, 50(3), 747.
12. Feng, F.; Ye, L. Morphologies and mechanical properties of polylactide/thermoplastic polyurethane elastomer blends. *J. Appl. Polym. Sci.* 2011, 119(5), 2778.

13. Bitinis, N.; Verdejo, R.; Cassagnau, P.; Lopez-Manchado, M. A. Structure and properties of polylactide/natural rubber blends. *Mater. Chem. Phys.* 2011, 129(3), 823.

14. Gupta, A. P.; Kumar, V. New emerging trends in synthetic biodegradable polymers – polylactide: a critique. *Eur. Polym. J.* 2007, 43(10), 4053.

15. Wu, C.-S. Polylactide-based renewable composites from natural products residues by encapsulated film bag: characterization and biodegradability. *Carbohydr. Polym.* 2012, 90(1), 583.

16. Lopez-Rodriguez, N.; Lopez-Arraiza, A.; Meaurio, E.; Sarasua, J. R. Crystallization, morphology, and mechanical behavior, of polylactide/poly(ε-caprolactone) blends. *Polym. Eng. Sci.* 2006, 46(9), 1299.

17. Na, Y.-H.; He, Y.; Shuai, X.; Kikkawa, Y.; Doi, Y.; Inoue, Y. Compatibilization effect of poly(ε-caprolactone)-b-poly(ethylene glycol) block copolymers and phase morphology analysis in immiscible poly(lactide)/poly(ε-caprolactone) blends. *Biomacromolecules*. 2002, 3(6), 1179.

18. Bartczak, Z.; Galeski, A.; Kowalczuk, M.; Sobota, M.; Malinowski, R. Tough blends of poly(lactide) and amorphous poly([R,S]-3-hydroxy butyrate) – morphology and properties. *Eur. Polym. J.* 2013, 49(11), 3630.

19. Han, L.; Han, C.; Zhang, H.; Chen, S.; Dong, L. Morphology and properties of biodegradable and biosourced polylactide blends with poly(3-hydroxybutyrate-co-4-hydroxybutyrate). *Polym. Compos.* 2012, 33(6), 850.

20. Jiang, L.; Wolcott, M. P.; Zhang, J. Study of biodegradable polylactide/poly(butylene adipate-co-terephthalate) blends. *Biomacromolecules*. 2006, 7(1), 199.

21. Jozziasse, C. A. P.; Topp, M. D. C.; Veenstra, H.; Grijpma, D. W.; Pennings, A. J. Supertough poly(lactide)s. *Polym. Bull.* 1994, 33(5), 599.

22. Ma, X.; Yu, J.; Wang, N. Compatibility characterization of poly(lactic acid)/poly(propylene carbonate) blends. *J. Polym. Sci., Part B: Polym. Phys.* 2006, 44(1), 94.

23. Wu, C.-S.; Liao, H.-T. A new biodegradable blends prepared from polylactide and hyaluronic acid. *Polymer*. 2005, 46(23), 10017.

24. Bonilla, J.; Fortunati, E.; Vargas, M.; Chiralt, A.; Kenny, J. M. Effects of chitosan on the physicochemical and antimicrobial properties of PLA. *J. Food Eng.* 2013, 119(2), 236.

25. Li, L.; Ding, S.; Zhou, C. Preparation and degradation of PLA/chitosan composite materials. *J. Appl. Polym. Sci.* 2004, 91(1), 274.

26. Nanda, R.; Sasmal, A.; Nayak, P. L. Preparation and characterization of chitosan-polylactide composites blended with Cloisite 30B for control release of the anticancer drug paclitaxel. *Carbohydr. Polym.* 2011, 83(2), 988.

27. Xiao, Y.; Li, D.; Chen, X.; Lu, J.; Fan, H.; Zhang, X. Preparation and cytocompatibility of chitosan-modified polylactide. *J. Appl. Polym. Sci.* 2008, 110(1), 408.

28. Xie, F.; Li, Q. F.; Gu, B.; Liu, K.; Shen, G. X. In vitro and in vivo evaluation of a biodegradable chitosan-PLA composite peripheral nerve guide conduit material. *Microsurgery*. 2008, 28(6), 471.

29. Baiardo, M.; Frisoni, G.; Scandola, M.; Rimelen, M.; Lips, D.; Ruffieux, K.; Wintermantel, E. Thermal and mechanical properties of plasticized poly(L-lactic acid). *J. Appl. Polym. Sci.* 2003, 90(7), 1731.

30. Chieng, B. W.; Ibrahim, N. A.; Yunus, W. M. Z. W.; Hussein, M. Z. Plasticized poly(lactic acid) with low molecular weight poly(ethylene glycol): mechanical, thermal, and morphology properties. *J. Appl. Polym. Sci.* 2013, 130(6), 4576.
31. Cuenoud, M.; Bourban, P.-E.; Plummer, C. J. G.; Manson, J.-A. E. Plasticization of poly-L-lactide for tissue engineering. *J. Appl. Polym. Sci.* 2011, 121(4), 2078.
32. Gumus, S.; Ozkoc, G.; Aytac, A. Plasticized and unplasticized PLA/organoclay nanocomposites: short- and long-term thermal properties, morphology, and nonisothermal crystallization behavior. *J. Appl. Polym. Sci.* 2012, 123(5), 2837.
33. Sungsanit, K.; Kao, N.; Bhattacharya, S. N. Properties of linear poly(lactic acid)/polyethylene glycol blends. *Polym. Eng. Sci.* 2012, 52(1), 108.
34. Hu, Y.; Hu, Y. S.; Topolkaraev, V.; Hiltner, A.; Baer, E. Crystallization and phase separation in blends of high stereoregular poly(lactide) with poly(ethylene glycol). *Polymer.* 2003, 44(19), 5681.
35. Piorkowska, E.; Kulinski, Z.; Galeski, A.; Masirek, R. Plasticization of semicrystalline poly(L-lactide) with poly(propylene glycol). *Polymer.* 2006, 47(20), 7178.
36. Pluta, M. Morphology and properties of polylactide modified by thermal treatment, filling with layered silicates and plasticization. *Polymer.* 2004, 45(24), 8239.
37. Prut, E. V.; Zelenetskii, A. N. Chemical modification and blending of polymers in an extruder reactor. *Russ. Chem. Rev.* 2001, 70(1), 65.
38. Rogovina, S. Z.; Vikhoreva, G. A.; Akopova, T. A.; Gorbacheva, I. N.; Zelenetskii, S. N. Reaction of chitosan with solid carbonyl-containing compounds under shearing deformation conditions. *Mendeleev Commun.* 1998, 8(3), 107.

CHAPTER 7

INORGANIC POLYOXIDE THERMOPLASTICS: THEIR HYBRIDS AND BLENDS

A. A. SHAULOV and A. A. BERLIN

Semenov Institute of Chemical Physics, Russian Academy of Sciences, Moscow, Russia

Email: ajushaulov@yandex.ru

CONTENTS

ABSTRACT

Synthesis, structure, and thermal and mechanical properties of hybrids and blends of inorganic polymers, such as boron oxides, phosphates, sulfo-phosphates, and fluorophosphates with organic thermoplastics, were investigated and analyzed. The use of these hybrids presents a very promising line in the development of polymer materials science. It was shown that the blends of inorganic polyoxides with organic thermoplastics are of specific interest due to various valuable properties, such as high thermal and radiation stability, oxidation resistance, inflammability, and the absence of volatiles at thermal degradation.

INTRODUCTION

Inorganic, metallorganic, and organic polymers as nonmetallic materials can be subdivided into polymers of hydrocarbon and inorganic nature, considering inorganic polymers as compounds without hydrocarbon groups.

Their behavior similar to macromolecular compounds has been well proved, although the concepts of the polymer nature of inorganic compounds are not widely believed.[1–9]

Every class of polymers is characteristic of specific chemical, physical, and technological features. Hydrocarbon polymers with their low relaxation times are characteristic of plasticity and elasticity, relatively low processing temperatures, and low thermal stability due to the tendency of hydrocarbon groups to oxidation, combustibility, low radiation resistance, and relatively low elastic parameters.

The intensive development of polymer materials science allowed one to synthesize materials based on organic polymers with characteristics close to the theoretical values. However, their service under extreme conditions, in particular, at high temperatures, cannot be realized fundamentally. In other words, their potential development is limited to some extent.

Thus, the use of inorganic polymers and their hybrids and blends with polyhydrocarbons presents a natural evolution line in the development of polymer materials science.

By analyzing inorganic compounds from the point of view of polymer structure, more than 10 classes of polymers can be distinguished: mono-elemental polymers (C_n, Si_n, Ge_n, P_n, B_n, As_n, Sbn, Bi_n, Te_n, Se_n, and S_n),[10] polyoxides, borides, nitrides, silicides, carbides, phosphides, chalcogenides (S, Se, and Te copolymers), fluorocarbon polymers, and others. Inorganic polyoxides (IPO; silicates, phosphates, aluminates, borates, germanates, titanates, complex polyoxides − clays, etc.) with the properties determined by the nature of the basic elements are of the highest interest for the synthesis and application of hybrid polymers.

IPO, comprising up to 80% of the Earth's crust, demonstrates a variety of space structures characteristic of polymers: a wide range of softening temperatures, high thermal and radiation stability, stability to oxidation, inflammability, the absence of volatiles at thermal degradation, and low vapor pressure on processing. Moreover, IPOs are ecologically safe and can be synthesized under mild conditions.

Polyoxides are also of interest due to further peculiarity: a number of polyoxides (e.g., silica, phosphates, titanates, aluminates, borides, boron nitride, graphite, complex layered polyoxides, clays, etc.) have planar (two-dimensional [2D]) structure, which is typical for low- and high-molecular inorganic compounds, unlike organic compounds, for which the planar structure occurs only in a number of low-molecular compounds. At the same time, IPOs require elevated synthesis and processing temperatures and are characteristic of low fracture toughness and brittleness.

Unlike organic polymers with purely covalent bonds between atoms or a minor portion of ionic bonds, in polyoxides, the fraction of ionic bonds is as high as 50% that is related to a great difference in the electric negativity of oxygen and basic element atoms. This determines the high energy of internal interactions − the energy of chemical bonds and intermolecular interactions and, consequently, high melting and glass transition temperatures.

RESULTS AND DISCUSSION

When considering the properties of IPOs in the frame of the energy theory, the equilibrium interatomic distances and chemical bond energy may be related to thermal and mechanical characteristics of polymers.[10] The comparative estimation of the chemical bond energy in monoelemental polymers, in which the ionic component of the bond energy is absent, and in polyoxides showed that the chemical bond energy in oxides is two times higher due to the contribution of dipole component

$$E_k \text{ monoel} \approx 2E_k \text{ oxide.}$$

The objects of investigations were chosen on the basis of the energy theory related to the energy of internal interactions in solids with their thermal and elastic characteristics (Figs. 7.1 and 7.2).

It was shown that the melting temperature depends on the chemical bond energy only for 3D cross-linked polymers, that is, the melting of these polymers occurs under conditions of thermal degradation, whereas the melting of linear and branched polymers is determined by intermolecular physical interactions with much lower energy.

The presented data allow one to make a conclusion that boron polyoxide and phosphorus polyoxide have the minimum softening temperatures and may be used as objects of investigation.

Low-softening thermoplastic polyoxides and their hybrids and hybrid resins can be prepared via the synthesis of oxides in oligomer and polymer forms, copolymerization, chemical and physical modification by organic and metallorganic compounds using synthetic procedures typical of the synthesis and processing technology of hydrocarbon polymers.

They can be used in the preparation of thermoplastic polymer blends, low- and non-flammable block, and reinforced materials with high heat resistance and a reduced amount of gaseous products evolved during thermal exposure.

This approach may provide a number of advantages:

i)   Co-processing with an organic polymer providing the degree of filling up to 95 vol%, which allows one to avoid rheology problems accompanying the polymer filling with dispersed solid particles.

ii) Preparation of hybrid polymer blends with a combination of both classes of components at the molecular level.

iii) Chemical modification of inorganic polymers and synthesis of hybrid polymers without the use of solvents.

iv) Changes in the morphology of organic polymer in blends.

v)   Fiber formation in mineral components during processing through extrusion and molding technology.

vi) A significant reduction in the content of the hydrocarbon groups, which determine the flammability of material and present a source of volatile degradation products that allows one to obtain incombustible materials without flame-retardants.

Ultraphosphates,[11,12] fluorophosphates,[13,14] sulfophosphates,[15] boron oxides [16] are used as such inorganic polymers.

It was shown that the melting temperature of different organic polymers in compositions with fluorophosphates decreases, up to full amorphization,[17] whereas the glass transition temperature of some polymers reduces by 14°.[18]

It was also demonstrated that a variety of morphologies, such as droplets and fibrils, could be achieved, [19,20] which was previously observed in mixtures of immiscible organic polymers.

The mechanical properties of various hybrids have also been reported. It has been shown that sulfophosphate exhibits higher Young's modulus and flexural strength than equivalently filled conventional composites,[21] and fluorophosphate increases creep resistance of composites and provides the dimensional stability at high temperatures.[18]

The rheology of sulfophosphate/polyetherimide (PEI) blends was qualitatively examined through evaluation of their processing behavior. It was found that the inorganic polymer modified the flow behavior of PEI and enhanced the formability of the hybrid and its resultant properties.[21]

The elongation flow behavior (elongation viscosity at the strain rate of 1.0 s[-1]) of fluorophosphate/low-density polyethylene (LDPE) [22] and fluorophosphate/polyamide blends [23] was studied as well.

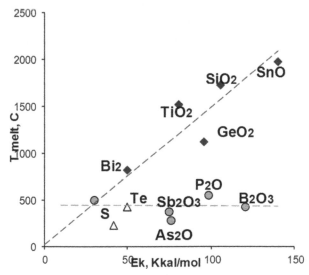

FIGURE 7.1   Melting Temperature of Polyoxides Versus Chemical Bond Energy.

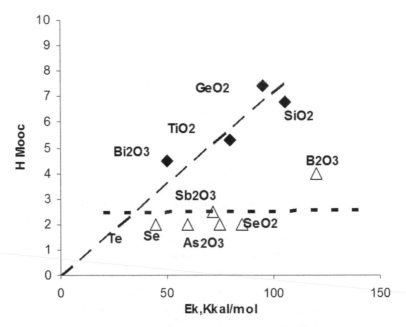

FIGURE 7.2   Microhardness of Polyoxides Versus Chemical Bond Energy.

Low-softening polyoxides are not exotic objects. The literature describes a wide range of such glasses of different compositions and content of the basic elements, which can be considered as a new class of thermoplastic oligomers. The main basic elements used in the preparation of glasses with $T_{soft} < 300°°C$ are B, P, Sn, Pb, F, Bi, and V (Table 7.1).

TABLE 7.1   Low-softening Polyoxides ($T_g \leq 250°C$)

Compositions	$T_g, °C$	$T_{soft.}, °C$	Ref.
Boron oxides	102–253	140–302	Our data
Phosphates			
P_2O_5-K_2O-Na_2O-B_2O_3-MgO-ZnO	122–300	113–180	Our data
SnO-SnF$_2$-P$_2$O$_5$	95–145	160–175	24–27
SnO-SnF$_2$-P$_2$O$_5$-B$_2$O$_3$	–	68.5–181	27
SnF$_2$-SnCl$_2$- PbO-P$_2$O$_5$	58.5–164	–	28
Lead glasses			
PbO-B$_2$O$_3$-SiO$_2$-Nd$_2$O$_3$-MgO- Bi$_2$O$_3$	–	300	29,30

TABLE 7.1 *(Continued)*

Compositions	T_g, °C	$T_{soft.}$, °C	Ref.
Bismuth glasses			
Bi_2O_3-PbO-ZnO-SiO_2-B_2O_3	–	270–300	31
Oxyfluoride glasses [18]			
$MnNbOF_5$- PbF_2	189	–	32
$MnNbOF_5$-BiF_2-BaF_2	209	–	
Phenyl phosphate glasses			
SnO-SnF_2-P_2O_5-$PhPO_3$	25–164	–	33

The feature of polyoxides is not only low-temperature processing, but also the ability of chemical synthesis and modification at mild conditions (T < 300°°C). This was the basis for the synthesis of a number of low-softening polyoxides: boron oxides, Si, Ti, Sb, Al, Mg, Zr borates, Al, Zn, Mg phosphates, and cross-linked phosphates (ultraphosphates) and fluorinated phosphates of tin and boron in order to choose the objects for investigation.

BORON OXIDE OLIGOMERS

Boron polyoxides, which are the products of polycondensation of orthoboric acid, are formed by the following scheme:

The process of polycondensation of orthoboric acid − a three-functional monomer of planar structure − includes the stage of synthesis of metaboric acids with the general formula $(HBO_2)_n$. Among them, b-form of metaboric acid constitutes a linear oligomer with the crystalline structure composed of six-membered boroxol rings connected by oxygen atoms with a hydroxyl group at each boroxol ring ($T_m = 200.9$°°C).[34-36]

The final product of boric acid (BA) polycondensation − boric anhydride − has $T_g = 300.4$°°C.

Oligomer boron oxides (OBOs) used as components of blends with polyhydrocarbons were obtained via polycondensation of BA at 220−300°°C, reaction time $\tau = 1$−3 h (Figs. 7.3 and 7.4).

The study of the conditions of synthesis of oligomers with $T_g < 150$°C resulting in oligomers in the glassy state, which can be characterized as thermoplastics with rubbery properties (Figs. 7.5 and 7.6), allows one to choose an oligomer, which meets the basic requirements of the inorganic thermoplastic as a component of blends. The obtained oligomers are stable in the atmosphere.

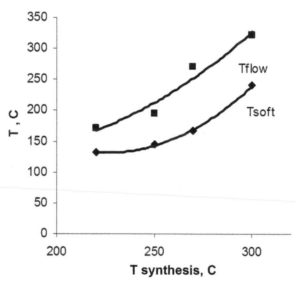

FIGURE 7.3 T_{soft} and T_{flow} of OBO Versus Synthesis Temperature ($\tau = 3$ h).

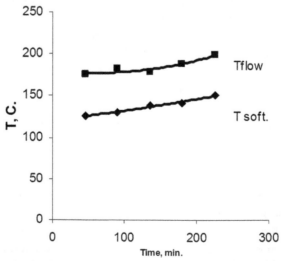

FIGURE 7.4 T_{soft} and T_{flow} of OBO Versus Synthesis Time ($T = 220°C$).

The ideal structure of boric anhydride can be presented as a planar macromolecule consisting of macrocycles from six-membered boroxol rings connected by oxygen bridges.[37] By the experimental data, the boron fraction in cycles comprises,[38] 82 ± 8 [39] and 70%.[40]

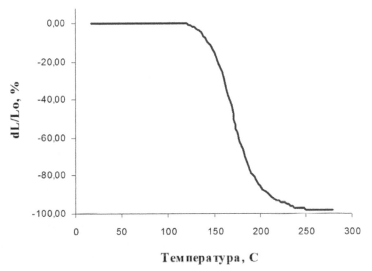

FIGURE 7.5   Thermomechanical Curve of Thermoplastic Boron Oxide Oligomer.

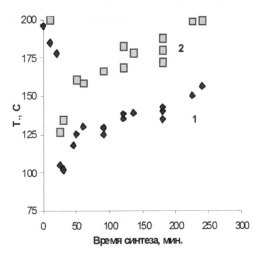

FIGURE 7.6   Temperatures of (1) Softening and (2) flow of OBOs Versus Synthesis Time ($T = 220°C$).

Crystalline boron polyoxide is characteristic of a layered supramolecular structure characteristic of weak van der Waals interactions between layers,[41,42] whereas the glassy polymer has a weakly branched disordered structure with folded and chaotically distributed branched macromolecules.[43]

We carried out the computer simulation of the structure of boron polyoxides with different degree of branching. The calculation was performed using a program package CHARMM v.30b2 under the control of OS Linux and a computer based on Intel Pentium IV 2.4 GHz processors.

The structure consideration was based on a fragment of continuous network of the Zakhariasen ideal model 300   300 nm.

The generation of macromolecules via opening of a fraction of boroxol rings of the continuous network by random law was carried out. The energy minimization and geometry optimization were performed at 0 K, as well as the structure simulation by molecular dynamics method at 300 and 500 K, with the step of 0.001 ps in the range of 10,000–100,000 steps, in vacuum.

The calculation was carried out for structures with different degrees of branching and molecular mass close to the value obtained by the measurement of melt viscosity of boric anhydride [44] with 100 (the Zakhariasen model), 70, 50, 30, and 15% boroxol rings opened. Interatomic potentials of B_2O_3 were fitted using the data [45–47] based on the Morse and Buckingham equations for calculation of valence bond parameters and van der Waals interactions, respectively, with account for the Coulomb forces and valence angle deviation (Fig. 7.7).

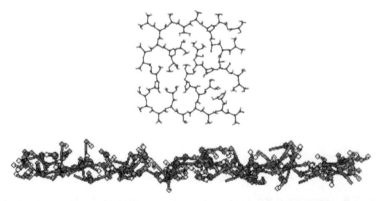

FIGURE 7.7 Computer Simulation Model of Fragment of BORon Polyoxide Macromolecule with 70% Boroxol Rings Opened ($T = 300$ K).

The borates of different metals were considered. The observed T_g are related to high temperature synthesis of polymers and allow one to determine the modifier concentration and possible changes in T_g (Fig. 7.8).

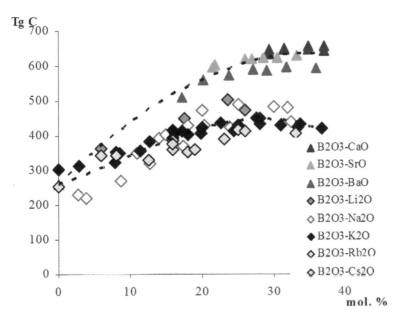

FIGURE 7.8 T_g of Borates of Uni- and Bivalent Elements (The Literature Data).

It should be noted that the modification allows the control of the relaxation properties of boron polyoxides to improve the hydrolytic stability of polymers. Aluminum borate may be considered as an example.

ALUMINUM BORATE

Thermoplastic based on BA and aluminum hydroxide was prepared by thermal treatment of powdered homogeneous oxide mixture at B_2O_3/Al_2O_3 ratios of 91.9/8.1 and 83.6/16.4 mol% ($T = 230°C$, $\tau = 3$ h)%. It was shown that the thermomechanical characteristics of aluminum borate can be changed by the addition of a third component – magnesium hydroxide. The reaction products are partially soluble in water. The yield of insoluble fraction is as high as 74.5%.

Figure 7.9 shows the thermomechanical curves for aluminum borates at the contents of aluminum hydroxide of 10 wt% (8.2 mol%), as well as for a mixture $BA/Al(OH)_3/Mg(OH)_2 = 80/19/1$ wt%.

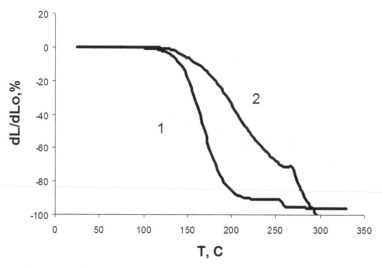

FIGURE 7.9 Thermomechanical Curves of the Product of BA and $Al(OH)_3$ Interaction and the Mixture.

$BA/Al(OH)_3/Mg(OH)_2 = 80/19/1$ wt% (T = 220°C, t = 3h): (1) $BA/Al(OH)_3 = 90/10$ wt% and (2) $BA/Al(OH)_3/Mg(OH)_2 = 80/19/1$ wt%.

The obtained aluminum borate has $T_{soft} = 140°C$, $T_{flow} = 198°C$, $\Delta T = 58°C$ and is stable in air. By comparing these data with the analogous parameters for OBO synthesized under the same conditions ($T_{soft} = 112$, $T_{flow} = 159$, and $\Delta T = 47°C$), it can be concluded that the pristine reagent changes considerably and a product with significantly higher temperature characteristics and molar mass distribution width is formed.

The data of IR and Raman spectra of the initial reagents and reaction product are also indicative of the transformation of aluminum hydroxide: the disappearance of absorption bands after thermal treatment of aluminum hydroxide in the region of high wave numbers 1018.6, 891.8, 710.4, 570 cm^{-1} and low wave numbers 147.2, 198.6, 293.2, 476.6 cm^{-1}. The comparison of the obtained data with the Raman spectra of suggested reaction products – alumina (263.5, 391.5, 429.4 cm^{-1}) and boehmite [– O – Al(OH) –]n (355.2, 460.2, 494.8, 629.9, 933.9 cm^{-1}) allows one to make a conclusion on the absence of these compounds in the reaction products (Fig. 7.10).

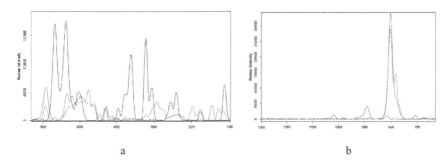

FIGURE 7.10   (a) IR and (b) Raman spectra of Al(OH)$_3$, BA, and the Product of Their Interaction.

The thermal analysis showed that the dehydration of Al(OH)$_3$ begins at $T = 220°C$. By comparing the heat of absorption of Al(OH)$_3$ equal to 859.6 J/g and the reaction product (6.4 wt%) in the same temperature range (55.5 J/g) gives the conversion of Al(OH)$_3$ of 95.6% (Fig. 7.11).

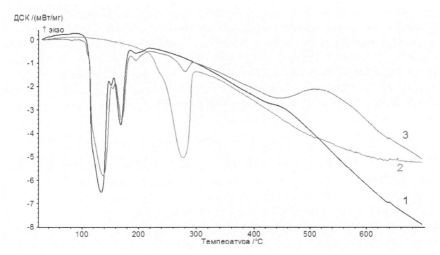

FIGURE 7.11   DSC Curves of Initial Reagents and Reaction Product: (1) BA, (2) Al(OH)$_3$, and (3) Reaction Product; B$_2$O$_3$/Al$_2$O$_3$ = 91.9/8.1 mol%.

The possibility of solid-phase synthesis of aluminum borate under the joint action of mechanical mixing and shear stresses in the Brabender mixer was studied. Figure 7.12 shows the thermomechanical curves of the products obtained in the Brabender mixer.

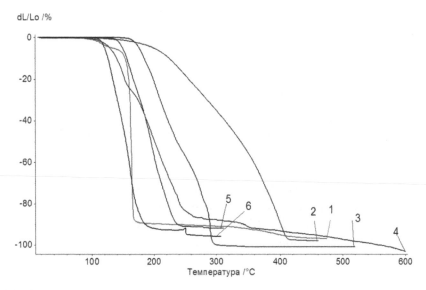

FIGURE 7.12   Thermomechanical Curves of Aluminum Borate $(B_2O_3/Al_2O_3 = 92/8$ °mol%) Prepared Under Different Conditions: (1) Mixing With Blade Stirrer ($T = 220$°C, 180 min), (2) Mixing in Brabender Mixer of ICP RAS ($T = 220$°C, 50 min), (3) Mixing in Brabender Mixer of ICP RAS ($T = 220$°C, 30 min), (4) Mixing in Brabender GmBh & Co. KG ($T = 250$°C, 6 min), (5) Mixing in Brabender GmBh & Co. KG ($T = 250$°C, 12 min), and (6) Mixing in Brabender GmBh & Co. KG ($T = 250$°C, 60 min).

The resulting thermoplastic compositions are of heterogeneous nature. It was shown that polyphosphates can be synthesized at simple technological conditions.

PHOSPHATES: LOW-TEMPERATURE SYNTHESIS

Di-and trivalent metal phosphates were synthesized under mild conditions: molar ratio $Al_2O_3/ZnO/P_2O_5 = 1/1/2$, $Al_2O_3/ZnO/P_2O_5 = 3/1/8$ (90°°C, 9 h) (Fig. 7.18), $MgO/P_2O_5 = 1/1$ (250°°C, 1.5 h) (Figs. 13 and 14). T_{soft} of the products as high as 230–235°°C and ΔT = 20°°C suggest the oligomeric character of the products, as well as high viscosity of aqueous solutions.

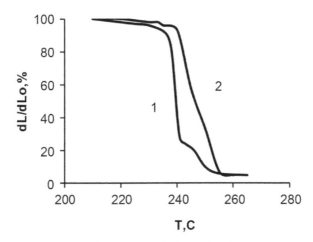

FIGURE 7.13   Thermomechanical Curves of Al-Zn-Phosphates.
1 – Al$_2$O$_3$ / ZnO / P$_2$O$_5$ 1 / 1 / 2 mol, 2 – Al$_2$O$_3$ / ZnO / P$_2$O$_5$ = 3 / 1 / 8 mol.

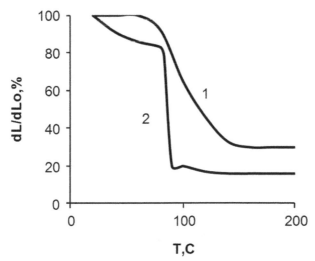

FIGURE 7.14   Thermomechanical Curves of Mg Phosphates (MgO/P$_2$O$_5$ = 1/1 mol/mol).

The dried samples are hygroscopic because of which their practical application is difficult.

ULTRAPHOSPHATES: HIGH-TEMPERATURE SYNTHESIS

The basic structure of cross-linked polyphosphates presents a modification of polyphosphoric acid, which allows one to control the temperature of the phase transition and hydrolytic stability of the polymer (Fig. 7.15). The methods that reduce glass transition temperature of ultraphosphates are based on the optimization of R_2O_5/OMe, simultaneous modification by two alkali metals, alkaline earth metals, and boron oxide (see Table 7.2). The synthesis was performed at 800°C, 2 h. Thus colorless glasses are obtained.

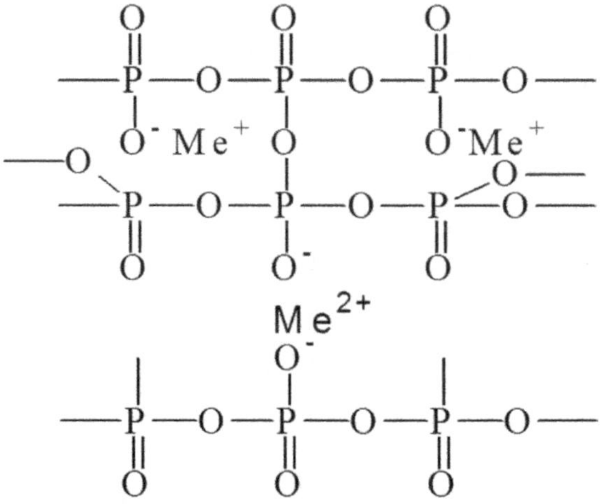

FIGURE 7.15   The Structure of Low-Softening Ultraphosphates. (Me$^+$- K, Li, Na; Me^{2+} - Mg, Ba, Zn).

TABLE 7.2   Compositions of Ultraphosphates, mol%

P$_2$O$_5$	Li$_2$O	K$_2$O	Na$_2$O	B$_2$0$_3$	Mgo	Zno	BaO
70	10	–	10	5	5	–	–
70	10	–	10	5	–	5	–
70	10	–	10	5	–	–	5

Hydrolytic stability of ultraphosphates was determined by the weight loss rate of 1 cm³ sample in boiling water for 1 h (Fig. 7.16).

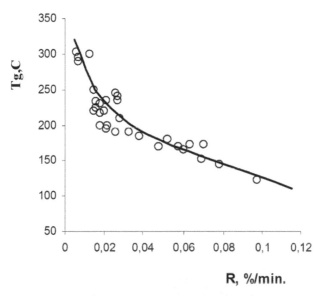

FIGURE 7.16 Hydrolytic Stability of Ultraphosphates (at T = 100°C).

FLUORINATED PHOSPHATES

One of the ways for stabilizing the hydrolytic stability of polyoxides is getting polyoxides fluorine derivatives. In this connection, fluorinated tin phosphate $[(-Sn-O)_n-(PO-(O)_d(F)_m]_k$ [24] and tin boron phosphate $[(-Sn-O)_n-(PO-(O)_d-B-O(O)_p(F_m]_k$ were synthesized at the reactant ratio $SnF_2/SnO/P_2O_5 = 40/30/30$ mol% and $SnF_2/SnO/P_2O_5 = 40/30/30 + 7\% B_2O_3$ [27] The synthesized polymers are durable thermoplastics c $T_g = 127^{\circ\circ}C$, $T_{soft} = 162^{\circ}C$, $T_{flow} = 232^{\circ\circ}C$, and $T_g = 166^{\circ\circ}C$, $T_{soft} = 181^{\circ\circ}C$, $T_{flow} = 247^{\circ\circ}C$. The synthesis was performed at 500°C for 3 h.

CHEMICAL MODIFICATION

The high activity of OH groups of polyoxides determining their chemical transformations allows their modification, which is accompanied by a decrease in the number of cross-links and by weakening of the inter-molecular interactions. The modification may be a means for controlling relaxation properties and organophilization.

The analysis of the literature data on the effect of chemical modification of SiO_2, Ge_2O_3, and B_2O_3 by univalent elements on T_g of polymer showed the dependence of T_g change on the macromolecular structure of modified polymer at low modifier contents[16] (Fig. 7.17).

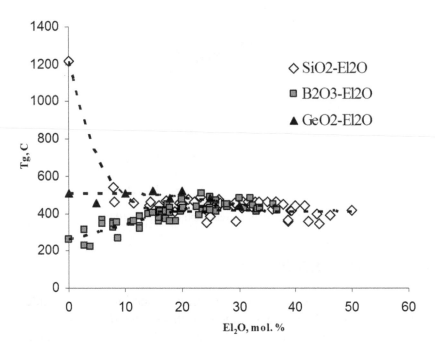

FIGURE 7.17 T_g of Silicon, Germanium, and Boron Polyoxides Versus Modifier Concentration (Li_2O, Na_2O, K_2O, Rb_2O, and Cs_2O).

Most desirable modification is the use of phenyl groups characteristic of high heat resistance and low-energy intermolecular interaction. Thus, Si, B polyoxides, and Zn phosphate phenyl derivatives were obtained.

Chemical modification was carried out via reaction of tetraethoxysi-loxane with its phenyl derivative

$$\text{Si}\left(OC_2H_5\right)_4 + +\text{Ph}-\text{Si}\left(OC_2H_5\right)_3 \rightarrow \left[-O-\text{Si}\left(Ph\right)\left(OH\right)-O-\right]_n\left[-O-\text{Si}\left(OC_2H_{5)}-O-\right]_m.$$

The presence of residual chemically active OH and C_2H_5O groups allows one to realize further reactions with the formation of cross-linked polymer by thermal treatment of the product, the modified silica with T_g 260°C can be obtained (Fig. 7.18).

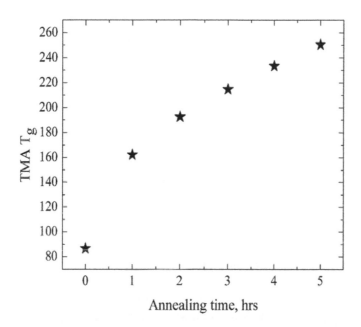

FIGURE 7.18 T_g of [Ph-Si-O$_{1.5}$-]$_n$ as a Function of Time (Synthesis Temperature 250°C).

The product of the interaction of BA with tetrafluorohydroquinone (TFHQ) was synthesized by the following reaction: (T = 140 − 200°C, 1 − 15 h).

$$B\left(OH\right)_3 + +HO-PhF_4-OH®\left(HO\right)_2-B-O-\left(PhF_4\right)-O-B-\left(OH\right)_2$$

Thermomechanical curves of reaction products with softening temperatures ranged from 50 to 160°C (Fig. 7.19).

The resulting polymers are colored black, insoluble in water, soluble in organic solvents, and incombustible.

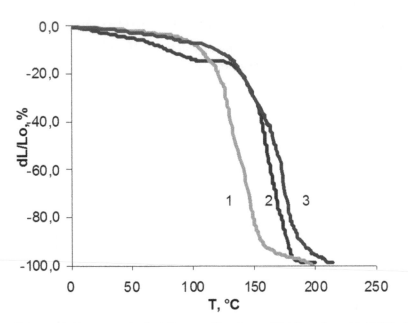

FIGURE 7.19 Thermomechanical Curves of Products of BA Reaction with TFHQ. (1: 1, mol/mol, T = 160°C; τ = (1) 3, (2) 10, and (3) 15 h.

ZINC PHENYLPHOSPHONATE

[-O-Zn-O-P(O)(Ph)-O-]$_n$ was synthesized via interaction of an aqueous solution of phenylphosphonic acid with ZnO (2: 1 mol/mol, 90°C, 1 h) under intensive stirring. The product yield was as high as 62%. The resulting zinc phenylphosphonate is water-resistant compound with T_{soft} = 481°C.

By the data of TGA and calorimetry, temperature of the onset of weight loss constitutes 508°C, and the intensive degradation of phenyl rings begins at 590°C (Fig. 7 20).

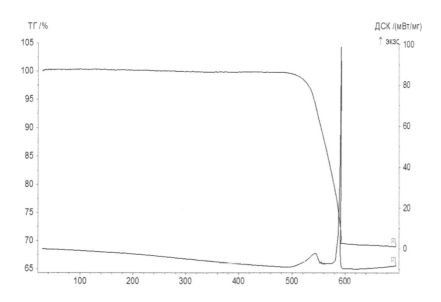

FIGURE 7.20 TG and DSC Curves of Zinc Phenylphosphonate Obtained from ZnO and Phenylphosphonic Acid.

HYBRID RESIN

Among the major disadvantages of organic resins are the flammability and toxicity of the degradation products. An improvement in these properties can be achieved by minimizing the content of hydrocarbon groups in the resin composition. For this purpose, a mixture of miscible reactive organosilicon oligomers and boron oxide was used. Hybrid resin was prepared from mutually soluble boron oxide oligomer and silicone oligomer $[CH_3SiO_{0.75}(CH_3O)_{1.5}]_n/H_{0.3}BO_{1.7}$ at a ratio of 1/0.37 mol/mol for 3 h without air access. The mixture is a clear colorless liquid (viscosity is 130 cps at 20°C, lifetime in the absence of air of 5 days at 20°C).

The resin forms polymers with $T_{soft} = 300 - 450°C$ depending on curing temperature (50 – 150°C), 5% weight loss at heating up to 500°C, and the oxygen index of 81 (Fig. 7.21).

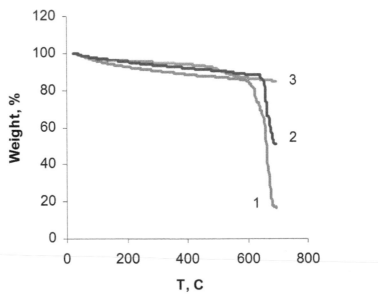

FIGURE 7.21 Thermogravimetric Data for Samples Cured at Different Temperatures: (1) 50, (2) 100, and (3) 150°C.

Based on the results of ^{13}C NMR spectroscopy, it was shown that the curing mechanism is associated with the formation of =B – O – Si bonds.

The cured samples are transparent to visible light and moisture resistant (Fig.7.22).

FIGURE 7.22 Cured Hybrid Resin (T = 120°C, 30 min); Plate is 2 mm Thick. The Signature is Located Under the Plate.

The hybrid resin was used as a binder of the composite reinforced by basalt fabric (0.15 mm, 5 layers). The noncombustible composite has flexural strength of 140 MPa, elastic modulus of 8.5 GPa, and relative elongation at break of 0.9%.

HYBRID COPOLYMERS OF BORON OXIDE OLIGOMERS WITH NITROGEN-CONTAINING COMPOUNDS

For obtaining hybrid polymers, the sol-gel method based on hydrolysis of the alkoxy compounds in acetone or alcohol solution followed by polycondensation of polyoxide nanoparticles and chemical grafting to the organic polymer is used. Using this method, the synthesis of hybrid and chemically modified polyoxides and their compositions was carried out.

The synthesis of the hybrid and chemically modified polyoxides and their compositions can be also performed in suspensions and melts of miscible compounds. The hybrid copolymers were synthesized via reaction of BA with imidazole, polyethylenepolyimine (PEPA), and caprolactam (CL) forming donor-acceptor bonds between electrophilic boron and nitrogen atoms (T = 25 − 250°C; Fig. 23). Weight loss of samples during the reaction of BA polycondensation and the softening temperature of the resulting products were determined (Fig. 7.24).

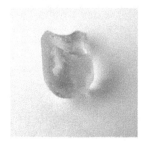

FIGURE 7.23   Samples of CL/BA = 60/40 vol%, Imidazole/BA = 70/30 vol%, and PEPA/ BA = 45/55 vol% (Temperature of Syntheses 250, 220, and 150°C, Respectively, 3 h).

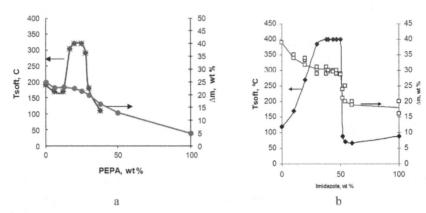

FIGURE 7.24 Softening Temperatures and Weight Loss of Hybrid Copolymers (a) BA/PEPA (150°C, 3 h) and (b) BA/Imidazole (220°C, 3 h).

The data of IR spectroscopy confirmed the formation of BN⁻ bonds and the chemical identity of the products with the maximum and minimum T_{soft}. It was shown that the stoichiometric ratio of BA and imidazole is equal to 55/45 wt%.

Abnormal changes in the relaxation properties of the resulting supramolecular structures at different ratios of reagents supposedly are related to their structure.

The formation of donor–acceptor bonds between boron and nitrogen atoms are also found in the product of BA and CL polycondensation.

The resulting polymers are environmentally stable, water soluble, and light transparent.

COPOLYCONDENSATION OF MONOMERS

The polycondensation of reagents was carried out in a homogeneous melt of BA in CL ($T_m = 62°C$) at a ratio of BA/CL = 70/30 wt% at 200 – 250°C. It was shown that during heat treatment of the reaction mixture, the polycondensation of BA takes place followed by hydrolysis of CL by released water, formation of ε-aminocapronic acid and its polymerization.

$$\text{(CL)} \quad \underset{-H_2O}{\overset{H_2O}{\rightleftarrows}} \quad NH_2\text{-}(CH_2)_5\text{-COOH} \quad \overset{-H_2O}{\longrightarrow} \quad (\text{-NH-}(CH_2)_5\text{-OC-})_n$$

The product of reaction (T = 200°C, 1 h) has T_{soft} = 200.7°C.

The structure of the reaction products was determined by IR and ^{13}C, ^{11}B, ^{1}H solid-state NMR spectroscopy.[48]

The comparison of the IR spectra of the reaction products obtained at 200°C and a mechanical mixture showed the presence of bands due to amino acid groups of starting reagents (1715, 1607, and 1317 cm^{-1}) indicative of the opening of the CL cycle. Moreover, there is an intense donor–acceptor band due to B^{+}...N^{-} at 1622 cm^{-1}. This conclusion is confirmed by the ^{11}B NMR spectra shifted upfield explained by increasing coordination number of boron.

The NMR magic-angle-spinning spectra display bands due to the CL ring opening at 200°C. The comparison of the ^{13}C NMR spectra with ^{1}H cross polarization (CP) of the initial crystalline CL and the products of thermal treatment of BA–CL mixture at 200°C showed a slight change in chemical shifts of aliphatic carbon atoms that results in the appearance of an additional line and the shift of the carbonyl C line to a stronger field (from 181.3 to 177.7 ppm).

The 2D spectrum of [^{11}B – ^{1}H] heteronuclear correlation shows that all boron atoms contact with all protons of the products obtained at 200 and 255°C.

The full contact of boron atoms with nitrogen atoms and protons indicate that the aforementioned compositions are solid solutions.

POLYMER BLENDS

A new type of composites can be obtained via blending in molten hydrocarbon and oxide polymers with similar thermomechanical and rheological properties using the conventional technologies of plastics processing.

The melt processing has several advantages: avoiding the rheological difficulties encountered in filling of solid dispersions and carrying out the synthesis of hybrid polymers using the chemical activity of the components.

The compositions of BA, OBO, boron polyoxide, and aluminum borate with PE of different molecular masses, PP, ethylene–vinyl acetate copolymers, and polyurethane were prepared by blending in Brabender mixers and a mixer of new type – RMX (elongation flow Reactor and MiXer; Scamex, France), and two-screw extruder.

During the preparation of the composition, the following effects were observed:

i)   An increase in the flow rate of polymer melts in mixtures of polyhydrocarbons with boron oxide explained by the planar structure of boron polyoxide, which is confirmed by computer simulation of its macromolecular structure.

ii) The decrease of the pressure on the walls of the extruder and torque of the screw with an introduction of OBO into the PE melt. These effects were explained by two factors: i) the larger rigidity of OBO molecules in comparison with PE chains, and ii) slippage of planar OBO on the extruder walls, as well as on the interfaces between OBO and PE. The latter one is in accordance with the tribological properties of the self-lubricated boric acid.[49,50] Further increase in OBO content leads to the abnormal drop of the pressure at volume fraction of OBO by 25 vol% (Fig. 7.25). This threshold lies in the concentration range corresponding to the interpenetrating structures of incompatible polymer blends. The sharp decrease in pressure apparently testifies a spontaneous change of the blend structure due to the violation of connectivity of the inorganic component.

Note the absence of a marked effect for PE–PMMA blends of nonplanar structure and PE–BA mixtures having the planar structure at the molecular level.

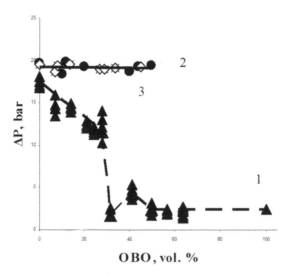

FIGURE 7.25 Pressure Drop on the Extruder Walls at Melt Mixing of PE/OBO (10, 15, 20, 30 min). (1) BA and (2, 3) PMMA (15 min).

As a result of the orientation of boron polyoxide in the melt flow, fibrils are formed at the component ratios of phase inversion accompanied by changes in the mechanical properties of molded samples of blends (Fig. 7.26).

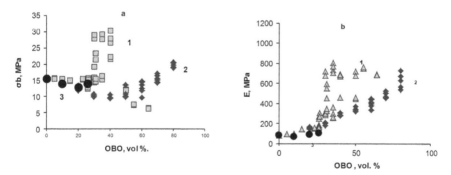

FIGURE 7.26 Dependence of (a) Tensile Strength and (b) Elastic Modulus of Blends (1) PE/OBO, (2) PE/PMMA, and (3) Composite PE/TiO$_2$ Versus Composition.

The fibril formation in the injection molding samples is observed on electron microscopic images (Fig. 7.27).

PE++10 vol% OBO PE++40 vol% OBO PE++64 vol% OBO

PE +10 vol % OBO   PE + 40 vol % OBO   PE + 64 vol % OBO

FIGURE 7.27   SEM Images of PE/Boron Oxide Blends.

VISCOMETRIC TESTS

By analyzing the flow curves of the compositions prepared at 150°C (Fig. 7.28), it was shown that the addition of HBO_2 to PE results in the enhancement of the maximum Newtonian viscosity of the two-component system, at HBO_2 concentrations up to ~30 vol%, without any qualitative changes in the shape of flow curves. However, at a higher HBO_2 content, the flow pattern changes abruptly (curve *6*): a clearly pronounced yield stress is observed, and the blend becomes viscoplastic.

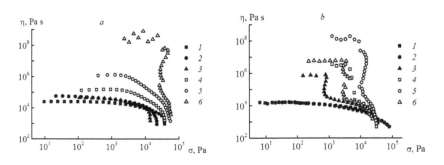

FIGURE 7.28   Viscosity Versus Shear Stress at (a) 150°C and (b) 180°C for (1) Pure PE and PE/HBO_2 Blends with (2) 6.8, (3) 14.1, (4) 21.9, (5) 30.4, and (6) 39.6 vol% HBO_2.

As the temperature increases to 180°C, the transition from the Newtonian flow of PE melt to the viscoplastic behavior of blends is observed, even for samples with 14.1 vol% HBO_2 (Fig. 7.27b). It should be emphasized that, at this temperature, the yield stress of the composition with 39.6 vol% HBO_2 is lower compared with the composition with 30.4 vol% HBO_2. Moreover, Figure 7.10b shows that, at the shear stresses above the yield point, the viscosity of these blends becomes lower compared with the PE viscosity.[51]

A similar result was obtained for PP/aluminum borate blends containing a significant amount of boron polyoxide (Fig. 7.29)[1].

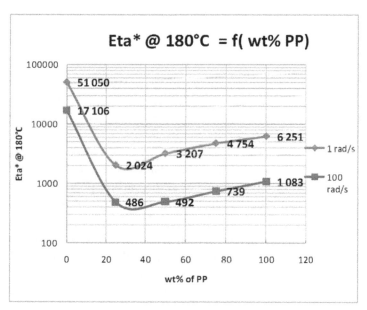

FIGURE 7.29 Viscosity of Aluminum Borate–PP Blends Prepared on RMX Mixer (Scamex).

BLENDS OF FLUORINATED POLYMERS

Other interesting systems are blends of polyoxides with fluorinated polymers characteristic of inflammability and hydrolytic and chemical stability.

---

[1]The data were obtained in collaboration with prof. R. Muller, ECPM, Strasburg University, France.

The blends were obtained by extrusion with three types of polyoxides: oligomers of boron oxide, fluorinated tin and phosphorus oxides $[(SnO)_n\text{-}(P(O)(F)_m]_k$ (FOOP), and fluorinated tin, phosphorus, and boron oxides $[(\text{-}Sn\text{-}O)_n\text{-}(P\text{-}O\text{-}(O)_d\text{-}B\text{-}O(O)_p (F)_m]_k$ (FOOPB). As fluorocarbon polymers, polyvinylidenes fluorides with different flow and processing temperatures (T_{mix}) F2 $(T_{flow} = 280°°C, T_{mix} = 280°°C, 15$ min) and F2M $(T_{flow} = 215°C, T_{mix} = 230°°C, 15$ мин) and F2MB $(T_{mix} = 280°°C, 15$ мин) were used (Fig. 7.30).

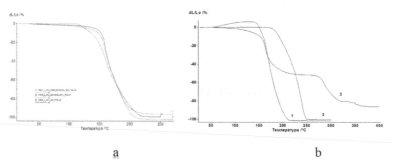

a   b

FIGURE 7.30   Thermomechanical Data of (a) F2M, OBO and F2M/OBO Blends and (b) (1) F2MB, (2) FOOPB, (3) F2MB/FOOPB = 50/50 vol% (3).

In the former case, the thermomechanical data are indicative of the unchanged softening temperature of fluorocarbon polymers and the incompatibility of components, and, in the latter case, of a new product of interaction[2].

The mechanical properties of all mixtures were measured and the stress–strain curves of molded samples were obtained. Figure 7.31 shows the mechanical properties of F2MB/FOOPB blends.

a   b   c

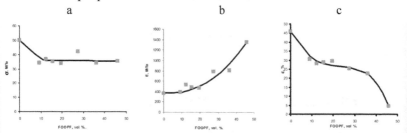

FIGURE 7.31   Dependence of (a) Tensile Strength, (b) Young's Modulus, and (c) Elongation at Break of F2MB–FOOPB Blends on FOOPB Content.

[2]The data were obtained in collaboration with V.M. Buznik, L.M. Ignatieva, and N.N. Loginova.

Fluorinated polyoxides in mixtures with F2MB exhibit chemical and catalytic activity.

STRUCTURE OF BLENDS

The degree of mixing of the components is an important characteristic of blends. The degree of mixing of the components was studied by different methods. The level of mixing of PE and PP with OBO was estimated by the paramagnetic probe method.[52]

Based on the ESR spectra of a paramagnetic molecule–iminoxyl radical incorporated into the composition, correlation times (τ) inversely proportional to the rotation frequency and characterizing the molecular mobility of the radical environment may be measured.

For this purpose, τ values in PE and PP (the paramagnetic probe is insoluble in boron oxides) were measured and compared with the ESR spectral patterns and τ values for the compositions.

It was found that the ESR spectra of probe in the compositions represent a superposition of the spectra with different τ values (Fig. 7.32).

Previously, no selective adsorption of the stable radical from heptane solution on a solid surface of boron oxide oligomer was observed, unlike the adsorption of the particles of Al $(OH)_3$.

This means that the blends contain the zones of a new phase with the molecular mobility significantly lower as compared with hydrocarbons. The new phase fraction increases with the oxide concentration up to 45% (Table 7.3).

TABLE 7.3   Spin Probe Correlation Times in New Phases of BA/LDPE Blends%

Composition, wt %	T, °C	τ_1, 10^{10}, c	%	τ_2, 10^{10}, c	%
	40	27	–	–
LDPE	60	16,3	100	–	–
	80	6,9		–	–
	40	26	84	130	16
OBO/LDPE 46/54	60	14	87	138	13
	80	6,9	85	130	15

TABLE 7.3   *(Continued)*

Composition, wt %	T, °C	τ_1, 10^{10}, c	%	τ_2, 10^{10}, c	%
	40	27	65	140	35
OBO/LDPE 69/31	60	13	55	110	45
	80	3,9	58	100	42

Vac_PE_R15

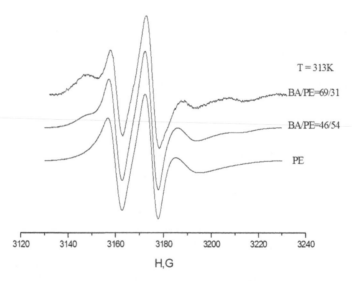

FIGURE 7.32   ESR Spectra of Spin Probe in OBO/LDPE Compositions.

The degree of blending of boron oxides with polyolefins was also studied by electron probe X-ray spectral microanalysis.[53]

The measurements were carried out with polymer films prepared by pressing at 170°C. The PE–BA–PP samples in the "sandwich" form were pressed from BA powder between the polymer films at 190°C for 15 min. The samples were annealed in the range of 190–260°C for 1–6 h.

Along with three-layer films, the phase structure of LDPE and PP blends with boron oxides prepared by the condensation of orthoboric acid, which were obtained by blending in a laboratory Brabender mixer at 190°C for 10 min followed by pressing at the same temperature was studied. The blend composition was ranged from 30 to 100 wt% BA.

The distribution profiles of polyolefins and BA in the contact zone of phases were determined by electron probe X-ray spectral microanalysis at different stages of thermal annealing.

The concentrations of C, O, and B elements in the compositions were analyzed along the line perpendicular to the contact boundary of films. The profiles of the distribution of the blend components in the contact zone of phases and their changes in the course of thermal treatment were obtained.

In the OBO phase, the intensity of the characteristic X-ray radiation of carbon corresponds to the background level, and as it moves to the polyolefin phase, the intensity increases. In monolithic phases of LDPE and PP contacting with BA particles, an extended transition zone characteristic of a smooth profile of composition change is observed (Figs. 7.33 and 7.34).

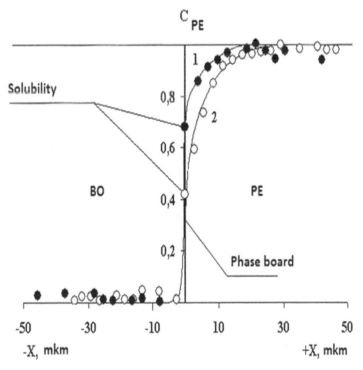

FIGURE 7.33   Evolution of Concentration Profiles in the Zone of OBO and PE Phase Conjugation at (1) 220 and (2) 260°C in the Annealing Time of 60 min.

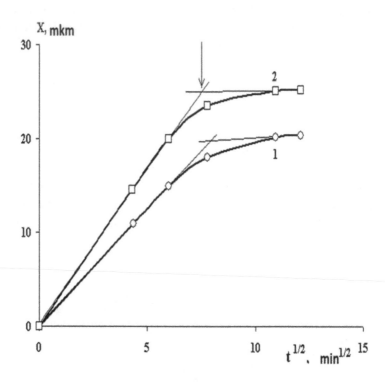

FIGURE 7.34 Kinetics of BA Diffusion Front Motion into Melt of (1) PE and (2) PP; T = 210°C.

It was found that the zone of phase conjugation increases with temperature and time of annealing, and at 260°C and annealing time of 60 min, its length achieves ~9 μm and ~10 μm for LDPE–BA and PP–BA blends, respectively.

The treatment of concentration profiles in the frame of the classical solution of the second Fick's law for semi-infinite media showed that the composition change in the transition zone is of diffusion character. By extrapolation of the concentration profiles on the interface, the solubility of BA in polyolefin melts was estimated. As indicated earlier, at 210 and 260°C, the solubilities of BA in LDPE are as high as 30 and almost 50 wt%, respectively.

The BA diffusion coefficients into the polyolefin matrix estimated by the initial portions of kinetic curves change from 1×10^{-9} cm^2/s for PP and 6×10^{-10} cm^2/s for LDPE up to 1×10^{-13} cm^2/s with the annealing time.

The results obtained are indicative of a partial molecular compatibility of components, and a diffusion-chemical model of the complex structure of composites may be proposed.

Another example of the deep mixing is the change of T_{soft} of amorphous phase of PP ($T_m = 162°C$, $T_g = 48°C$, $T_{soft = 75°C}$) in blends with boron oxides (Table 7.4).

The change in T_g of polymer can be used as a parameter reflecting the degree of mixing of the components.

The increase in T_{soft} can be explained by the formation of a new blend with the other relaxation properties dissimilar to the properties of the initial components. In both cases, such significant changes are associated with a deep interpenetration of the components.

TABLE 7.4 T_g of amorphous phase of PP in blends with boron oxide

Concentration of Boron Oxide, vol%	0	8	16	19	27
T_g, °C		48	58	70	73	77

It has been reported that T_g of polymers can change by as much as ±30°C due to the addition of a nanofiller.[54,55] The decrease of T_g may occur during heterogeneous plasticization of organic polymers by nanoparticle filler with a low surface energy, which prevents the interaction between the polymer chains. This is shown with an example of graphite,[54] alumina,[56] clay,[57] and phosphate.[58]

It was shown that some blends, namely, samples of boron oxide/polyolefin and boron oxide/ethylene–vinyl acetate copolymer blends, are transparent in the visual light that suggests a high level of mixing (the lettering is located beneath the plate, 2 mm thick; Fig. 7.35).

Taking into account that the diffraction of light waves is possible only when the size of the obstacle is 10^2–10^3 nm, one can conclude that the size of the oxide particles is less than these values.

PP + boron oxide, PE + boron oxide PE + ethylene and vinylacetate copolymer
57/43 wt% 47/53 wt% 80/20 wt%

FIGURE 7.35   Transparent Organic/Inorganic Blends.

BORON OXIDE OLIGOMERS AS INHIBITORS OF OXIDATION PROCESSES

It was shown that in PE/boron oxide blends, the thermal stability of PE
increases along with the changing direction of the reactions of thermal
oxidative degradation of the polymer.[59] The temperature of 1% weight loss
for PE–boron oxide blends increased by 140 K at the maximum oxide content up to 64 vol% (Fig. 7.36).

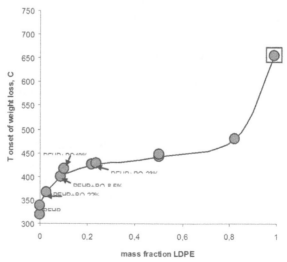

FIGURE 7.36   Temperature of Onset of Weight Loss for LDPE–Boron Oxide Blend
Versus Oxide Weight Fraction Calculated on Assumption that the Activation Energy of
Degradation is Equal to the Energy of C–C Bond Scission in PE (19.7 kJ/mol).

The following scheme of the inhibition of catalytic oxidation of poly-hydrocarbons can be presented.

Hydroperoxide formation

$$R - H + +O_2 \rightarrow HO_2 + +R^* + +O_2 \rightarrow ROO^* + +RH \rightarrow RO - OH + -$$

The stage of inhibition

$$RO - OH + +HO - B \ll\rightarrow RO^* + +^*\hat{I} - \hat{A} \ll$$

$$RO^* \rightarrow R == \hat{I}, \quad RO^* + +HO - B \ll\rightarrow R - OH.$$

It was shown that the retardation of the degradation rate is related to the effective inhibition of the process via interaction of chemically active –B–OH groups with hydroperoxides inevitably forming upon hydrocarbon oxidation. This is corroborated by the formation of oxygen-containing products of flash pyrolysis of PE in the presence of boron oxides, which are virtually absent in the products of pure PE pyrolysis that is in an accordance with the classical scheme of thermal degradation of hydrocarbons. The scheme under consideration correlates with the mechanism of liquid-phase oxidation of paraffins in the presence of catalytic amounts of boric anhydride.[60, 61]

In all obtained mass spectra of gaseous products of degradation of LDPE–OBO compositions, no boron-containing compounds were found.

The degradation of PE in the presence of boron oxides is characteristic of the presence of an additional feature, namely, more intensive formation of double bonds as a result of PE chain scission that is accompanied by conjugation and cross-linking processes and, hence, enhanced coke formation.

Thus, a conclusion can be made on a significant decrease in the rate of thermooxidative degradation of PE in the presence of boron oxides, which act as high-temperature antioxidants. The effect of thermal stabilization of polyhydrocarbon by boron oxide was also found for blends with ethylene–vinyl acetate copolymer.

LOW-TEMPERATURE CARBONIZATION

Powdered blends of BA and poly(vinyl alcohol) (PVA) with the component ratio of 50/50 wt% were studied (T = 200–600°C, 3 h). An intensive carbonization process beginning at 200°C was observed, with the shape and color of carbonized sample being retained at further heating, whereas the analogous thermal treatment of PVA does not result in the coke residue[62] (Fig. 7.37).

FIGURE 7.37   Carbonized Sample of BA and Polyvinyl Alcohol Mixture Obtained at T = 200°C.

On the examination of processes in powdered compositions, it should be taken into account that all reactions begin at the interface at temperature above T_m of BA; in this case, the polycondensation accompanied by water release with the formation of oligomer oxides.

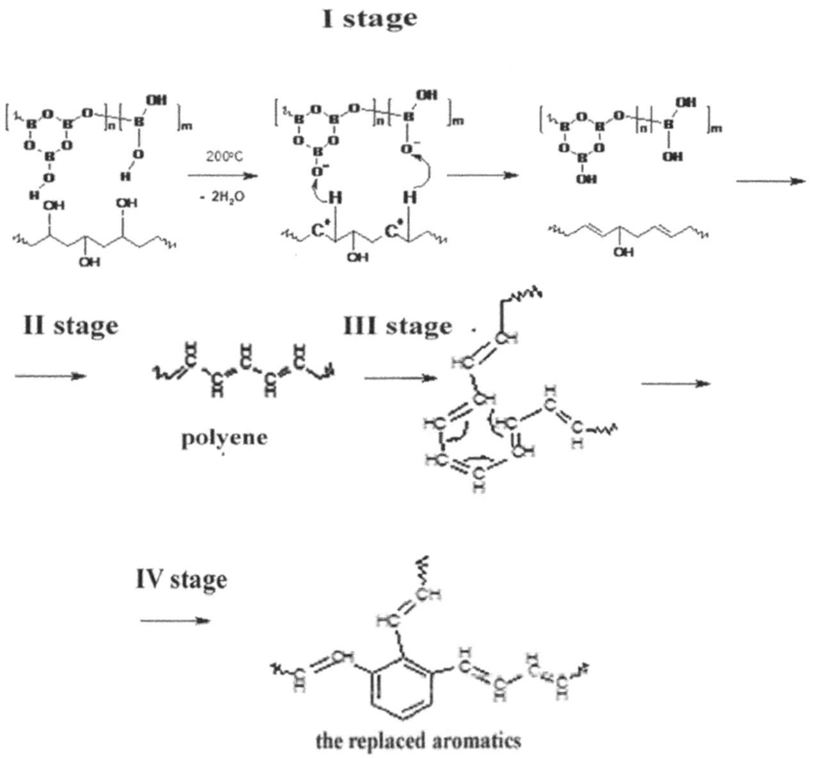

I stage

II stage   **III stage**

polyene

IV stage

the replaced aromatics

FIGURE 7.38   Reactions of Catalytic Dehydration and Dehydrogenation of PVA in the Presence of Boron Oxide.

Supposedly, in the course of thermal treatment, two parallel processes take place: catalytic degradation and dehydrogenation of PVA in the presence of boron polyoxide with the formation of diene bonds followed by their aromatization (Fig. 7.38) and chemical interaction of –B–OH groups with hydroxyl groups of PVA "retaining" carbon atoms in materials at high temperatures.

CONCLUSION

In conclusion, it should be noted that the relevant field of research is an attempt to extend the range of polymeric materials constructing an intermediate region between the hydrocarbon and inorganic polymers.

Low-softening inorganic polymers and their hybrids and blends obtained under mild conditions have a wide promise not only as inflammable block polymers but also as binders for reinforced materials.

Furthermore, the study in this field opens the way to endow inorganic polymers with the deformability.

KEYWORDS

- **Organic-inorganic hybrids**
- **polymer blends**
- **polyoxides**
- **phosphates**
- **thermal stability**
- **inflammability**

REFERENCES

1. Stone, F. G. A.; Graham, W. A. G.; Eds. *Inorganic Polymers*; Academic: New York, 1962.
2. Bartenev, G. M. *The Structure and Mechanical Properties of Inorganic Glasses*; Stroiizdat (in Rus.) Publishing House: Moscow, 1966.
3. Ray, N. H. *Inorganic Polymers*; Academic: London, 1978.
4. Tarasov, V. V. *Problems of Glass Physics*; Stroiizdat (in Rus.) Publishing House: Moscow, 1979.
5. Sanditov, D. S.; Bartenev, G. M. *Physical Properties of Disordered Structures*; "Nauka" ("Science", in Rus.): Novosibirsk, 1982.
6. Saifulin, R. S. *Inorganic Polymer Materials*; "Khimiya" ("Chemistry", in Rus.) Publishing House: Moscow, 1983.
7. Feltz, A. *Amorphe und glasartige anorganische Festkorper*; Akademie: Berlin, 1983.
8. Gerasimov, V. V. *Inorganic Polymer Materials based on Silicon and Phosphorus Oxides*; Stroiizdat (in Rus.) Publishing House: Moscow, 1993.
9. Urusov, V. S. *Energetic Crystal Chemistry*; "Nauka" ("Science", in Rus.) Publishing House: Moscow, 1975.
10. Shaulov, A.Yu. Monoelemental polymers: structure and properties. *Polym. Sci., Ser. B.* 2006, 48, 308.
11. Shaulova, A. B.; Shaulov, A.Yu.; Shalumov, B. Z. Synthesis of low-softening point ultraphosphates as component of composites materials. The Problem Statement. Abstracts of XVI International Congress on Glass, Madrid, 1992.

12. Shaulova, A. B.; Shaulov, A.Yu.; Shalumov, B. Z. A method of producing composite materials based on polymer-polymer mixtures. RF Patent 2,048,488, 1995. Bull. No. 32. (in Rus.)

13. Otaigbe, J. U. Relationship of polymer engineering with glass technology. *Trends Polym. Sci.* 1996, 4, 70.

14. Otaigbe, J. U.; Urman, K. New phosphate glass/polymer hybrids–current status and future prospects. *Prog. Polym. Sci.* 2007, 32, 1462.

15. Young, R. T.; Baird, D. G. Processing and properties of injection molded thermoplastic composites reinforced with melt. *Process. Glasses Polym. Compos.* 2000, 21, 645.

16. Shaulov, A.Yu; Berlin, A. A. Low-softening inorganic polyoxides as polymer components of materials. *Recent Res. Dev. Polym. Sci.* 2012, 11, 21.

17. Guschl, P. C.; Otaigbe, J. U. The crystallization kinetics of low-density polyethylene and polypropylene melt-blended with a low-Tg tin-based phosphate glass. *J. Appl. Polym. Sci.* 2003, 90, 3445.

18. Urman, K.; Otaigbe, J. U. Novel phosphate glass/polyamide 6 hybrids: miscibility, crystallization kinetics, and mechanical properties. *J. Polym. Sci., Part B: Polym. Phys.* 2006, 44, 441.

19. Otaigbe, J. U.; Quinn, C. J.; Beall, G. H. Processability and properties of novel glass-polymer melt blends. *Polym. Compos.* 1998, 19, 18.

20. Guschl, P. C.; Otaigbe, J. U. An experimental study of morphology and rheology of ternary glass-PS-LDPE hybrids. *Polym. Eng. Sci.* 2003, 43, 1180.

21. Young, R. T.; McLeod, M. A.; Baird, D. G. Extensional processing behavior of thermoplastics reinforced with a melt processable glass. *Polym. Compos.* 2000, 21, 900.

22. Urman, K.; Schweizer, T.; Otaigbe, J. U. Uniaxial elongation flow effects and morphology development in LDPE/phosphate glass hybrid. *Rheol. Acta.* 2007, 46, 989.

23. Urman, K.; Schweizer, T. Rheology of fluorophosphate glass/polyamide-12 hybrids in the low concentration regime. *J. Rheol.* 2007, 51, 1171.

24. Tick, P. A. Water durable glasses with ultra low melting temperatures. *Phys. Chem. Glasses.* 1984, 25, 149.

25. Shaw, C. M.; Shelby, J. E. The effect of stannous oxide on the properties of stannous fluorophosphates glasses. *Phys. Chem. Glasses.* 1988, 29, 87.

26. Xu, X. J.; Day, D. E. Properties and structure of Sn-P-O-F glasses. *Phys. Chem. Glasses.* 1990, 31, 183.

27. Xu, X. J.; Day, D. E.; Brow, R. K.; Callahan, P. M. Structure of tin fluorophosphates glasses containing PbO or B_2O_3. *Phys. Chem. Glasses.* 1995, 36, 264.

28. Dima, V.; Balta, P.; Eftimie, M.; Ispas, V. Low melting glasses in the system SnF_2-$SnCl_2$-PbO-P_2O_5. *J. Optoelectron. Adv. Mater.* 2006, 8, 2126.

29. Schepochkina, Yu.A., Low-softening glass. RF Patent 2,326,072, 2008. (in Rus.).

30. Pavlushkin, N. M.; Juravlev, A. K. *Low-Softening Glasses*; ("Energy", in Rus.) Publishing House: Energia, Moscow, 1970.

31. Rachkovskaya, G. E.; Zacharevich, G. B.; Poliakov, V. B.; Semenkova, O. S.; Poliakov, A. V. Lowsoftening glass. RF Patent 2,237,623, 2004. (in Rus.).

32. Polischuk, S. A.; Ignatieva, L. N.; Marchenko, Yu.V; Buznik, V. M. Oxyfluoride glass. *Phys. Chem. Glasses.* 2011, 37(1), 3. (in Rus.).

33. O'Reilly, J. M. O.; Papadopoulos, K. Novel organic-inorganic glasses. *J. Mater. Sci.* 2001, 36, 1595.

34. Urusov, V. V. Thermographic study of boric acid the dehydration process. *Reports of Academy of Sciences of USSR (in Rus.)*. 1957, 116, 97.
35. Freyhardt, C. C.; Wiebcke, M.; Felsche, J. The monoclinic and cubic phases of metaboric acid (precise redeterminations). *Acta Crystallogr*. 2000, 56, 276.
36. Zachariasen, W. H. The crystal structure of monoclinic metaboric acid. *Acta Crystallogr*. 1963, 16, 380.
37. Zachariasen, W. H. The atomic arrangement of glass. *J. Am. Chem. Soc*. 1932, 54, 3841.
38. Sakowski, J.; Herms, G. Structure of vitreous and molten B_2O_3. *J. Non-Cryst. Solids*. 2001, 293, 304.
39. Jellison, G. E. Jr.; Panek, L. W.; Bray, P. J.; Rouse, G. B. Jr. NMR study bonding in some solid boron compounds. *J. Chem*. 1977, 66, 802.
40. Joo, C.; Werner-Zwanziger, U.; Zwanziger, J. W. The ring structure of boron trioxide glass. *J. Non-Cryst. Solids*. 2000, 271, 265.
41. Krebs, H. *Grundzüge der anorganischen Kristallchemie*; Ferdinand Enke: Stuttgart, 1968.
42. Brazhkin, V. V.; Katayama, Y.; Inamura, Y.; Kondrin, M. V.; Lyapin, A. G.; Popova, S. V.; Voloshin, R. N. Structural transformations in liquid, crystalline and glassy B_2O_3 under high pressure. *Lett. JETP*. 2003, 78(6), 845.
43. Krohg-Moe, J. The structure of vitreous and liquid boron oxide. *J. Non-Cryst. Solids*. 1969, 1(4), 269.
44. Volarovich, M. P.; Leontieva, A. A. *J. Phys. Chem. (in Rus.)*. 1937, 10, 439.
45. Takada, A.; Catlow, C. R. A.; Price, G. D. Computer modelling of B_2O_3. Part 1. *J. Phys.: Condens. Matter*. 1995, 7, 8659.
46. Takada, A.; Catlow, C. R. A.; Price, G. D. Computer modelling of B_2O_3 Part 2. *J. Phys.: Condens. Matter*. 1995, 7, 8693.
47. Halgren, T. A. J. Merck molecular force. *Comput. Chem*. 1996, 17, 490.
48. Shaulov, A.Yu; Skachkova, V. K.; Salamatina, O. B.; Rudnev, S. N.; Shchegolikhin, A. N.; Lomakin, S. M.; Eihoff, U.; Stoernagel, S.; Samoilenko, A. A.; Berlin, A. A. Synthesis of an inorganic-organic polymer blend from orthoboric acid and caprolactam. *Polym. Sci., Ser. A*. 2006, 48, 228.
49. Erdemir, A. Tribological properties of boric acid and boric-acid-forming surfaces. Part 1: Crystal chemistry and mechanism of self-lubrication of boric acid. *Lubr. Eng*. 1991, 47, 168.
50. Ma, X. D.; Unertl, W. H.; Erdemir, A. The boron oxide-boric acid system: Nanoscale mechanical and wear properties. *J. Mater. Res*. 1999, 14, 3455.
51. Ilyin, S. O.; Malkin, A.Ya; Filippova, T. N.; Kulichikhin, V. G.; Shaulov, A.Yu; Stegno, E. V.; Patlazhan, S. A.; Berlin, A. A. Rheological properties of polyethylene boron oxide oligomer thermoplastic blends. *Rheol. Acta*. 2008, 63, 193–200.
52. Shaulov, A.Yu; Aliev, I. I.; Lyumpanova, A.Yu; Zarkhina, T. S.; Barashkova, I. I.; Vasserman, A. M.; Berlin, A. A. Structure of boron oxide/polyethylene polymer blends. *Dokl. Phys. Chem*. 2007, 413, 63.
53. Chalykh, A. E.; Gerasimov, V. K.; Aliev, A. D.; Lyumpanova, A.Yu; Shaulov, A.Yu; Berlin, A. A. Solubility, diffusion, and phase structure of hybrid composites of polyolefins – boric acid. *Dokl. Chem*. 2008, 423, 286.

54. Bansal, A.; Yang, H.; Li, C.; Schandler, L. C. Controlling the thermomechanical properties of polymer nanocomposites by tailoring the polymer–particle interface. *J. Polym. Sci., Part B: Polym. Phys.* 2006, 44, 2944.

55. Long, D.; Lequeux, F. *Heterogeneous dynamics at the glass transition in van der Waals liquids, in the bulk and in thin films. Eur. Phys. J. E. 2001, 4, 371.*

56. Ash, B. J.; Siegel, R. W.; Schadler, L. S. Glass-transition temperature behavior of alumina/PMMA nanocomposites. *J. Polym. Sci., Part B: Polym. Phys.* 2004, 42, 4371.

57. Lee, Y.-H.; Bur, A. J.; Roth, S. C.; Start, P. R. Accelerated alpha relaxation dynamics in the exfoliated nylon 11/clay nanocomposite observed in the melt and semi-crystalline state by dielectric spectroscopy. *Macromolecules.* 2005, 38, 3828.

58. Otaigbe, J. U.; Urman, K. Polyphosphate glasses as a plasticizer for nylon and making nylon composites. Patent USA 2007 - 717934 2007290405 20070314, 2007.

59. Shaulov, A.Yu; Lomakin, S. M.; Rakhimkulov, A. D.; Koverzanova, E. V.; Glushenko, P. B.; Shchegolikhin, A. N.; Shilkina, N. G.; Berlin, A. A. Thermal stability of polyethylene in composites with boron oxide. *Dokl. Phys. Chem.* 2004, 398, 231.

60. Wood, W. G. In *Progress in Boron Chemistry*; Wood, W. G., Brotheron, R. J., Eds.; Pergamon: London-New-York-Paris, 1970; Vol. 3.

61. Novak, F. I.; Bashkirov, AP.; Kamzolkin, V. V.; Talizenkov, Yu.A. Oxidation of n-paraffin hydrocarbons in presence of boric anhydride. *Rep. Acad. Sci. USSR (in Rus.)*. 1971, 196, 149.

62. Shaulov, A.Yu; Lomakin, S. M.; Zarkhina, T. S.; Rakhimkulov, A. D.; Shilkina, N. G.; Muravlev, Yu.B; Berlin, A. A. Carbonization of polyhydrocarbon in composite with boron oxide. *Dokl. Phys. Chem.* 2005, 403, 154.

CHAPTER 8

HYPERBRANCHED 1,4-CIS+1,2-POLYBUTADIENE SYNTHESIS USING NOVEL CATALYTIC DITHIOSYSTEMS

SHAHAB HASAN OGLU AKHYARI[1], FUZULI AKBER OGLU NASIROV[1,2], EROL ERBAY[2], and NAZIL FAZIL OGLU JANIBAYOV[1]

[1]Institute of Petrochemical Processes of National Academy of Sciences of Azerbaijan, Baku, Azerbaijan; [2]Petkim Petrokimya Holding, Izmir, Turkey

Email: j.nazil@yahoo.com, fnasirov@petkim.com.tr

CONTENTS

INTRODUCTION

Branching degree of polydienes (polybutadiene, polyisoprene, and so on) is a very important parameter with the necessary specifications such as stereoregularity, molecular mass, and molecular mass distribution. Branches in the structure of polydienes strongly reduce solution viscosity (SV) by improving its processability and increasing operating ability of vulcanizates. On the other hand, the branched polymers easily decompose via light and by biological factors, which solves the problem of the elimination of waste polymers from the environment.

By using known catalytic systems, we can achieve high molecular mass, linear 1,4-cis–polybutadienes of high SV. These easily prevent the polymer processing and limit using them in preparation of impact resistant polystyrenes. At the same time, as it is known,[1] because of their high crystallization temperature, high stereo, regular 1,4-cis–polybutadienes cannot be used in the production of tires and technological rubbers, which are employed in cold climatic conditions. Therefore, it is of critical importance to obtain cold resistant, hyperbranched, and high molecular 1,4-cis+1,2-polybutadiene with 20–40% of 1,2-vinyl bonds.[1]

Earlier, in the Institute of Petrochemical Processes of Azerbaijan National Academy of Sciences, we have developed the new bifunctional nickel- and cobalt-containing catalytic dithiosystems, which have shown high catalytic activity and stereo selectivity in butadiene polymerization process. Particularly, by using the cobalt-dithiocarbamate catalytic system (diethylditiocarbamate cobalt + alkylaluminummonochloride), it was possible to obtain linear high molecular 1,4-cis+1,2-polybutadiene of high SV.[1]

For the synthesis of hyperbranched 1,4-cis+1,2-polybutadienes, the new dialkyldithiocarbamate cobalt + alkylaluminumdichloride (or alkylaluminumsesquichloride) catalytic dithiosystems have been prepared and used in the butadiene polymerization process. It has been shown that by using the dialkyldithiocarbamate cobalt + alkylaluminumdichloride (or alkylaluminumsesquichloride) catalytic dithiosystems, it is possible to synthesize hyperbranched, low SV, and high molecular weight 1,4-cis–polybutadiene with 20–40% 1,2-vinyl bond content.

In this chapter, the main results of butadiene polymerization to high molecular weight hyperbranched 1,4-cis+1,2-polybutadienes using new dialkyldithiocarbamate cobalt + alkylaluminumdichloride (or alkylalumi-numsesquichloride) catalytic dithiosystems are given.

EXPERIMENTAL PROCEDURE

Butadiene (99.8 wt.%) and aluminum organic compounds (85.0 wt.% – in benzene) were obtained from Sigma-Aldrich (St. Louis, MO, USA). Organic dithioderivatives (dithiocarbamates, dithiophosphates, and xanthogenates) of cobalt were synthesized according to the previously described procedures.[1-3]

After drying over metallic sodium for 24 h, polymerization was conducted in toluene, which was distilled and preserved over sodium. Where necessary, manipulations were carried out in 50–200 mL Schlenk-type glass reactors under dry, oxygen-free argon or nitrogen with appropriate techniques and gas-tight syringes. The usual order of addition was solvent, cobalt component, aluminum organic compound (at −78°C), and finally, the monomer. Polymerizations were conducted at 25–60°C, and then the polymerizate was poured to ethanol (or methanol) followed by the termination. The obtained polymer was washed several times with ethanol (or methanol) and was dried at 40°C in a vacuum to constant weight and stored under argon (or nitrogen).

Solution viscosities of polybutadienes were measured using an Ubbelohde viscometer in toluene solution at 30°C at a concentration of 0.2 °g/dL.[4] The molecular weight of 1.4-cis+1.2-polybutadienes were determined by viscometric method[1,4] using the following formula:

$$[\eta]_{30(toluene)} = 15.6 \cdot 10^{-5} M^{0.75}$$

The molecular weights (M_w and M_n) and molecular weight distribution (M_w/M_n) of polybutadienes were measured by a Gel Permeation Chromatograph (GPC) constructed in the Czech Republic with a 6000 A pump, original injector, R-400 differential refractive index detector, and styragel columns with nominal exclusions of 500, 10^3, 10^4, 10^5, and 10^6. The GPC instrument was calibrated according to the universal calibration method by using narrow molecular-weight polystyrene standards.[5]

The microstructure of the polybutadiene was determined using FT-IR-spectrometry ("Nicholet NEXUS 670" with spectral diapason from

400 cm^{-1} till 4000 cm^{-1}, as a film on KBr, obtained from toluene solution).[6,7]

Mooney viscosity (MV) was determined by the means of a SMV-201 Mooney viscometer (Shimadzu Co., Ltd) in accordance with ASTM D 1646.[8,9] Branching degree was calculated based on the viscosity of polybutadiene at 25°C in 5% toluene solution. SV in proportion to MV at 100°C (SV/MV) and its branching index (g_M) were calculated from proportional literature sources.[8,9] Branching index for linear polybutadiene is $g_M = 1.0$ and for branched polybutadienes is $g_M < 1.0$.

RESULTS AND DISCUSSION

In this study, we have used dialkyldithiocarbamate cobalt as the main catalyst component with the structure as shown below:

$$\underset{R}{\overset{R^1}{\diagdown}} N - C \underset{\diagdown S}{\overset{\diagup S}{\diagup\diagdown}} Co \underset{\diagdown S}{\overset{\diagup S}{\diagdown\diagup}} C \underset{\diagdown R}{\overset{\diagup R^1}{=}} N \underset{\diagdown R}{\overset{\diagup R^1}{}}$$

where R and R^1 are alkyl, aryl, and alkylaryl radicals, and co-catalysts are the aluminum organic compounds (dialkylaluminumchlorides, alkylaluminumdichlorides, alkylaluminumsesquichlorides, and aluminumoxanes) with formulas AlR^2Cl_2 and/or $Al_2R^2_3Cl_3$, AlR^2_2Cl (where R^2 is methyl, ethyl, isobutyl radicals, and so on).

We have investigated the peculiarities of dialkyldithiocarbamate cobalt + alkylaluminumdichloride (or alkylaluminumsesquichloride) catalytic dithiosystems in the polymerization of butadiene to highly branched and high molecular weight 1,4-cis+1,2-polybutadiene. These catalysts were studied in comparison with a known cobalt-containing catalytic dithiosystems for butadiene polymerization dependence of process outcomes, catalyst activity and selectivity on the nature of ligands of cobalt compounds, catalyst component concentration and ratios, as well as the influence of temperature and polymerization time. As may be, in polymerization, the ligand nature of the transition metal is also crucial for the catalyst activity and stereoselectivity. Ligands used in this work were mainly dithiocarbamates, dithiophosphates, and xanthogenates. As can be seen in Table 8.1., Exp. 28–31, that cobalt-dithiophosphates and cobalt-dialkylxanthogenates in combination with diethylaluminumchloride (DEAC), ethylaluminumdichloride (EADC), and methylaluminoxane (MAO) allow yield of linear high molecular weight 1,4-cis polybutadiene with polymer yield of 80.0–98.0%, catalyst activity of 12.0–34.0 kg PBD/g Co h, intrinsic viscosity [η] of 1.8–2.5 dL/g, and 1,4-cis content of 90.0–97.0%. Cobalt-dialkyldithiocarbamate catalytic system–cobalt-diethyldithiocarbamate (DEDTC-Co) + DEAC (or MAO) gives linear high-molecular weight 1,4-cis+1,2-polybutadiene with polymer yield of 85.0–90.0%, catalyst activity of 20.0–22.0 kg PBD/g Co h, intrinsic viscosity [η] of 2.3–2.8 dL/g, 1,4-cis content of 63.0–66.0%, and 1,2-vinyl content of 20.0–25.0% (Table 8.1, Exp.24–27).

As seen from Table 8.1, Exp. 28–31, cobalt-dithiophosphates and cobalt-dialkylxanthogenates in combination with DEAC, EADC, and MAO allow yield of linear high molecular weight 1,4-cis polybutadiene with a polymer yield of 80.0–98.0%, catalyst activity of 12.0–34.0 kg PBD/g Co h, intrinsic viscosity [η] of =1.8–2.5 dL/g, and 1,4-cis content of 90.0–97.0%.

TABLE 8.1 Results of Butadiene Polymerization to Hyper Branched and Linear Polybutadienes with Bifunctional Cobalt-containing Catalytic Dithiosystems

№	Cobalt Compound	Aluminum Organic Compounds (AOC)	$[Co] \times 10^4$, mol/L	Al:Co	[M], mol/L	T,°C	Reaction Time, min.	Yield of PBD,%	Catalyst activity, Kg of PBD/g Co hour	[η], dL/g	Branching		M_w/M_n	Microstructure of PBD%		
											Index (g_M)	Degree (SV/MV)		1,4-cis	1,4-trans	1,2-
1	2	3	5	6	7	8	9	10	11	12	13	14	15	16	17	18
1	DEDTC	EADC	1.0	100	5.0	25	60	96.0	44.0	2.5	0.35	2.5	2.3	70	5	25
2	DMDTC	EADC	1.0	100	5.0	25	60	92.0	42.0	3.5	0.55	2.7	2.5	60	5	35
3	DBenz-DTC	EADC	1.0	100	5.0	25	60	97.0	45.0	3.3	0.40	2.6	2.6	60	10	30
4	DPhDTC	EADC	1.0	100	5.0	25	60	98.0	55.0	3.0	0.52	2.7	2.3	60	7	33
5	DEDTC	EASC	1.0	100	5.0	25	60	94.0	33.0	2.9	0.60	3.0	2.6	72	3	25
6	DBDTK	EASC	1.0	100	5.0	25	60	92.0	22.0	2.8	0.70	3.4	2.7	71	7	22
7	DEDTK	EADC	1.0	100	5.0	25	60	92.0	42.0	2.5	0.60	3.1	2.6	70	5	25
8	DMDTK	EADC	1.0	100	5.0	25	60	97.0	44.0	2.7	0.65	3.25	2.3	72	6	22
9	DBDTK	EADC	1.0	100	5.0	25	60	88.0	40.0	2.3	0.55	2.7	2.8	75	5	20

10	DEDTC	EADC	0.5	100	5.0	25	90	90.0	50.0	3.5	0.65	3.5	2.5	75	5	20
11	DEDTC	EADC	2.5	100	5.0	25	60	95.0	44.9	2.9	0.55	2.7	2.3	68	7	25
12	DEDTC	EADC	5.0	100	5.0	25	30	98.0	43.0	2.5	0.49	2.5	2.1	65	5	30
13	DEDTC	EADC	10.0	100	5.0	25	10	99.0	40.0	2.0	0.42	2.4	2.0	55	5	40
14	DEDTC	EADC	1.0	10	5.0	25	60	87.0	39.8	3.0	0.75	3.7	2.4	72	2	25
15	DEDTC	EADC	1.0	50	5.0	25	60	95.0	43.5	2.8	0.55	2.8	2.5	72	1	27
16	DEDTC	EADC	1.0	100	1.0	25	60	98.0	22.0	2.0	0.22	2.5	2.2	60	10	30
17	DEDTC	EADC	1.0	100	3.0	25	60	97.0	26.6	2.8	0.28	3.0	2.3	65	7	28
18	DEDTC	EADC	1.0	100	8.0	25	60	90.0	50.0	3.5	0.60	3.6	2.8	70	10	20
19	DEDTC	EADC	1.0	100	5.0	45	60	95.0	43.5	2.9	0.40	2.3	2.4	70	3	27
20	DEDTC	EADC	1.0	100	5.0	60	60	90.0	41.2	2.7	0.35	2.3	2.3	71	2	27
21	DEDTC	EADC	1.0	100	5.0	25	10	86.0	36.1	2.8	0.70	3.6	2.2	62	3	35
22	DEDTC	EADC	1.0	100	5.0	25	90	99.0	30.2	3.5	0.45	2.4	2.5	68	4	28
23	DEDTC	EADC	1.0	100	5.0	25	120	90.0	20.0	2.5	0.95	9.2	3.2	62	13	25
24	DMDTC	DI-BAC	1.0	100	5.0	25	120	85.0	19.5	2.3	0.96	10.0	3.2	65	15	20
25	DBDTC	DEAC	1.0	100	5.0	25	120	92.0	22.0	2.7	0.94	8.8	3.5	66	12	22
26	DBenz-DTC	MAO	1.0	100	5.0	25	120	88.0	20.0	2.7	0.95	9.0	3.6	65	13	22
27	DPhDTC	DI-BAC	1.0	100	5.0	25	120	90.0	21.0	2.8	0.96	9.5	3.5	63	12	25
28	DCDTPh	DEAC	1.0	100	3.0	25	60	98.0	12.0	1.8	0.92	8.6	1.8	90	8	2
29	DEDTPh	MAO	1.0	100	3.0	25	60	97.0	30.0	2.2	0.93	8.7	2.6	97	2	1
30	DPh-DTPh	EADC	1.0	100	3.0	25	60	94.0	34.0	2.3	0.98	8.5	3.2	95	3	2
31	BuXh	EASC	2.0	100	3.0	25	60	80.0	28.0	2.5	1.0	10.0	2.6	95	2	3
32	EtXh	TEA	2.0	100	3.0	25	60	93.0	30.0	2.5	1.0	-	-	2	2	96
33	i-PrXh	TEA	2.0	100	3.0	25	60	98.0	35.0	2.3	1.0	-	-	1	1	98

Notes: In experiments 7–9, benzene, chlorobenzene, and hexane were used as solvent, respectively; In experiments 24, 25, 27, alkylaluminummonochlorides (such as DEAC and DIBAC) were used as co-catalysts, respectively; In experiments, 28–33 O,O-dialkylsubstituted dithiophosphates and alkylxanthogenates were used as cobalt compounds, respectively.

Abbreviations: DCDTPh-Co, Cobalt-O,O'-di-4-methylphenyl dithiophosphate; DEDTPh-Co, Cobalt-Diethyldithiophosphate; DMDTC-Co, Cobalt-Dimethyldithioicarbamate; DBDTC-Co, Cobalt-Dibutyldithiocarbamate; DBenzDTC-Co, Cobalt-Dibenzyldithiocarbamate; DPhDTC-Co, Cobalt-Diphenyldithiocarbamate; DPhDTPh-Co, Cobalt-Diphenyldithiophosphate; EtXh-Co, Cobalt-Ethylxanthogenate; DCDTPh-Co, Cobalt-O,O'-di-4-methylphenyl dithiophosphate; EASC, Ethylaluminumsesquichloride; BuXh-Co, Cobalt-Butylxanthogenate; AlkXh-Co, Cobalt-Alkylxanthogenate; TEA, Triethylaluminum; iPrXh-Co, Cobalt-iso-Propylxanthogenate.

The cobalt-dialkyldithiocarbamate catalytic system DEDTC-Co + DEAC (or MAO) gives linear high molecular weight 1,4-cis+1,2-polybutadiene with polymer yields of 85.0–90.0%, catalyst activity of 20.0–22.0 kg PBD/g Co h, intrinsic viscosity [η] of 2.3–2.8 dL/g, 1,4-cis content of 63.0–66.0%, and 1,2-vinyl content of 20.0–25.0% (Table 8.1, Exp.24–27).

Cobalt-dialkylxanthogenates + TEA catalytic dithiosystems give high molecular and high crystalline syndiotactic 1,2-PBD (1,2-SPBD). In their presence, polymer yield was 93.0–98.0%, catalyst activity was 30.0–35.0 kg PBD/g Co h, and the sythesized 1,2-SPBD had an intrinsic viscosity (135°C, tetralin) of 2.3–2.5 dL/g, and 1,2-vinyl content of 96–98% (Table 8.1, Exp. 32 and 33).

By using only the dialkyldithiocarbamate cobalt + EADC (or EASC) as catalyst results in obtaining of hyperbranched high molecular weight 1,4-cis+1,2-polybutadiene with the yields of 92.0–98.0%, intrinsic viscosity [η] = 2.8–3.5 dL/g, 1,2-vinyl content of 22.0–35.0%, branching index of 0.35–0.70, and with the capability of regulation of polymers chain branching in wide intervals. Productivity of these catalysts is 22.0–55.0 kg PBD/g Co h (Table 8.1, Exp. 1–23).

The influence of the organic solvents (toluene, benzene, chlorobenzene, and hexane) on the activity and stereoselectivity of the diethyldithiocarbamate cobalt + ethylaluminumdichloride (DEDTC-Co + EADC) catalytic dithiosystem was studied at [Co] = 1.0 10^{-4} mol/L, [M] = 5.0 mol/L, Al:Co = 100:1. The butadiene polymerization reactions were conducted at 25°C for an hour.

From the results shown in Table 8.1, Exp.1 and 7, it seems that when toluene and benzene were used as the solvent, the polymer yield was 96.0% and 92.0%, and the catalyst activity was 44.0 kg PBD/g Co h and 42.0 kg PBD/g Co h, respectively. The obtained polybutadiene had 1,2-vinyl content of 30.0% and 25.0% and branching index of 0.35 and 0.60, respectively. The used catalyst showed the same results in terms of activity and stereoselectivity when chlorobenzene and hexane were used as the process solvent (Table 8.1, Exp. 8 and 9).

For future investigations, toluene, a largely used industrial polymerization process solvent, has been chosen as an optimal solvent in the polymerization of butadiene to hyperbranched polybutadiene. In these experiments, polymerization of butadiene was carried out at 25°C, and the cobalt-compound concentration was varied within the limits – [Co] = (0.5–10)×10⁻

[4] mol/L at Al:Co = 100:1. The results are shown in Table 8.1, Exp.1, 10–13. An increase in cobalt-compound concentration from 0.5×10^{-4} mol/L up to approximately 10.0×10^{-4} mol/L resulted in an increase in the yield of polymer from 90.0 to 99.0%. Increase in the concentration of cobalt compound resulted in a decrease in the branching index of polymer from 0.65 to 0.42 and intrinsic viscosity [η] from 3.5 to 2.0 dL/g. Content of the 1,2-vinyl chain increased in the range of 20.0–44.0%.

The influence of Al:Co ratio in the interval of (10–100):1 on the activity, stereoselectivity, and productivity of catalyst was investigated at [Co] = 1.0×10^{-4} mol/L and t = 25°°C. From the results of Table 8.1, Exp.1, 14, 15, it seems that an increase of Al:Co ratio from 10:1 to 100:1 resulted in an increase of polybutadiene yield of 87.0–95.0% and 1,2- vinyl content of 25.0–30.0%. It was also observed that the intrinsic viscosity decreased to 3.0–2.5 dL/g and branching index of polymer decreased to 0.75–0.35.

The influence of monomer concentration in a range of 1.0 to 8.0 mol/L was studied at [Co] = 1.0×10^{-4} mol/L, t = 25°C (Table 8.1, Exp.1, 16–18). Increase in the monomer concentration resulted in a decrease in the polymer yield of 98.0–90.0% and 1,2-vinyl content of 30.0–20.0 with an increase in the catalyst activity to 22.0–50.0 kg PBD/g Co h and branching index to 0.22–0.60.

The influence of temperature on the yield of the butadiene polymerization reaction was investigated between 25°C and 60°C. As it can be seen in Table 8.1, Exp. 1, 19, and 20, an increase in temperature resulted in a decrease in polybutadiene yield of 96.0–90.0% and intrinsic viscosity [η] of =2.9–2.5 dL/g. But this has no influence on the 1,2-vinyl content (25.0–27.0%) and branching index (0.35–0.40) of the product.

Experimental results show that only the new cobalt alkyldithiocarbamate + alkylaluminumdichloride (or alkylaluminumsesquichloride) catalytic dithiosystems allow us to obtain high molecular weight 1,4-cis+1,2-polybutadienes with high activity and stereoselectivity, high cold resistance, and various branching index.

The presence of known catalytic systems, such as cobalt-dithiophosphate + AOC, cobalt-xanthogenate + AOC, and cobalt-dithiocarbamate + dialkylaluminumchloride, did not yield branched polybutadienes. Obtaining of branching polymers in the presence of alkyldithiocarbamate cobalt + alkylaluminumdichloride (or alkylaluminumsesquichloride) could be explained by the cationic nature of an alkylaluminumdichloride and alkylaluminumsesquichloride co-catalysts. These co-catalysts act as catalysts in the polymerization process of butadiene (I. Direction) and in the co-polymerization of the obtained polymer with a new molecule of monomer (II. Direction) or with another molecule of polymer (III. Direction), resulting in the formation of branching with various lengths in the polybutadiene chain as shown in scheme 8.1.

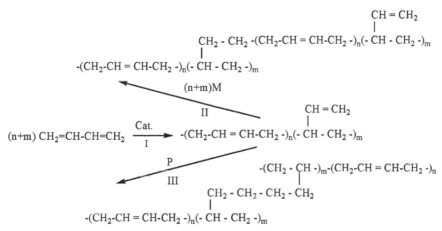

SCHEME 8.1   Formation of Branching in 1,4-cis+1,2-Polybutadiene Molecule.

CONCLUSIONS

High activity and stereo selectivity cobalt-dithiocarbamate catalytic dithiosystems for butadiene polymerization process have been developed. Based on alkyldithiocarbamate cobalt as the main catalyst compound in combination with alkylaluminumdichloride (or alkylaluminumsesquichloride) as the co-catalyst, they provide for the preparation of high molecular hyperbranched 1,4-cis+1,2-polybutadienes with 20–40% 1,2-vinyl bond content.

Productivity of these catalysts is 20.0–40.0 kg PBD/g Co h with the yields of polybutadiene 85.0–99.0%. The obtained hyperbranched polybutadienes have a branching index of 0.70–0.20 versus the branching index of linear polybutadienes, which is between 0.96–0.94. Known catalytic systems based on cobalt-dithiophosphates or cobalt-xanthogenates as the main cobalt compound and aluminum organic compounds (such as dialkylaluminumchlorides, alkylaluminumdichloride, alkylaluminumsesquichloride, and trialkylaluminum) and cobalt-dithiocarbamate + dialkylaluminumchloride do not lead to the formation of hyperbranched poly-butadienes.

Synthesized hyperbranched high molecular 1,4-cis+1,2-polybutadiene can be used in the production of tires, technological rubbers, and impact resistant polystyrenes.

KEYWORDS

- **Polydienes**
- **molecular mass**
- **waste polymers**
- **environment**
- **catalytic systems**
- **binders**

REFERENCES

1. Nasirov, F. A. Bifunctional Nickel- or Cobalt containing catalyst-stabilizers for poly-butadiene production and stabilization (Part I): kinetic study and molecular mass stereoregularity correlation. *Iran. Polym. J.* 2003, 12(4), 217–235.
2. Nasirov, F. A. Organic dithioderivatives of metalls – components and modificators of petrochemical processes. *Petrochemistry.* 2001, 6, 403–416. (in Russian).
3. Byrko, V. M. *Dithiocarbamates*; Nauka: Moscow, 1984; p 342. (in Russian).
4. Rafikov, S. P.; Pavlova, S. A.; Tvyordokhlebova, I. I. *Methods of Determination Molecular Weight and Polydispersity of High Molecular Materials*; Academy of Sciences of USSR: Moscow, 1963; p 336. (in Russian).
5. Deyl, Z.; Macek, K.; Janak, J.; Eds. *Liquid Column Chromatography*; Elsevier, Amsterdam Scientific: Amsterdam, 1975; Vol. 1, p 2.
6. Haslam, J.; Willis, H. A. *Identification and Analysis of Plastics*; Iliffe Books: London/D. Van Nostrand Co: Princeton, NJ, 1965; pp 172–174.
7. Bellami, L. J. *The Infra-red Spectra of Complex Molecules*; Methuen and Co: London/J. Wiley: New York, 1957; p 592.
8. Grechanovsky, V. A. Branching in polymer chains. Uspekhy Khimii, т. 38, 12, c.2194-2219, 1969 (Russian).
9. Jang, Y.-C.; Kim, P.-S.; Kwag, G.-H.; Kim, A.-J.; Lee, S.-H. Process for controlling degree of branch of high 1,4-cis polybutadiene. US Patent 20020016423 A1, 2002.

CHAPTER 9

A STUDY ON THE FORMATION OF THE PHASE STRUCTURE OF SILANOL-MODIFIED ETHYLENE COPOLYMERS WITH VINYL ACETATE AND VINYL ACETATE AND MALEIC ANHYDRIDE IN A WIDE RANGE OF TEMPERATURES AND COMPOSITIONS

N. E. TEMNIKOVA[1], O. V. STOYANOV[1], A. E. CHALYKH[2], V. K. GERASIMOV[2], S. N. RUSANOVA[1], and S. YU SOFINA[1]

[1]Kazan National Research Technological University, K. Marx street, 68, Kazan, 420015, Tatarstan, Russia

[2]Frumkin Institute of Physical Chemistry and Electrochemistry, Russian Academy of Sciences, Leninskii pr. 31, Moscow, 119991, Russia

CONTENTS

ABSTRACT

The solubility of components has been studied in a wide range of temperatures and compositions in the systems of copolymers of ethylene – aminoalkoxysilane. Phase diagrams have been constructed. The solubility of the components in different temperatures and concentration areas has been identified, and structure of the modified copolymers has been studied.

INTRODUCTION

Copolymers of ethylene are widely used for the synthesis of materials and products for various purposes including coatings and adhesives. In this connection, there is a need for continuous improvement in the properties of existing materials because synthesis of a new polymer is difficult. Thus, to extend the scope of industrially produced copolymers of ethylene is possible by their modification. One of the effective ways of modification is the introduction of organosilicon compounds. Introduction of such additives allows to achieve various changes in polymer properties],[1-3] including adhesive characteristics.[4,5]

So γ-aminopropyltriethoxysilane (AGM-9) is used in fiberglass and paint industries to improve the adhesion of different polymers and coatings (acrylates, alkyds, polyesters, and polyurethanes) to inorganic substrates (glass, aluminum, steel, and others) and to increase water resistance and corrosion stability of paint materials. AGM-9 is also used as pigmenting additives (to enhance the interaction of the pigment with the polymeric matrix of composite material or paint material).

In order to optimize the composition of the polymer compounds and conditions of the structure formation of their mixtures, the information about the phase organization of these systems is of considerable interest.[5]

Despite the fact that there are studies focused on improving adhesion to various substrates when modifying copolymers of ethylene by aminosilanes,[6,7] the information about the influence of monoaminofunctional silane on phase balance and phase structure of the polyolefin compositions in the scientific literature is not available.

The aim of this work was to study the formation of the phase structure of silanol-modified ethylene copolymers with vinyl acetate and vinyl acetate and maleic anhydride in a wide range of temperatures and compositions.

SUBJECTS AND METHODS

In this study, copolymers of ethylene with vinyl acetate Evatane 2020 (EVA20) and Evatane 2805 (EVA27) with vinyl acetate content of 20 and 27 wt%, respectively, and copolymers of ethylene with vinyl acetate and maleic anhydride brand Orevac 9307 (EVAMA13) and Orevac 9305 (EVAMA26) with vinyl acetate content of 13 and 26 wt% were used. Main characteristics of the copolymers are given in Table 9.1.

TABLE 9.1 Characteristics of the Copolymers of Ethylene

Polymer	Symbol	VA Content,%	MA Content, %	Melting Temperature, °C	M_v	MFR, g/10 min 125°C	Density, g/cm³
Evatane 2020	EVA20	20	–	80	44,000	2.23	0.936
Evatane 2805	EVA27	27	–	72	57,000	0.74	0.945
Orevac 9305	EVA-MA26	26	1.5	47	20,000	11.13	0.951
Orevac 9307	EVA-MA13	13	1.5	92	73,000	1.1	0.939

Silane containing an amino group gAGM-9 was used as a modifier. It is a transparent, colorless liquid with a molecular weight of 221. Density is 962 kg/m³. Refractive index n^d_{20} =is 1.4178, and the content of amine groups is 7–7.5%. The melting temperature is -70°°C.

Determination of the composition of coexisting phases and the inter-diffusion coefficients were carried out by processing of series of interfero-grams obtained by microinterference method. Interferometer ODA-3 was used for the measurements. Measurements were performed at a range of temperatures from 50 to 150°°C. To construct profiles of concentrations by interference patterns, the temperature dependencies of the refractive index of the components are required.[6–8] Refractive index measurements were carried out by an Abbe refractometer IRF-454 BM at a range of temperatures from 20 to 150°°C.

The structure of the modified copolymers was investigated by the transmission electron microscopy. Identification of the phase structure of the samples was carried out by etching of the surface in high-oxygen plasma discharge with the subsequent preparation of single-stage carbon-platinum replicas. Samples were viewed using PEM EM-301 (Philips, Holland).

RESULTS AND DISCUSSION

KINETICS OF MIXING OF THE COMPONENTS

Typical interferograms of interdiffusion zones of the systems EVA (EVA-MA) – modifier are shown in Figure 9.1.

Preliminary studies have shown that at high temperatures, the interdiffusion process is completely reversible, that is, the phase structures, which occurred when the temperature was lowered, dissolved again when the temperature was increased. This means that the net of the diffusion experiment is not formed or is formed but broken during the diffusion of the modifier, which is uncommon.

It is known that during the interaction of polymers with a modifier, transitional zones appear, within which the structure, composition, and properties vary continuously during the transition from one phase to another.

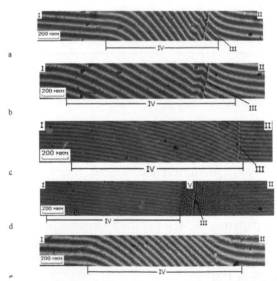

FIGURE 9.1   The Interferograms of Interdiffusion Zones of the Systems: (a) EVAMA26–AGM-9 (100°°C); (b) EVAMA26–AGM-9 (60°°C – cooling); (c) EVA20–AGM-9 (120°°C); (d) EVA20–AGM-9 (90°°C – cooling), (e) EVAMA26–AGM-9 (140°°C).

For all systems, the general picture is characteristic of partially compatible systems with a primary dissolution of alkoxysilanes in the melt of copolymers.

Phase boundary (III), a region of diffusion dissolution of modifiers in the melt of copolymers (IV), and phases of a pure copolymer (I) and the modifier (II) are clearly expressed on interferograms. Situation in the systems changes with decrease in the temperature: near the interface, there appears a region of opacity in between the dispersed phase in the melt of the copolymer (V) and the modifier. However, when the temperature increases again, the region of opacity disappears, indicating the reversibility of phase transformations occurring in the systems.

However, the compatible systems and the temperatures corresponding to this dissolution have been registered. AGM-9 is compatible in the systems EVA27 – AGM-9 and EVAMA26 – AGM-9 (see Fig. 9.1e) at a temperature above 100°°C and in the systems EVA20 – AGM-9 (at a temperature above 120°C) and EVAMA13 – AGM-9 (at a temperature above 150°°C).

Typical profiles of the concentration distribution in these systems are shown in Fig. 9.2.

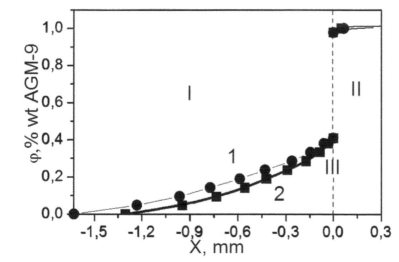

FIGURE 9.2 Profiles of the Concentration Distribution in the System EVAMA13 – AGM-9 at 135°C. Diffusion Time: (1) 11 min, (2) 7 min. (I) Diffusion Zone of AGM-9 in EVAMA13, (II) Diffusion Zone of EVAMA13 in AGM-9, and (III) Phase Boundary.

The size of the diffusion zone is influenced by several factors: the temperature and the time of observation.

It can be seen that the sizes of the diffusion zones on both sides of the phase boundary increase in time, whereas the values of concentrations near the interphase boundary in isothermal process conditions do not change.

The influence of temperature has a number of characteristic features. The higher the temperature, the greater the distance the molecules of the modifier diffuse for equal periods of time and larger the diffusion zone. This distribution of the concentration profiles when the temperature changes indicates that the system belongs to a class of systems with upper critical point of mixing.[8]

Figure 9.3 shows the kinetic curves of isoconcentration planes moving at different temperatures for the systems studied.

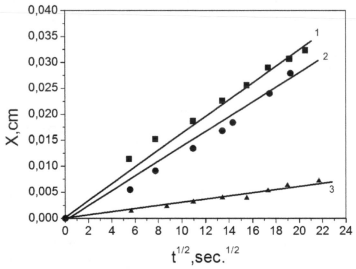

FIGURE 9.3   The Time Dependence of the Size of the Diffusion Zone EVAMA26 – AGM-9 at Various Temperatures: (1) 140°°C, (2) 120°°C, (3) 100°C.

Despite the fact that the kinetics of movement of the modifier front in the matrix of the copolymer has a linear dependence, IR-spectroscopy showed that all of the systems react chemically.[9] We can assume that the method of interferometry was unable to fix a chemical reaction under the given observation time. The rate of chemical reaction is comparable or slightly higher than the rate of diffusion, and the movement of the modifier already occurs in a chemically modified matrix, which is the reason for the lack of bending motion of the modifier front in the copolymer. The resulting matrix is soluble in the copolymer; hence, the phase decomposition by reheating cannot be observed.

As the temperature increases, the nature of the concentration distribution and mixing of the components is maintained in the zone of diffusion. Only the velocity of the movement of the isoconcentration planes changes. The angle of inclination of these relationships varies with temperature: the higher the temperature, the greater the angle of inclination of the line in the coordinates $X-t^{1/2}$. The slope of the kinetic lines is proportional to the coefficient of the modifier diffusion into the matrix. Therefore, the greater the angle of inclination, the higher the numerical value of the diffusion coefficient.

PHASE EQUILIBRIA

Consideration of diffusion zones of interacting copolymers and modifiers allows us to obtain not only the concentration profiles, but also the phase diagrams of the studied systems by quantitative analysis of interferograms obtained at different temperatures.

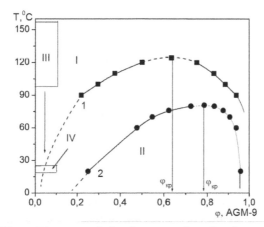

FIGURE 9.4 Phase Diagrams of the Systems of a Copolymer of Ethylene with Vinyl Acetate – AGM-9: (1) EVA20; (2) EVA27. (I, II) The Areas of True Solutions, Heterogeneous Condition; (III) The Area of Preparation of the Compositions; and (IV) The Area of Study of the Structure and Physical Properties.

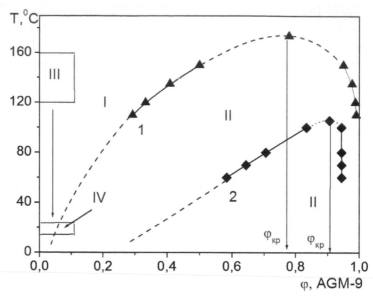

FIGURE 9.5 Phase Diagrams of the Systems of a Copolymer of Ethylene with Vinyl Acetate and Maleic Anhydride – AGM-9: (1) EVAMA13; (2) EVAMA26: (I, II) the Areas of True Solutions, Heterogeneous Condition; (III) The Area of Preparation of the Compositions; and (IV) the Area of Study of the Structure and Physical Properties.

There are two binodal curves on all phase diagrams: the right branch of the binodal corresponds to the solubility of the copolymer in the modifier and is located in the area of infinitely dilute solutions. The second binodal curve represents the solubility of the modifier in the copolymers and is located in a fairly wide concentration area. The solubility of the modifier in the copolymer is increased as the temperature increases.

In all phase diagrams (see Figs. 9.4 and 9.5), there are the areas corresponding to temperature-concentration areas of the components mixing (area III) and the areas corresponding to the structure of the compounds and their physical properties (area IV). The diagrams show that the preparation of the mixtures takes place in the single-phase area (area I in the diagrams). As the temperature decreases, the figurative point of the systems crosses the binodal curve and the system goes into a heterogeneous area (area II in the diagrams). The phase decomposition takes place, and it is uniquely fixed in electron microscopic images (see Figs. 9.6 and 9.7). Judging from the microphotographs, in the mixtures with EVAMA13, precipitated phases have a size that ranges from 0.1 to 1 micron, whereas for EVAMA26, the dispersed particles have a size that ranges from 50 to 100 micron.

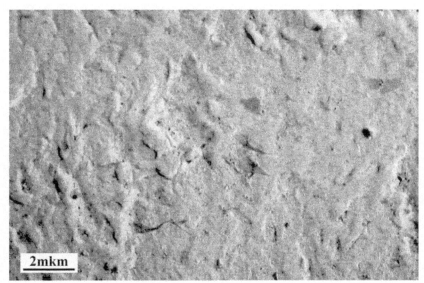

FIGURE 9.6 Microphotograph of EVAMA13 – AGM-9 (10%).

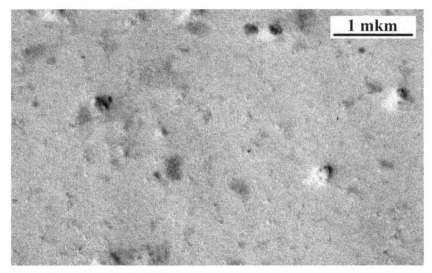

FIGURE 9.7 Microphotograph of EVAMA26 – AGM-9 (10%).

Particles, protruding from the surface, etched in plasma of high-oxygen discharge have smaller etching rate compared with the dispersion medium. It has been shown previously[10] that the lowest etching rate among carbo- and heterochain polymers have polysiloxanes. Thus, it can be concluded that the dispersed phase is enriched by siloxanes.

With the increase of the content of vinyl acetate groups both in EVA and in EVAMA, solubility of AGM-9 increases. In this case, the tendency in the change of solubility is the same for high and low temperature areas.

CONCLUSION

Thus, comprehensive studies of diffusion, phase, and structural-morphological characteristics of the compositions allow us to identify the contribution of chemical reactions for the change of phase equilibrium and the phase structure formation.

ACKNOWLEDGMENT

This work was financially supported by the Ministry of Education and Science of Russia in the framework of the theme No. 693 "Structured composite materials based on polar polymer matrices and reactive nanostructured components."

KEYWORDS

- **Copolymers of ethylene**
- **Aminoalkoxysilane**
- **Phase diagrams**

REFERENCES

1. Stoyanov, O. V.; Rusanova, S. N.; Khyzakhanov, R. M.; Petykhova, O. G.; Chalykh, A. E.; Gerasimov, V. K. Modificatsya promyshlennykh etilenvinilatsetatnykh sopo-

limerov predelnymi alkoksisilanami. *Vestnik Kazanskogo tekhnologicheskogo universiteta.* 2002, 1-2, 143–147.

2. Chalykh, A. E.; Gerasimov, V. K.; Rusanova, S. N.; Stoyanov, O. V.; Petykhova, O. G.; Kulagina, G. S.; Pisarev, S. A. Formirovanie fazovoi structury silanolno-modifitsirovannykh polimerov etilena s vinilatsetatom. *VMS seriya A.* 2006, 48(10), 1801–1810.

3. Stoyanov, O. V.; Rusanova, S. N.; Khuzakhanov, R. M.; Petuhova, O. G.; Deberdeev, T. R. Structure-mechanical characteristics of ethylene with vinylacetate copolymers modified by saturated alkoxysilanes. *Russ. Polym. News.* 2002, 7(4), 7.

4. Temnikova, N. E.; Rusanova, S. N.; Sof'ina, S.Yu; Stoyanov, O. V.; Garipov, R. M.; Chalykh, A. E.; Gerasimov, V. K. The effect of aminoalkoxy and glycidoxyalkoxy silanes on adhesion characteristics of double and triple copolymers of ethylene. *Polym. Sci. Ser. D.* 2014, 7(3), 84–187.

5. Temnikova, N. E.; Rusanova, S. N.; Sofina, S.Yu; Stoyanov, O. V.; Garipov, R. M.; Chalykh, A. E.; Gerasimov, V. K.; Zaikov, G. E. Influence of aminoalkoxy- and glycidoxyalkoxysilanes on adhesion characteristics of ethylene copolymers. *Polym. Res. J.* 2014, 18(4), 305–310.

6. Osipchik, V. S.; Lebedeva, E. D.; Vasilets, L. G. Razrabotka i issledovanie svoistv silanolnosshitogo polietilena. *Plast. Massy.* 2000, 9, 27–31.

7. Kikel, V. A.; Osipchik, V. S.; Lebedeva, E. D. Sravnitelnyi analiz structury i svoistv sshitykh razlichnymi metodami polietilenov. *Plast. Massy.* 2005, 8, 3–6.

8. Polimernye smesi. V 2-kh tomakh/Pod red. D.R. Pola i K.B. Baknella/Per. s angl. pod red. Kulezneva V.N. – Spb.: Naychnye osnovy i tekhnologii, 2009, 1224 s.

9. Rusanova, S. N.; Temnikova, N. E.; Stoyanov, O. V.; Gerasimov, V. K.; Chalykh, A. E. IK-spektroskopichskoe issledovanie vzaimodeistviya glitsidoksisilana i sopolimerov etilena. *Vestnik Kazanskogo tekhnologicheskogo universiteta.* 2012, 22, 95–96.

10. Temnikova, N. E. Vliyanie amino- i glitsidoksialkoksisilanov na formirovanie fazovoi structury i svoistva etilenovykh sopolimerov: dis. kand. tekh. nauk. Kazan. 2013;154 s.

CHAPTER 10

QUANTUM CHEMICAL CALCULATION OF MOLECULE 3,4,5,6,7-6,7- PENTADIMETHYLIN-DENE BY METHOD AB INITI

V. A. BABKIN[1], D. S. ANDREEV[1], YU. A. PROCHUKHAN[2], K. YU. PROCHUKHAN[2], and G. E. ZAIKOV[3]

[1]Sebrykov Department, Volgograd State Architect-build University, Volgograd, Russia

[2]Bashkir State University, Ufa, Russia

[3]N. M. Emanuel Institute of Biochemical Physics, Russian Academy of Sciences, 4, Kosygin Street, Moscow 119991, Russia

CONTENTS

ABSTRACT

For the first time, quantum chemical calculation of a molecule of 3,4,5,6,7-pentamethylindene by the method AB INITIO with optimization of geometry on all parameters is executed. The optimized geometrical and electronic structure of this compound is obtained. Acid power of 3,4,5,6,7-pentamethylindene is theoretically appreciated. It is observed that this compound pertains to a class of very weak H-acids (pKa was equal to +33, where pKa is the universal index of acidity).

AIM AND BACKGROUND

The aim of this work is to study the electronic structure of the molecule 3,4,5,6,7-6,7-pentadimethylindene[1] and theoretical estimation of its acid power by quantum-chemical method AB INITIO in base 6-311G**. The calculation was carried out with optimization of all parameters by standard gradient method using built-in PC GAMESS[2]. The calculation was carried out by the method of the insulated molecule in gas phase. Program Mac-MolPlt was used for the visual presentation of the model of the molecule.[3]

EXPERIMENTAL PROCEDURE

Geometric and electronic structures and general and electronic energies of the molecule 3,4,5,6,7-6,7-pentadimethylindene were obtained by the method AB INITIO in base 6-311G** and are shown in Figure 10.1 and in Table 10.1. The universal factor of acidity was calculated by the following formula[4,5]:

$$pKa = 49.04 - 134.6 * q_{max}^{H+},$$

where q_{max}^{H+} is a maximum positive charge on atom of the hydrogen $q_{max}^{H+} = +0.12$ (for 3,4,5,6,7-6,7-pentadimethylindene q_{max}^{H+} alike in Table 10.1). This same formula is used by Babkin and Zaikov,[6] and the pKa was calculated to be 33.

Quantum-chemical calculation of the molecule 3,4,5,6,7-6,7-penta-dimethylindene by method AB INITIO in base 6-311G** was executed for the first time. Optimized geometric and electronic structure of this compound was obtained. Acid power of the molecule 3,4,5,6,7-6,7-pentadimethylindene was theoretically evaluated (pKa = 33). This compound pertains to the class of very weak H-acids (pKa > 14).

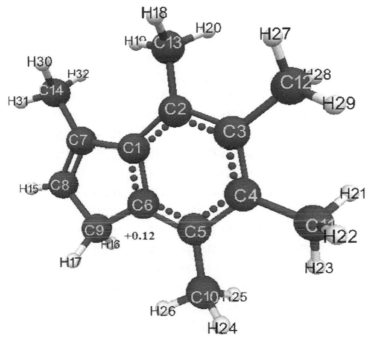

FIGURE   10.1 Geometric   and   Electronic   Molecule   Structure   of 3,4,5,6,7-6,7-Pentadimethylindene.

$(E_0 = -1419471$ kDg/mol, $E_{el} = -3740911$ kDg/mol$)$

TABLE 10.1 Optimized Bond Lengths, Valence Corners, and Charges on Atoms of the Molecule 3,4,5,6,7-6,7-Pentadimethylindene

Bond Lengths	R,A	Valence Corners	Grad	Atom	Charges on Atoms
C(2)–C(1)	1.39	C(5)–C(6)–C(1)	122	C(1)	-0.02
C(3)–C(2)	1.41	C(9)–C(6)–C(1)	110	C(2)	-0.09
C(4)–C(3)	1.4	C(1)–C(2)–C(3)	118	C(3)	-0.10
C(5)–C(4)	1.4	C(13)–C(2)–C(3)	119	C(4)	-0.12
C(6)–C(5)	1.38	C(2)–C(3)–C(4)	121	C(5)	-0.09

C(6)–C(1)	1.4	C(12)–C(3)–C(4)	119	C(6)	-0.15
C(6)–C(9)	1.51	C(3)–C(4)–C(5)	120	C(7)	-0.10
C(7)–C(1)	1.5	C(11)–C(4)–C(5)	120	C(8)	-0.16
C(8)–C(7)	1.33	C(9)–C(6)–C(5)	128	C(9)	-0.06
C(9)–C(8)	1.5	C(4)–C(5)–C(6)	118	C(10)	-0.20
C(10)–C(5)	1.52	C(2)–C(1)–C(6)	120	C(11)	-0.21
C(11)–C(4)	1.52	C(8)–C(9)–C(6)	102	C(12)	-0.21
C(12)–C(3)	1.52	C(10)–C(5)–C(6)	120	C(13)	-0.20
C(13)–C(2)	1.52	C(2)–C(1)–C(7)	133	C(14)	-0.18
C(14)–C(7)	1.5	C(1)–C(7)–C(8)	109	H(15)	+0.08
H(15)–C(8)	1.07	C(14)–C(7)–C(8)	123	**H(16)**	**+0.12**
H(16)–C(9)	1.09	C(7)–C(8)–C(9)	112	H(17)	+0.12
H(17)–C(9)	1.09	C(4)–C(5)–C(10)	122	H(18)	+0.11
H(18)–C(13)	1.09	C(3)–C(4)–C(11)	120	H(19)	+0.10
H(19)–C(13)	1.08	C(2)–C(3)–C(12)	119	H(20)	+0.10
H(20)–C(13)	1.08	C(1)–C(2)–C(13)	123	H(21)	+0.10
H(21)–C(11)	1.08	C(1)–C(7)–C(14)	128	H(22)	+0.11
H(22)–C(11)	1.09	C(7)–C(8)–H(15)	125	H(23)	+0.10
H(23)–C(11)	1.08	C(8)–C(9)–H(16)	112	H(24)	+0.11
H(24)–C(10)	1.09	C(8)–C(9)–H(17)	112	H(25)	+0.11
H(25)–C(10)	1.09	C(2)–C(13)–H(18)	112	H(26)	+0.10
H(26)–C(10)	1.08	C(2)–C(13)–H(19)	112	H(27)	+0.10
H(27)–C(12)	1.08	C(2)–C(13)–H(20)	111	H(28)	+0.11
H(28)–C(12)	1.09	C(4)–C(11)–H(21)	111	H(29)	+0.10
H(29)–C(12)	1.08	C(4)–C(11)–H(22)	112	H(30)	+0.11
H(30)–C(14)	1.09	C(4)–C(11)–H(23)	112	H(31)	+0.10
H(31)–C(14)	1.08	C(5)–C(10)–H(24)	112	H(32)	+0.11
H(32)–C(14)	1.09	C(5)–C(10)–H(25)	112		
		C(5)–C(10)–H(26)	111		
		C(3)–C(12)–H(27)	111		
		C(3)–C(12)–H(29)	112		
		C(7)–C(14)–H(30)	112		
		C(7)–C(14)–H(31)	110		
		C(7)–C(14)–H(32)	112		

KEYWORDS

- **Quantum chemical calculation**
- **Method AB INITIO**
- **3,4,5,6,7-pentamethylindene**
- **acid power**

REFERENCES

1. Kennedi, J. *Cationic Polimerization of Olefins*. Nauka (Science) Publisher: Moscow, 1978; p 431.
2. Shmidt, M. W.; Baldrosge, K. K.; Elbert, J.A.; Gordon, M. S.; Enseh, J. H.; Koseki, S.; Matsvnaga, N.; Nguyen, K. A.; Su, S. J.; and anothers. Advances in electronic structure theory: GAMESS a decade later. *J. Comput. Chem.* 1993, 14, 1347–1363.
3. Bode, B. M.; Gordon, M. S. *J. Mol. Graphics Mod.* 1998, 16, 133–138.
4. Babkin, V. A.; Fedunov, R. G.; Minsker, K. S et al. *Oxid. Commun.*, 2002, 25(1), 21–47.
5. Babkin, V. A.; and others. *Oxid. Commun.*, 1998, 21(4), 454–460.
6. Babkin, V. A.; Zaikov, G. E. *Nobel laureates and nanotechnology of the applied quantum chemistry*. Nova Science Publisher: New York, USA, 2010; p 351.

CHAPTER 11

COBALT ALKYLXHANTHOGENATE + TRIALKYLALUMINUM CATALYTIC DITHIOSYSTEMS FOR SYNTHESIS OF SYNDIOTACTIC 1,2-POLYBUTADIENE

NEMAT AKIF OGLU GULIYEV[1], FUZULI AKBER OGLU NASIROV[1,2], and NAZIL FAZIL OGLU JANIBAYOV[1]

[1]Institute of Petrochemical Processes of National Academy of Sciences of Azerbaijan, Baku, Azerbaijan

[2]Petkim Petrokimya Holding, Izmir, Turkiye

E-mail: fnasirov@petkim.com.tr; j.nazil@yahoo.com

CONTENTS

INTRODUCTION

Syndiotactic 1,2-polybutadiene is a photodegradable polymer with molecular irregularities – double bonds in its polymeric backbone. Studies have shown a 95% loss of properties after photodegradable materials have been exposed to direct sunlight over a year.

It is used in the manufacturing of tires, packing polymer-film materials, microcapsules for the medical purposes, ceramics, semi-permeability membranes, adhesives, synthetic leather, oil-resistant tubes, coatings for semi-conductor devices, non-woven materials, and carbon fibers, and so on.[1,2]

Syndiotactic 1,2-polybutadienes have been synthesized using various catalysts based on compounds of Ti, Cr, Pd, Co, V, Fe, and Mo. The results, relating to the catalysts and processes of syndicotactic 1,2-polybutadiene synthesis, have been described in many studies[3–11] and patents.[12]

Ashitaka H. et al., in a series of articles,[3–6] have described the $Co(acac)_3 + AIR_3 + CS_2$ catalyst for the polymerization of butadiene to high stereoregularity and high melting point syndiotactic 1,2-polybutadiene. This catalyst had very low catalytic activity that polymer yield is only 10–20% in toluene solution (the mostly used industrial polymerization processes as solvent).

Chinese researchers have recently focused on the synthesis and characterization of syndiotactic 1,2-polybutadiene,[7–11] but catalytic systems have shown very low catalytic activity and stereoregularity.

Earlier, we have developed the new bifunctional nickel- and cobalt-containing catalytic dithiosystems for the polymerization of butadiene.[13,14] In this article, the results of the polymerization of butadiene to the photo-degradable highly crystalline syndiotactic 1,2-polybutadiene in the presence of a new cobalt alkylxhantogenate (AlkXh-Co) + trialkylaluminum (TAA) catalytic dithiosystems is shown.

EXPERIMENTAL PROCEDURES

Butadiene (99.8% wt) and aluminum organic compounds (85.0% wt in benzene) were used as obtained from Aldrich.

Organic dithioderivatives (dithiophosphates, dithiocarbamates, and xhantogenates) of cobalt were synthesized according to procedures explained by Nasirov and Djanibekov.[13–16]

Polymerization was conducted in toluene, which, after predrying over metallic Na for 24 h, was distilled and preserved under Na. Where necessary, manipulations were carried out under dry, oxygen-free argon or nitrogen in 50–200 mL glass reactors. The desired volumes of toluene, monomer, triethylaluminum (TEA; or DEAC, MAO), and cobalt-compound solutions from the calibrated glass reservoir were added to the reactor under stirring at controlled temperature. The usual order of addition was solvent, cobalt component, aluminum organic compound (at −78°C), and finally, the monomer. All polymerizations were conducted at a temperature range of 0–100°C. After polymerization, the polymerizate was poured into ethanol or methanol and the polymerization reactions were terminated. The precipitated polymer was washed several times with ethanol (or methanol). Polybutadiene was dried at 40°C in a vacuum to constant weight and stored under argon or nitrogen.

The viscosity of dilute solutions of syndiotactic 1,2-polybutadienes was measured using a Ubbelohde viscometer in tetralin at 135°C or in o-dichlorobenzene (o-DCB) at 140°C at a concentration of 0.2 g/dL.[2,3] The molecular mass of high molecular weight and high crystalline syndiotactic polybutadiene was determined by viscometric method[17] using the following relationship:

$$[h\eta]_{135(\text{tetralin})} = [h\eta]_{140(\text{o-DCB})} = 9.41 \cdot 10^{-5} \cdot M^{0.854}$$

The molecular masses (M_w and M_n) and molecular mass distribution (M_w/M_n) were measured by a Gel Permeation Chromatograph (GPC), which was constructed in Czech Republic, consisting of a 6000 A pump, original injector, R-400 differential refractive index detector, and styragel columns with nominal exclusion of 500, 10^3, 10^4, 10^5, and 10^6. The GPC was operated at a flow rate of 0.8 mL/min with o-dichlorobenzene as solvent. The sample concentration was kept at about 0.3–0.6% with a sample volume of 100–200 mL. The GPC instrument was calibrated according to the universal calibration method by using narrow molecular weight polystyrene standards.[18]

The microstructure of the polybutadiene was determined by a FTIR spectrometer (Nicholet NEXUS 670, with spectral diapason from 400 cm^{-1} to 4000 cm^{-1}, as a film on KBr, received from THF or o-dichloroben-zene solution).[19,20] Tactisity and crystallinity of polymer were determined accordingly.[21] Melting point (mp) was determined under nitrogen by differential scanning calorimeter (DSC Q20 V23.4).[21,22]

RESULTS AND DISCUSSION

We have investigated the peculiarities of cobalt alkylxhantogenate + tri-alkylaluminum catalytic dithiosystems in the polymerization of butadi-ene to syndiotactic 1,2-polybutadiene.[23–26] These catalysts were studied in comparison with known cobalt-containing catalytic dithiosystems for bu-tadiene polymerization dependence of process outcomes on the nature of ligands of cobalt-compounds, catalyst components concentration and ra-tio, as well the influence of temperature on catalyst activity and selectivity.

Apart from the metal chosen for polymerization, the ligand nature is also of high importance. Ligands that have been used for polymerization of butadiene were mainly dithiophosphates, dithiocarbamates, and xhan-thogenates. The results are given in Table 11.1.

It can be observed from Table 11.1 that co-dithiophosphates and co-al-kylxhantogenates in combination with diethylaluminumchloride (DEAC) allow obtaining of high molecular mass 1,4-cis polybutadiene with 1,4-cis contents of 90.0–96.0%. Cobalt alkyldithiocarbamate catalytic system (DEDTC-Co + DEAC) yields high molecular mass 1,4-cis + 1,2-polybu-tadiene with 1,4-cis content of 58.0% and 1,2-content of 34.0%.

Only the cobalt alkylxhantogenate catalytic dithiosystems (Cobalt al-kylxhantogenates + TEA) yield high molecular mass and high crystalline syndiotactic 1,2-PBD (1,2-SPBD). In their presence, polymer yields are 93.0–99.0%, and the obtained 1,2-SPBD has intrinsic viscosity (135°C, in tetralin) of 2.2–3.5 dL/g, 1,2-content of 94.0–99.0%, crystallinity of 86.0–95.0%, and mp of 175–208°C.

The effect of the organic solvents (toluene, benzene, chlorobenzene, methylene chloride, and hexane) on the activity and selectivity of the iso-propylxhantogenate (iPrXh-Co) + TEA catalytic dithiosystem was studied at [Co] = 1.0×10^{-4} mol/L, [M] = 3.0 mol/L, Al:Co = 100:1. For that, the necessary amount of a particular solvent was mixed with iPrXh-Co followed by the addition of TEA and butadiene into reactor at −78°C. The butadiene polymerization reactions were conducted at 40°C for 120 min.

From the results shown in Table 11.2, it appears that when toluene and benzene were used as the solvent, the polymer yield was 96.0–99.0%. The obtained polybutadiene had 1,2-vinyl contents of 96.0–98.0%, mp of 203–208°C, crystallinity of 93.0–95.0%, and syndiotacticity of 95.0–97.5%. The known catalytic systems for butadiene 1,2-polymerization showed very low catalytic activity and stereoregularity in toluene and benzene solutions. For future investigations, toluene, a commonly used industrial polymerization process solvent, has been chosen as an optimal solvent in the 1,2-polymerization of butadiene.

An increase in cobalt concentration from 0.2×10^{-4} mol/L up to approximately 10.0×10^{-4} mol/L resulted in an increase in the initial reaction rate. As the concentration of cobalt compound was increased, the polybutadiene yield increased from 50.0 to 100.0%, 1,2-vinyl content decreased from 99.0 to 95.0%, and mp decreased from 212 to 187°C as shown in the results in Table 11.3.

The increase of Al:Co ratio from 10:1 to 200:1 resulted in an increase of polybutadiene yield to 65.0–99.0% and a decrease in the intrinsic viscosity to 2.9–1.5 dL/g, 1,2-content of 99.0–94.0%, crystallinity of 96.0–76.0%, and mp of 212–175°C (Table 11.3).

With an increase in temperature, there was an increase in polybutadiene yield to 26.0–100.0% as expected. This results in decreasing of intrinsic viscosity to 3.5–1.3 dL/g, 1,2-content of 99.0–94.0%, crystallinity of 98.0–63.0%, and mp of 215–177°C.

Experimental results show that the high activity and stereoregularity of the new cobalt alkylxhantogenate + trialkylaluminum catalytic dithiosystems in toluene solution allow for the formation of syndiotactic 1,2-polybutadienes of varying crystallinity, syndiotacticity, and molecular mass when experimental conditions are varied.

TABLE 11.1 Comparison of Efficiency of Different Cobalt-containing Catalytic Dithiosystems CoX_2 + AOC in the Butadiene Polymerization Process

Item No	Cobalt-containing Catalytic System (CoX_2 + AOC)	Polymer Yield% (mass)	Crystallinity%	Syndiotacticity%	Melting Point,°C	Intrinsic Viscosity [η], dL/g	Molecular Mass		Microstructure%			Yield of Polymer, kg PBD/g Co h
							$M_w \times 10^{-3}$	M_w/M_n	1,4-cis	1,4-trans	1,2-	
1	DCDTPh-Co + DEAC	95	–	–	–	2.62	570	2.1	92	6	2	57
2	TBDTPh-Co + DIBAC	98	–	–	–	2.6	610	2.3	90	7	3	109
3	4m-6-TBPh-Co + MAO	92	–	–	–	2.5	625	2.5	92	5	3	59
4	X-Co + DEAC	95	–	–	–	2.9	610	1.9	93	6	1	57
5	NGDTPh-Co + DIBAC	88	–	–	–	3.1	630	1.85	93	5	2	46
6	DEDTC-Co + DEAC	58	–	–	–	2.55	600	2.3	58	8	34	44
7	EtXh-Co + DEAC	95	–	–	–	2,8	540	1.85	96	2	2	127
8	BuXh-Co + DIBAC	90	–	–	–	2,5	410	1.65	95	2	3	105
9	EtXh-Co + TEA	95	92	98.5	208	2.2	270	2.1	2	1	99	96
10	i-PrXh-Co + TEA	99	95	97.0	205	2.3	255	1.82	1	1	98	115
11	BuXh-Co + TEA	98	93	95.5	200	2.5	240	1.6	3	1	96	125
12	HeXh-Co + TEA	95	88	94.0	185	2.9	330	1.75	5	1	94	108
13	OcXh-Co + TEA	93	86	94.5	175	3.5	485	2.2	6	2	95	112

The obtained experimental results allow us to establish the optimal parameters for the synthesis of high molecular mass and highly crystalline syndiotactic 1,2-polybutadiene[42–45]: [Co] = (1.0–2.0) × 10^{-4} mol/L; [M] = 3.0 mol/L; Al:Co = (50–100):1; T = 40–80°C.
Reaction conditions: [Co] = 2.0 × 10^{-4} mol/L; [M] = 3.0 mol/L; Al:Me = 100; T = 25˚C; τ = 60 min; solvent – toluene.
Notes: In experiments 9–13, an intrinsic viscosity was measured in tetraline at 135°C.

TABLE 11.2 The Influence of Organic Solvents Type on the Conversion of Butadiene, Selectivity and Productivity of iPrXh-Co + TEA Catalytic System [Co] = 1.0×10^{-4} mol/L, [M] = 3.0 mol/L, Al:Co = 100:1, T = 40°C, and τ = 120 min

Solvent	Yield of PBD,% (mass)	$[\eta]_{135},$dL/g	Crystallinity,%	Syndiotacticity,%	Melting point, °C	Microstructure%		
						1,4-cis	1,4-trans	1,2-
Toluene	99.0	2.5	95	97.5	208	1	1	98
Benzene	96.0	2.3	93	95.0	203	3	1	96
Chlorobenzene	92.0	1.75	91	94.0	194	3	2	95
Methylene chloride	90.0	1.65	90	93.8	185	3	2	95
Hexane	70.0	1.5	76	91.0	190	6	2	92
-	56.0	3.5	93	95.0	195	4	1	95

TABLE 11.3 The Influence of Butadiene Polymerization Parameters on the Conversion of Butadiene, Selectivity and Productivity of iPrXh-Co + TEA Catalytic System (Solvent–Toluene)

[Co]·10⁴, mol/L	[M] mol/L	Al:Co	Temperature, °C	Reaction tion, min	Yield of 1,2-SPBD, % (mass)	$[\eta]_{135},$ dL/g	Crystallinity,%	Syndiotacticity,%	Melting Point, °C	Microstructure,%		
										1,4-cis	1,4-trans	1,2-
0.2	3.0	100	40	180	50	3.0	68	98	212	1	–	99
0.5	3.0	100	40	120	65	2.8	72	98	212	1	–	99
1.0	3.0	100	40	120	95	2.5	76	97	210	1	1	98
2.0	3.0	100	40	60	99	2.3	85	98	205	1	1	98
5.0	3.0	100	40	15	99	2.1	90	96	197	2	1	97
10.0	3.0	100	40	15	99	1.7	95	95	187	3	2	95
2.0	1.5	100	40	60	97	2.0	84	96	207	2	1	97
2.0	6.0	100	40	45	90	3.2	87	98	203	1	–	99
2.0	3.0	10	40	180	65	2.9	96	98	212	1	–	99
2.0	3.0	25	40	90	83	2.7	93	98	208	1	–	99
2.0	3.0	50	40	60	91	2.4	78	97	206	1	1	98
2.0	3.0	150	40	45	99	2.0	73	96	186	3	1	96
2.0	3.0	200	40	30	99	1.5	76	94	175	4	2	94

2.0	3.0	100	0	180	26	3.5	98	97	215	1	–	99
2.0	3.0	100	10	120	45	2.8	93	97	212	1	1	98
2.0	3.0	100	25	60	95	2.4	90	96	209	2	1	97
2.0	3.0	100	55	45	99	2.0	81	95	194	4	1	95
2.0	3.0	100	80	45	99	1.5	68	93	180	4	2	94
2.0	3.0	100	100	45	99	1.3	63	93	177	5	1	94

CONCLUSIONS

In this work, novel, highly active, and stereoregular cobalt alkylxhanto-genate + trialkylaluminum catalytic dithiosystems have been developed. Activity and stereo regularity of these catalysts were studied in comparison with known cobalt-containing catalytic dithiosystems of butadiene polymerization. These experiments were conducted under varied conditions including catalyst concentration, ratio, and temperature. The developed catalytic systems allowed for the synthesis of high molecular weight and high crystalline syndiotactic 1,2-polybutadiene with yields of 50.0–99.0% and 1,2-contents of 94.0–99.0%, intrinsic viscosity between 1.3 and 3.5 dL/g, crystallinity between 63.0 and 98.0%, and mps between 175 and 212°C in toluene solution.

The optimal conditions for the synthesis of high molecular mass and highly crystalline syndiotactic 1,2-polybutadiene were established with the following parameters: $[Co] = (1.0–2.0) \times 10^{-4}$ mol/L; $[M] = 3.0$ mol/L; Al:Co = (50–100):1; $T = 40$–80°C.

KEYWORDS

- Syndiotactic 1,2-polybutadiene
- polymerization of butadiene
- microstructure
- cobalt alkylxhantogenate
- trialkylaluminum catalytic dithiosystems

REFERENCES

1. Obata, Y.; Ikeyama, M. Bulk properities of syndiotactic 1,2-polybutadiene. I. Thermal and viscoelastic properties. *Polym. J.* 1975, 7(02), 207–216.
2. Junji, K.; Shoko, S. Syndiotactic 1,2-polybutadiene rubber characteristics and applications. *JETI.* 1998, 46, 111–115.
3. Ashitaka, H.; Ishikawa, H.; Ueno, H.; Nagasaka, A. Syndiotactic 1,2-polybutadiene with Co-CS2 catalyst system. I. Preparation, properties, and application of highly crystalline syndiotactic 1,2-polybutadiene. *J. Polym. Sci., Polym. Chem. Ed.* 1983, 21, 1853–1860.
4. Ashitaka, H.; Ishikawa, H.; Ueno, H.; Nagasaka, A. Syndiotactic 1,2-polybutadiene with Co-CS2 catalyst system. II. Catalysts for stereospecific polymerization of butadiene to syndiotactic 1,2-polybutadiene. *J. Polym. Sci., Polym. Chem. Ed.* 1983, 21, 1951–1972.
5. Ashitaka, H.; Ishikawa, H.; Ueno, H.; Nagasaka, A. Syndiotactic 1,2-polybutadiene with Co-CS2 catalyst system. III. 1H-and 13C-NMR Study of highly syndiotactic 1,2-polybutadiene. *J. Polym. Sci., Polym. Chem. Ed.* 1983, 21, 1973–1988.
6. Ashitaka, H.; Ishikawa, H.; Ueno, H.; Nagasaka, A. Syndiotactic 1,2-polybutadiene with Co-CS2 catalyst system. IV. Mechanism of syndiotactic polymerization of Butadiene with Cobalt compounds-organoaluminum-CS2. *J. Polym. Sci., Polym. Chem. Ed.* 1983, 21, 1989–1995.
7. Cheng-zhong, Z. Synthesis of syndiotactic 1,2-polybutadiene with silica gel supported CoCl2-Al(I-Bu)₃-CS₂ Catalyst. *China Synth. Rubber Ind.* 1999, 22(04), 243.
8. Cheng-zhong, Z. Synthesis and morphological structure of crystalline syndiotactic 1,2-polybutadiene. *Chem. J. Chin. Univ.* 2003, 11.
9. Cheng-zhong, Z. Investigation on synthesis of high vinyl polybutadiene with iron-based catalysts. I. Effect of Triphenyl Phosphate. *Chin. J. Catal.* 2004, 08, 1219.
10. Cheng-zhong, Z.; Zhen, D.; Li-hong, N. Research progress of syndiotactic 1,2-polybutadiene. *Chem. Propell. Polym. Mater.* 2005, 04.
11. Lan-guo, D.; Weijian, H.; Cheng-zhong, Z. Preparation and characterization of high 1,2-syndiotactic polybutadiene/polystyrene in situ blends. *China Synth. Rubber Ind.* 2005, 28(03).
12. XXX. Patent USA 4751275 A, 1988; Patent USA 5239023 A, 1993; Patent USA 5356997 A, 1994; Patent USA 5677405 A, 1997; Patent USA 5891963 A, 1999; Patent USA 6720397 B2, 2004; Patent USA 6956093 B1, 2005; Patent USA7186785 B2, 2007.
13. Nasirov, F. A. Dissertation Prof. Doctor (Chemistry). IPCP, Azerbaijan National Academy of Sciences, Baku, 2003; 376 p (in Russian).
14. Nasirov, F. A. Bifunctional nickel- or cobalt containing catalyst-stabilizers for polybutadiene production and stabilization (part I): kinetic study and molecular mass stereo regularity correlation. *Iran. Polym. J.* 2003, 12, 217–235.
15. Djanibekov, N. F. Dissertation Prof. Doctor (Chemistry). IPCP, Azerbaijan Academy of Sciences, Baku, 1987; 374 p (in Russian).
16. Nasirov, F. A. Organic dithioderivatives of metals – components and modificators of petrochemical peocesses. *Petrochemistry.* 2001, 6, 403–416. (in Russian).

17. Rafikov, S. P.; Pavlova, S. A.; Tvyordokhlebova, I. I. Methods of determination molecular weight and polydispersity of high molecular materials. *M.: Acad. Sci. USSR.* 1963, 336. (in Russian).

18. Deyl, Z.; Macek, K.; Janak, J. *Liquid Column Chromatography*; Elsevier, Amsterdam Scientific: New York, 1975.

19. Haslam, J.; Willis, H.A. *Identification and Analysis of Plastics*; Iliffe Books: London; D.Van Nostrand Co: Princeton, NJ, 1965.

20. Bellami, L. J. *The Infra-red Spectra of Complex Molecules*; Methuen and Co.: London; J. Wiley: New York, 1957; p 592.

21. Rabek, J. F. *Experimental Methods in Polymer Chemistry: Physical Principles and Applications*; John Wiley & Sons: New York, 1982.

22. Huhne, G. W. H.; Hemminger, W. F.; Flammersheim, H.-J. *Differential Scanning Calorimetry*; Springer: Berlin Heidelberg, 2003.

23. Nasirov, F. A.; Novruzova, F. M.; Azizov, A. G.; Djanibekov, N. F.; Golberg, I. P.; Guliev, N. A. Method of Producing Syndiotactic 1,2-Polybutadiene. Patent 20010128, Azerbaijan, 1999 (in Azerbaijanian).

24. Nasirov, F. A.; Novruzova, F. M., Golberg, I. P.; Aksenov, V. I. New Bifunctional Catalyst for Obtaining of Syndiotactic 1,2-Polybutadiene, Proceedings of III. Baku International Mamedaliev Petrochemistry Conference, Baku, 289, 1998.

25. Nasirov, F. A.; Guliev, N. A.; Novruzova, F. M.; Azizov, A. G.; Djanibekov, N. F. Pecularities of Butadiene Polymerization to Syndiotactic 1,2-Polybutadiene in Toluene Solution, Proceedings of IV. Baku International Mamedaliev Petrochemistry Conference, Baku, 266, 2000.

26. Nasirov, F. A.; Azizov, A. H.; Novruzova, F. M.; Guliyev, N. A. Polymerization of Butadiene to Syndiotactic 1,2-Polybutadiene in the Presence of Co-xantogenate + AlEt3 Catalytic Systems, Proceedings of Polychar – 10 World Forum on Polymer Applications and Theory, Denton, USA, 226, 2002.

CHAPTER 12

A STUDY ON THE EFFECT OF THE PHASE STRUCTURE OF THE MODIFIED EVA (EVAMA) ON THEIR PROPERTIES

N. E. TEMNIKOVA[1], A. E. CHALYKH[2], V. K. GERASIMOV[2], S. N. RUSANOVA[1], O. V. STOYANOV[1], and S. YU. SOFINA[1]

[1]Kazan National Research Technological University, K. Marx street, 68, Kazan, 420015, Tatarstan, Russia

[2]Frumkin Institute of Physical Chemistry and Electrochemistry, Russian Academy of Sciences, Leninskii pr. 31, Moscow, 119991, Russia

Email: ov_stoyanov@mail.ru; vladger@mail.ru

CONTENTS

ABSTRACT

The mutual solubility of the components was investigated, and the phase diagrams in a wide range of temperatures and compositions in the systems EVA (EVAMA) – glycidoxyalkoxysilane – were constructed. The effect of structural heterogeneity of silanol-modified EVA (EVAMA), associated with the chemical interaction of the components, on the properties of the compositions was identified.

INTRODUCTION

Introduction of the reactive additives, which chemically interacts with macromolecules, to the polymer not only changes the chemical nature of the material, but also naturally affects the properties of the complex.[1-4] Graft structures formed in the matrix increase the molecular weight of the polymer, thereby affecting the process of the melt flow and solutions of the compositions.

Previously, it has been found[5] that the introduction of small amounts (up to 3%) of ethyl silicate (ETS) into the copolymers of ethylene with vinyl acetate (EVA) leads to the increase in intrinsic viscosity of EVA. Herewith further increase in the concentration of ETS did not affect the process of solutions flow. Dependence of melt flow index of modified EVA on the concentration of the modifier has an extreme character with a minimum. Minimum amount of the ETS significantly affects the proportion of vinyl acetate in the copolymer.

The aim of this work was to study the effect of the phase structure of the modified EVA (EVAMA) on their properties. Thus, the solubility of the components was studied, and the phase diagrams in a wide range of temperatures and compositions for the systems EVA (EVAMA) – glycidoxyalkoxysilane – were constructed.

SUBJECTS AND METHODS

Copolymers of ethylene with vinyl acetate Evatane2020 (EVA20) and Evatane2805 (EVA27) with a vinyl acetate content of 20 and 27 wt%, respectively, and copolymers of ethylene with vinyl acetate and maleic anhydride brand Orevac9307 (EVAMA13) and Orevac9305 (EVAMA26) with a vinyl acetate content of 13 and 26 wt% were used as the objects of the study. Main characteristics of the copolymers are given in Table 12.1.

TABLE 12.1　Characteristics of the Copolymers of Ethylene

Polymer	Symbol	VA Content,%	MA Content, %	Melting Temperature, °C	M_V	MFR, g/10 min 125°C
Evatane2020	EVA20	20	–	80	44,000	2.23
Evatane2805	EVA27	27	–	72	57,000	0.74
Orevac9305	EVAMA26	26	1.5	47	20,000	11.13
Orevac9307	EVAMA13	13	1.5	92	73,000	1.1

Silane containing glycidoxy group – (3-glycidoxypropyl)trimethoxysilane (GS) – was used as the modifier. It is a clear, colorless liquid with a molecular weight of 236. Density is 1070 kg/m^3, refractive index n^D_{20} is 1.4367, and content of glycidoxy groups is 31%. Melting point is -70°C, flashpoint is 135°C, and boiling point is 264°C.

Modification of the copolymers was carried out in the melt on laboratory micro-rollers for 10 min in the temperature range from 100 to 120°C. (The rotational speed of the rolls is 12.5 m/min, and friction is 1:1.2.)

The melt flow rate (MFR) was measured in accordance with GOST 11645–73 at 190°C and under a load of 2.16 kg.

Viscosity was measured by viscometric method by dissolving the compositions in carbon tetrachloride.

Determination of the composition of coexisting phases and the interdiffusion coefficients were carried out by processing of series of interferograms obtained by microinterference method. Measurements were performed at a range of temperatures from 50 to 150°C. To construct profiles of concentrations by interference patterns, the temperature dependencies of the refractive index of the components are required.[6] Refractive index measurements were carried out by an Abbe refractometer IRF-454 BM at a range of temperatures from 20 to 150°C.

The structure of the modified copolymers was investigated by the transmission electron microscopy. Identification of the phase structure of the samples was carried out by etching of the surface in high-oxygen plasma discharge with the subsequent preparation of single-stage carbon-platinum replicas. Samples were viewed using PEM EM-301 (Philips, Holland).

RESULTS AND DISCUSSION

The temperature dependencies of the solubility of (3-glycidoxypropyl) trimethoxysilane (GS) in the initial double and triple copolymers of ethylene were identified in diffusion experiments by directly bringing EVA (EVAMA) and GS into contact.

Studies of mutual solubility of the components are shown in Figures 12.1 and 12.2.

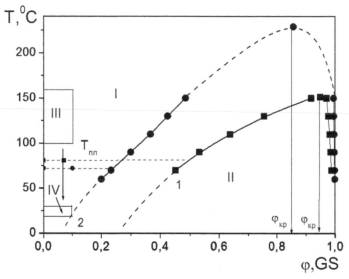

FIGURE 12.1   Phase Diagrams of the Systems Copolymer of Ethylene with Vinyl Acetate – GS: (1) EVA20; (2) EVA27. (I, II) The Areas of True Solutions, Heterogeneous Condition; (III) The Area of Preparation of the Compositions; (IV) The Area of the Study of the Structure and Physical Properties.

There are two binodal curves on all phase diagrams: the right branch of the binodal corresponds to the solubility of the copolymer in the modifier and is located in the area of infinitely dilute solutions. The second binodal curve represents the solubility of the modifier in the copolymers and is located in a fairly wide concentration area. The solubility of the modifier in the copolymer is increased as the temperature increases. In the systems EVA – GS, the solubility of the modifier reduces with an increase in VA content. However, for the systems EVAMA – GS, this is not observed. In these systems, the solubility of GS in the copolymer increases with the increase in VA content.

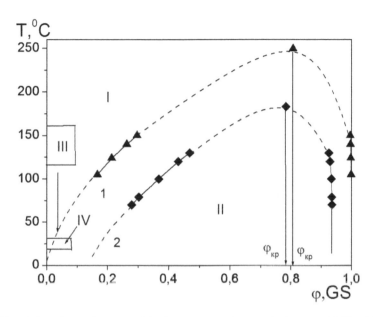

FIGURE 12.2 Phase Diagrams of the Systems Copolymer of Ethylene with Vinyl Acetate – GS: (*1*) EVAMA13; (*2*) EVAMA26. (I, II) The Areas of True Solutions, Heterogeneous Condition; (III) The Area of Preparation of the Compositions; (IV) The Area of the study of the Structure and Physical Properties.

It was found that the rate of a chemical reaction is greater than or comparable with the diffusion rate, and the movement of the modifier occurs in a chemically modified matrix, so we cannot see the appearance of the "hourglass" on the diagram.[5] The resulting matrix is soluble in the copolymer, and therefore, we do not observe the phase decomposition during reheating.[7]

Thus, it can be argued that these mixtures are characterized by diagrams with upper critical point of solubility (UCPS). This is also evidenced by the temperature dependencies of the pair interaction parameters (Fig. 12.3), calculated according to the equation, assuming that the right branch of the binodal is located on the axis $\varphi_2 = 1$:

$$\phi_{2,\hat{e}\hat{o}\hat{e}\hat{o}} = \frac{\sqrt{r_1}}{\sqrt{r_1} + \sqrt{r_2}},$$

$$\chi = \frac{1}{2}\left(\frac{1}{\sqrt{r_1}} + \frac{1}{\sqrt{r_2}}\right)^2,$$

where c is the pair interaction parameter; ϕ_1 and ϕ_2 are the concentration of EVA and GS, respectively; r_1, r_2 are their degrees of polymerization.

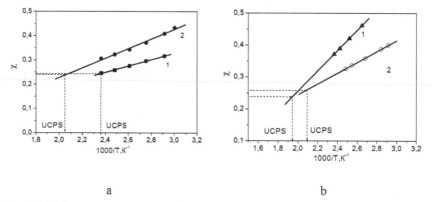

a   b

FIGURE 12.3   Temperature Dependence of the Pair Interaction Parameter of the Systems: (a) (*1*) EVA20 – GS; (*2*) EVA27 – GS; (b) (1) EVAMA13 – GS; (2) EVAMA26 – GS. Arrows show UCPS.

In all phase diagrams (see Figs. 12.1 and 12.2), there are the areas corresponding to temperature-concentration areas of the components mixing (area III) and the areas corresponding to the structure of the compounds and their physical properties (area IV). The diagrams show that the preparation of the mixtures takes place in the single-phase area (area I in the diagrams) and the study of the part of the system takes place in a heterogeneous area (area II in the diagram).

After etching of the cooled samples in plasma of high-oxygen discharge, there appear some particles, which have a lower etching rate compared with the dispersion medium. Earlier, it has been shown that the lowest etching rate among carbo- and heterochain polymers have polysiloxanes. Thus, it can be concluded that the dispersed phase is enriched by siloxanes (Fig. 12.4).

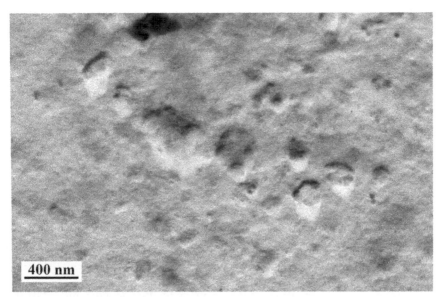

FIGURE 12.4   Microphotograph of EVAMA26 – GS (10%).

Since the reaction mixing was accompanied by intense shear impacts in the presence of oxygen and air moisture, the composite material obtained is the result of the processes of diffusion mixing and chemical interactions between the components. This may explain the presence of the dispersed phase, although based on the phase diagrams it should not be here.[8]

Formation of grafted siloxane units affects viscosity and rheological characteristics of the compositions. With the introduction of glycidoxysilanes, there is a decrease in the melt flow (see Fig. 12.5) of the modified polymers (by 30%) and an increase in the intrinsic viscosity (in 1.3–1.5 times), indicating that the preferential formation of branching and intermolecular bridges, increasing the length of the macromolecule, takes place.

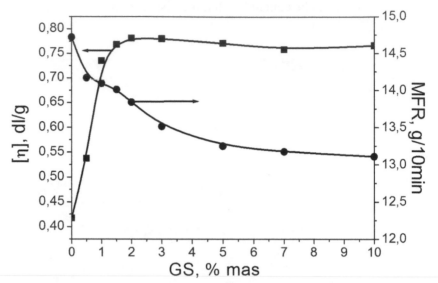

FIGURE 12.5   The Dependence of the Intrinsic Viscosity and MFR from the Modifier Content: EVAMA26 – GS (Viscosity at 40°C, MFR 125°C – 2.16kg).

Isolation of siloxane into a separate phase leads to an enrichment of the surface of the composition by them, and, consequently, to an increase of the adhesion characteristics.[9]

Glycidoxypropyltrimethoxysilane is used as an additive for polyesters, polyacrylates, polysulfides, urethanes, epoxy, and acrylic resins to improve their adhesion to glass, aluminum, steel, and other substrates. Aqueous and alcoholic solutions of glycidoxysilane are used to improve the adhesion of epoxy resin to aluminum plates.

Thus, the system EVAMA26 - GS has a good adhesion to PET. For this system, at the moment of the substrate rupture, an adhesion strength of the content of GS of 1.5 wt.% increased to 3.4 times (see Fig. 12.6). In this case, the rapture has a cohesion nature and is accompanied by the rapture of the substrate.[1]

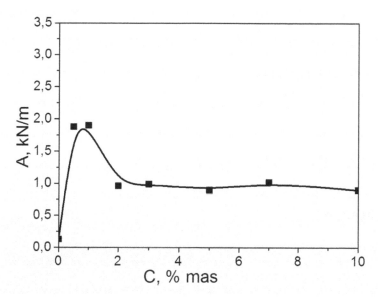

FIGURE 12.6   Adhesion Strength of the Polymer – PET: EVAMA26 – GS. Condition of the Formation: at 160°C for 10 min.

CONCLUSION

Thus, the data of these phase diagrams can be used in the study of the compositions, as well as selecting regimes for their preparation. From the diagram, we can define the aggregate state, the amount and chemical composition of the phases, as well as structural-phase state of the compositions according to the temperature and concentration of its constituent components.

The triple copolymer, modified by glycidoxysilane, can be used in the production of laminated multilayer films as it simultaneously has a good flowability and a high value of the adhesion strength.

ACKNOWLEDGMENT

This work was financially supported by the Ministry of Education and Science of Russia in the framework of the theme №693 "Structured composite materials based on polar polymer matrices and reactive nanostructured components".

KEYWORDS

- Copolymers of ethylene
- aminoalkoxysilane
- phase diagrams
- viscosity
- rheology
- adhesion strength

REFERENCES

1. Temnikova, N. E.; Rusanova, S. N.; Tafeeva, Yu. S.; Sof'ina, S. Y.; Stoyanov, O. V. The effect of an amino-containing modifier on properties of ethylene copolymers. *Polym. Sci. Ser. D*. 2012, 5(4), 259–265.
2. Temnikova, N. E.; Rusanova, S. N.; Sofina, S. Yu.; Stoyanov, O. V.; Garipov, R. M.; Chalykh, A. E.; Gerasimov, V. K.; Zaikov, G. E. Influence of aminoalkoxy- and glycidoxyalkoxysilanes on adhesion characteristics of ethylene copolymers. *Polym. Res. J.*. 2014, 8(4), 305–310.
3. Temnikova, N. E.; Rusanova, S. N.; Sof'ina, S. Yu.; Stoyanov, O. V.; Garipov, R. M.; Chalykh, A. E.; Gerasimov, V. K. The effect of aminoalkoxy and glycidoxyalkoxy silanes on adhesion characteristics of double and triple copolymers of ethylene. *Polym. Sci. Ser. D*. 2014, 7(3), 84–187.
4. Temnikova, N. E.; Rusanova, S. N.; Tafeeva, Yu. S.; Stoyanov, O. V. Study of modification of ethylene copolymers by aminosilanes by IR spectroscopy FTIR. *Bulletin of Kazan Technological University*. 2011, 19, 112–124.
5. Rusanova, S. N.; Stoyanov, O. V.; Gerasimov, V. K.; Chalykh, A. E. Influence of the phase structure of copolymers of ethylene with vinyl acetate, modified by ethyl silicate, on their rheological properties. *Bulletin of Kazan Technological University*. 2006, 1, 156–163.
6. Chalykh, A. E.; Gerasimov, V. K.; Petukhova, O. G.; Kulagina, G. S.; Pisarev, S. A.; Rusanova, S. N. Phase structure of silanol-modified ethylene-vinyl acetate copolymers. *Polym. Sci. Ser. A*. 2006, 48(10), 1058–1066.
7. Temnikova, N. E. Effect of Amino and Glycidoxyalkoxysilanes on the Formation of the Phase Structure and Properties of Ethylene Copolymers. Thesis PhD, Kazan, 2013, 154 p.
8. Rusanova, S. N.; Temnikova, N. E.; Stoyanov, O. V.; Gerasimov, V. K.; Chalykh, A. E. IR spectroscopic study of the interaction of glycidoxy silane and copolymers of ethylene. *Bulletin of Kazan Technological University*. 2012, 22, 95–96.
9. Chalykh, A. E.; Gerasimov, V. K.; Rusanova, S. N.; Stoyanov, O. V. Effect of structural heterogeneity of ethylene-vinylacetate copolymers modified by ethyl silicate on their stress-strain characteristics. *Polym. Sci. Ser. D*. 2011, 4(2), 85–89.

CHAPTER 13

SILOXANE MATRIX WITH METHYLPROPIONATE SIDE GROUPS AND POLYMER ELECTROLYTE MEMBRANES ON THEIR BASE

NATIA JALAGONIA, IZABELA ESARTIA, TAMAR TATRISHVILI, ELIZA MARKARASHVILI, DONARI OTIASHVILI, JIMSHER ANELI, and OMAR MUKBANIANI

Department of Chemistry, Iv. Javakhishvili Tbilisi State University, I. Chavchavadze Ave. 3, Tbilisi 0179, Georgia

Institute of Macromolecular Chemistry and Polymeric Materials, Iv. Javakhishvili Tbilisi State University, Faculty of Exact and Natural Sciences, I. Chavchavadze Ave. 13, Tbilisi 0179, Georgia

E-mail: omarimu@yahoo.com

CONTENTS

INTRODUCTION

Solvent-free polymer electrolytes may be formed by the interaction of polar polymers with metal ions. Ion transport in polymer electrolytes is extensively studied since Wright[1] discovered that polyethyleneoxide (PEO) can act as a host for sodium and potassium salts, thus producing a solid electrical conductor polymer/salt complex. The unique idea of employing these polymer electrolytes in battery applications belong to Armand et al.[2]

Transport mechanism models developed by Ratner et al.[3] indicated that polymers with low T_g have extremely high free volumes that favor the ion transport. Better results are obtained for polymers with highly flexible backbones, bearing oligo(ethylene glycol) side chains.

Interest in polysiloxane-based polymer electrolytes arose early in the 1980s. PEO-substituted polysiloxanes as ionically conductive polymer hosts have been previously investigated.[4–6] Their relatively high ionic conductivity was ascribed to the highly flexible inorganic backbone, which produced a totally amorphous polymer host. In recent years, improved battery performance has been observed for systems containing polymer electrolytes, with a Li$^+$ transference number close to unity.[7] Efforts have also been made to design and synthesize siloxane-based single-ion conductors.[8,9] Polysiloxanes are promising components for comb polyelectrolytes because they possess a flexible backbone that enhances the transports of ions. Their amorphous and highly flexible [Si–O]$_n$ backbone produces glass transition temperatures as low as -100°C and yields little or no crystallinity at room temperature. In addition, each monomer unit has two sites for cross-links or functional side chains through bond formation with silicon. Simulations indicate that comb polyelectrolytes should display higher conductivity values than their analogs to local motion of the bound anions in comb systems.[10–12]

EXPERIMENTAL PROCEDURE

MATERIALS

D_4^H (Aldrich), platinum hydrochloric acid (Aldrich), Karstedt's catalyst $(Pt_2[(VinSiMe_2)_2O]_3)$ or platinum(0)-1,3-divinyl-1,1,3,3-tetramethyldisiloxane complex (2% solution in xylene) (Aldrich), vinyltriethoxysilane (Aldrich), and methyl acrylate (Aldrich) were used as received. Lithium trifluoromethanesulfonate (triflate) and lithium bis(trifluoromethanesulfonyl) imide were purchased from (Aldrich). Toluene was dried over and distilled from sodium under an atmosphere of dry nitrogen. Tetrahydrofuran (THF) was dried over and distilled from K–Na alloy under an atmosphere of dry nitrogen.

CHARACTERIZATION

FTIR spectra were recorded on a Varian 660/670/680-IR series spectrometer. ^1H, ^{13}C NMR and ^{29}Si NMR spectra were recorded by a Varian Mercury 300VX NMR spectrometer, using DMSO and CCl_4 as the solvent and as an internal standard. Differential scanning calorimetric (DSC) investigation was performed on a Netzsch DSC 200 F3 Maia apparatus. Thermal transitions including glass transition temperatures T_g were taken as the maxima of the peaks. The heating and cooling scanning rates were 10 K/min.

Size-exclusion chromatographic (SEC) study was carried out with the use of Waters Model 6000A chromatograph with an R 401 differential refractometer detector. The column set comprised 10^3 and 10^4 Å Ultrastyragel columns. Sample concentration was approximately 3% by weight in toluene, and typical injection volume for the siloxane was 5 μL, and flow rate was 1.0 mL/min. Standardization of the SEC was accomplished by the use of styrene or polydimethylsiloxane standards with the known molar mass.

Wide-angle X-ray analysis was performed on a Dron-2 (Burevestnik, Saint Petersburg, Russia) instrument. A-CuK$_\alpha$ was measured without a filter; the angular velocity of the motor was $\omega \approx 2°/min$.

Determination of $\equiv Si–H$ content was calculated according to the method described by Iwahara et al.[13]

HYDROSILYLATION REACTION OF D_4^H WITH METHYL ACRYLATE AND VINYLTRIETHOXYSILANE

D_4^H (5.000 g and 0.0208 mol) were transferred into a 100 mL flask under nitrogen using standard Schlenk techniques. High vacuum was applied to the flask for half an hour before the addition of methyl acrylate (5.3636 g and 0.0624 mol) and vinyltriethoxysilane (3.9562 g and 0.0208 mol). The mixture was then dissolved in 7 mL of toluene, and 0.1 M solution of platinum hydrochloric acid in THF ($5/9 \times 10^{-5}$ g per 1.0 g of starting substance) was introduced. The homogeneous mixture was degassed and placed in an oil bath, which was previously set to 60°C and reaction continued at 60°C. The reaction was controlled by decrease of intensity of active $\equiv$Si–H groups. Then 0.1 wt.% activated carbon was added and refluxed for 12 h for deactivation of catalysts.

All volatile products were removed by rotary evaporation, and the compound was precipitated at least three times into pentane to remove side products. Finally, all volatiles were removed under vacuum for 24 h to isolate 13.7 g (95.6%) of colorless viscous compound I - 2.4.6.8-tetramethtyl-2.4.6-tri(methyl propionate)-8-ethyltriethoxysilanecyclotetrasiloxane ($D_4^{R,R'}$).

RING-OPENING POLYMERIZATION REACTION OF $D_4^{R,R'}$

The 1.1365 g (1.4046 mmol) of compound $D_4^{R,R'}$ was transferred into 50 mL flask under nitrogen. High vacuum was applied to the flask for half an hour. Then the compound was dissolved in 1.8 mL dry toluene, and 0.01% of total mass powder-like potassium hydroxide was added. The mixture was degassed and placed in an oil bath that was previously set to 60°C and was polymerized under nitrogen for 25 h. After the reaction, 7 mL of toluene was added to the mixture, and the product was washed using water. The crude product was stirred with $MgSO_4$ for 6 h, filtered, and evaporated, and the oligomer was precipitated at least three times into pentane to remove side products. Finally, all volatiles were removed under vacuum to isolate 1.06 g (93%) colorless viscous oligomer (II).

Ring-opening polymerization reaction of compound I at various temperatures had been carried out in the same manner.

GENERAL PROCEDURE FOR PREPARATION OF CROSS-LINKED POLYMER ELECTROLYTES

In a typical preparation, 0.75 g of polymer II was dissolved in 4 mL of dry THF and thoroughly mixed for half an hour before the addition of catalytic amount of acid (one drop of 0.1 M HCl solution in ethyl alcohol) to initiate the cross-linking process. After stirring for another 3 h, required amount of lithium triflate from the previously prepared stock solution in THF was added to the mixture and stirring continued for another 1 h. The mixture was then poured into a Teflon mould with a diameter of 4 cm, and solvent was allowed to evaporate slowly overnight. Finally, the membranes were dried in an oven at 70°C for 3 days and at 100°C for 1 h. Homogeneous and transparent films with average thickness of 200 μm were obtained in this manner. These films were insoluble in all solvents and only swollen in THF.

AC IMPEDANCE MEASUREMENTS

The total ionic conductivity of samples was determined by locating an electrolyte disk between two 10 mm diameter brass electrodes. The electrode/electrolyte assembly was secured in a suitable constant volume support, which allowed extremely reproducible measurements of conductivity to be obtained between repeated heating–cooling cycles. The cell support was located in oven, and the sample temperature was measured by thermocouple positioned close to the electrolyte disk. The bulk conductivities of electrolytes were obtained during a heating cycle using the impedance technique (impedance meter BM 507–TESLA for frequencies 50–500 kHz) over a temperature range between 20 and 100°C.

RESULTS AND DISCUSSION

Comb-type polymers for solvent-free solid polymer electrolytes usually are obtained via hydrosilylation or dehydrocondensation reactions of industrial linear polymethylhydrosiloxanes (PMHS) with donor group containing vinyl-, allyl- or hydroxyl-containing organic compounds. It should be noted that often these reactions proceed incompletely resulting in the formation of irregular defect structures.

The aim of our work is to synthesize organocyclotetrasiloxane with desired propionate donor side groups at silicon via hydrosilylation reaction of 2.4.6.8-tetrahydro-2.4.6.8-tetramethylcyclotetrasiloxane (D_4^H) with methylacrylate and vinyltriethoxysilane at 1:3:1 of initial compounds in the presence of platinum catalysts; polymerization reactions of organocyclotetrasiloxane in the presence of nucleophilic catalysts and obtaining comb-type polymers with regular arrangement of methyl propionate and ethoxyl group; obtaining solid polymer electrolyte membranes by the incorporation of lithium salt into polymer matrices; and investigation of ionic conductivity of membranes via impedance method.

Preliminary heating of initial compounds separately in the temperature range of 50–60°C in the presence of catalysts showed that under these conditions, polymerization of D_4^H or allyl acetoacetate and scission of siloxane backbone does not take place. No changes in the NMR and FTIR spectra of initial compounds were found. It was established that in melt condition, the hydrosilylation reaction proceeds vigorously with initiation of side reactions;[14] therefore, for obtaining fully addition product, hydrosilylation reaction have been carried out in dilute solutions.

Hydrosilylation reaction of D_4^H methylacrylate and vinyltriethoxysilane at 1:3:1 of initial compounds in the presence of platinum catalysts was carried out in 50% solution of dry toluene or THF at 60–70°C temperature. It was established that hydrosilylation in the presence of Karstedt's and platinum hydrochloride acid catalyst proceeds very slowly, so in hydrosilylation reaction, we did not use catalyst Pt/C (5%).

In order to increase the reaction rate, the reaction temperature at the final stage of the reaction mixture was increased up to 80°C.

The reaction generally proceeds according to the following Scheme 13.1:

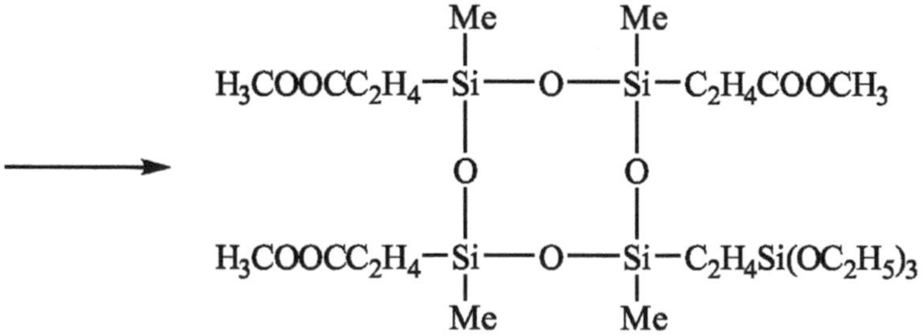

SCHEME 13.1 Hydrosilylation Reaction of D_4^H with Methyl Acrylate and Vinyltriethoxysilane.

From the literature, it is known that hydrosilylation with unsaturated bonds may be proceeded according to Markovnikov and anti-Markovnikov (Farmer) rules.[11,12]

In addition, methyl acrylate is a conjugated compound. As it is known from the literature, in conjugated systems, hydrosilylation might proceed not only in the direction of 1.2, but also in the direction of 1.4 (Scheme 13.2):

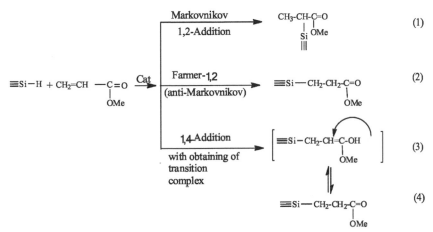

SCHEME 13.2   Possible Addition of ≡Si–H Bonds to Methyl Acrylate.

As it is seen from the possible Scheme 13.2.3, in the case of 1.4-hydride addition, reaction proceeds with obtaining of intermediate-transition complex. By regrouping of the intermediate product according to the Eltekov rule (Scheme 13.2.4), the products of 1.2-addition by anti-Markovnikov rule are obtained.

In addition, in hydrosilylation reaction, the mixture of isomeric cyclic compounds (cis– and trans–isomeric mixture) can take place. The obtained substance is a transparent viscous liquid, which is well soluble in common organic solvents. Compound I was identified by IR, ^{29}Si, ^{1}H and ^{13}C NMR spectral data, determining the molecular mass and molecular refractivity: $n_D^{20} = 1.4389$; $d_4^{20} = 1.1185$; calculated $M_{RD} = 153.36$; and found $M_{RD} = 152.67$.

In the FTIR spectra of compound I absorption bands at 1020 and 1197 cm^{-1}, characteristic for asymmetric valence oscillation of ≡Si–O–Si≡, Si–O–C, and CO–O–C bonds were observed. Also, one can observe absorption bands at 1257 and 2860–2970 cm^{-1} region, characteristic for valence oscillation of ≡C–H bonds. In the FTIR spectra of compound I at 2167 cm^{-1}, there is no absorption bands characteristic for unreacted ≡Si–H bonds. In the spectra absorption bands at 1736 cm^{-1}, characteristic for carbonyl groups is observed.

In ^{29}Si NMR spectra of compound I, one can see signal with chemical shift δ≈-18.98 and δ≈-26.86 ppm corresponding to the presence of RR'SiO (D) units in cyclotetrasiloxane fragment;[15,16] on the other hand, two signals δ≈-46.09÷-47.43 and δ≈-58.28÷-61.03 ppm corresponding to the presence of M$^{(OR)}_2$ ꝏ D^{OR} fragments and the signal with chemical shift δ≈-65.66 ppm T units at -55 and –65 ppm are observed,[16] which is confirmed reaction direction, with the formation of above structure. The direction of the reaction is confirmed by the structure formation.

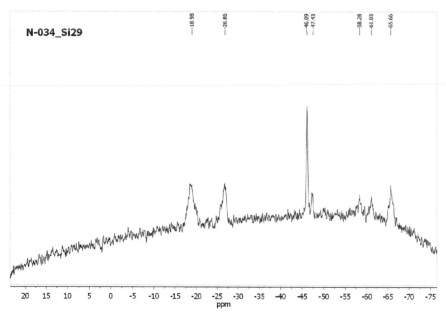

FIGURE 13.1 ^{29}Si NMR Spectra of Compound I.

In the ^{1}H NMR spectrum of compound I (Figs. 13.1 and 13.2), one can observe singlet signal with a chemical shift δ = 0.1–0.2 ppm characteristic for protons in ≡Si–Me groups in isomeric mixture of cyclotetra-

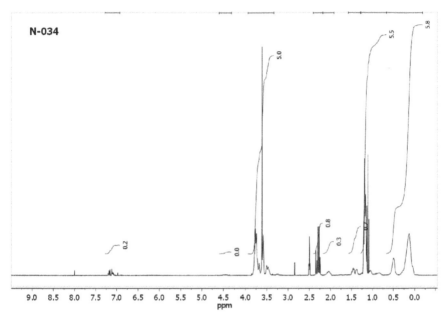

FIGURE 13.2 I ¹H NMR Spectra of Compound I.

siloxane fragments, triplet signal for methylene protons in ≡Si–CH₂- fragment with chemical shift δ = 0.50 ppm (during addition to anti-Markovnikov rule), and triplet signal for methyl protons in =CH–CH₃ group with chemical shift δ = 1.1 ppm (during addition to Markovnikov rule). Also, one can observe multiplet signals with center of chemical shifts δ = 2.3, 3.6, -3.8 ppm characteristics for proton in CH₂CH=, -CH₂CO-, and OCH₃ group, respectively.

In the ¹³C NMR spectra of isomeric mixture of compound I (Fig. 13.3), the resonance signals with chemical shifts δ≈-3.83, -2.45, 1.01, 17.75, 26.38, 30.13, 50.45, and 57.33 ppm characteristic for ≡Si–Me, CH₃ to CH=, ≡Si–CH₂-, -CH₂CO-, -CH(CH₃)- and OCH₃ accordingly ¹³C NMR spectra correspond to ¹H NMR spectra data.

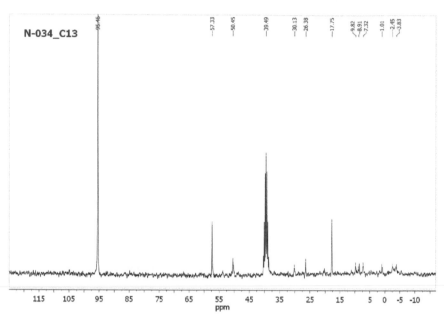

FIGURE 13.3 ^{13}C NMR Spectra of Compound I.

The synthesized compound I was used in the polymerization and co-polymerization reactions. From literature,[12] it is known that during polymerization reactions of ethoxyl group containing compound $D_4^{R,R'}$ proceeds with obtaining of cross-linking systems, which may be explained via intermolecular condensation reactions of Si–OH and -Si(OC$_2$H$_5$)$_3$ groups. Therefore, we have investigated copolymerization reactions of $D_4^{R,R'}$ with hexamethyldisiloxane as a terminating agent at 8:1 ratio of initial compounds. In most cases, terminating agent helps to regulate the molecular masses.[17]

The polymerization reactions of compound I in the presence of anhydrous powder-like potassium hydroxide (0.01 mass%) as a catalyst in dilute solution of dry toluene (C = 0.8606 mol/L) at 60–80°C temperature have been investigated. It was established that polymerization reactions proceed slowly during 80–100 h. The optimal condition of reaction temperature is 80-90°C. During polymerization reaction of $D_4^{R,R'}$ in toluene, the polymer precipitated from the solvent, but during polymerization in THF, the polymer remained in the solution. The copolymerization reaction generally proceeds according to the Scheme 13.3:

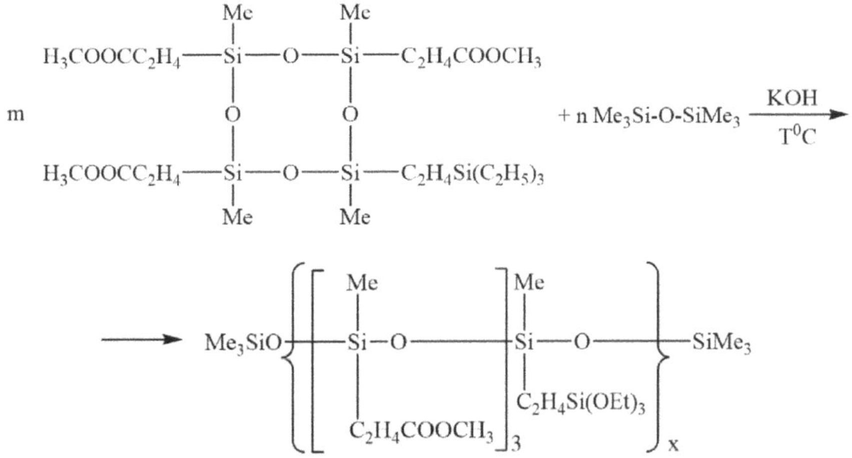

SCHEME 13.3 Copolymerization Reaction of Compound I.

Where, m:n = 8:1. II^1 (60°C), II (80°C).

The synthesized polymers are vitreous, viscous products, which are well soluble in ordinary organic solvents with specific viscosity $\eta_{sp} \approx$ 0.14–0.27. The structure and composition of polymers were determined by elemental analysis and molecular masses by FTIR and NMR spectra data. Elemental composition yields and some physical–chemical properties of oligomers were studied.

In FTIR spectra of polymers, the absorption band at 1080 cm^{-1} is characteristic for asymmetric valence oscillation of $\equiv$Si–O–Si$\equiv$ bonds. In the spectrum are reserved all the absorption bands characteristic of the initial monomeric compound I.

In ^{29}Si NMR spectra of polymers (Fig. 13.4), one can see resonance signal with chemical shifts $\delta \approx -19.10$ and $\delta \approx -26.97$ ppm corresponding to the presence of D to M$^{(OR)}_2$ and D^{OR} units; on the other hand, the signal at $\delta \approx -65.73$ ppm corresponds to T units.

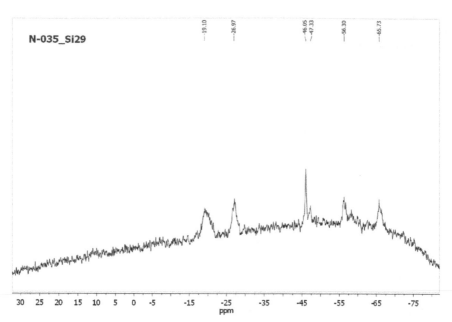

FIGURE 13.4 ^{29}Si NMR Spectra of Polymer II.

On the ^{1}H NMR spectra of polymers (Fig. 13.5), singlet signal with a chemical shift δ≈0.1–0.2 ppm characteristic for protons in ≡Si–Me groups, triplet signal for methylene protons in ≡Si–CH$_2$- fragment with che-mical shift δ≈0.50 ppm (during addition to anti-Markovnikov rule), triplet signal for methyl protons in =CH–CH$_3$ group with chemical shift δ = 1.1 ppm (during addition to Markovnikov rule) were observed. Also, one can observe multiplet signals with center of chemical shifts δ≈2.3, 3.6, -3.8 ppm characteristics for proton in C̲H$_2$CH=, -C̲H$_2$CO-, and OC̲H$_3$ group, respectively.

In Figure 13.6, ^{13}C NMR spectra of polymer II is presented. The signal with chemical shifts at δ≈ δ≈-3.83, -2.45, 1.01, 17.75, 26.38, 30.13, 50.45 and 57.33 ppm characteristic for ≡Si–Me, C̲H$_2$CH=, ≡Si–CH$_2$-, -C̲H$_2$CO-, -C̲H(CH$_3$)- and OCH$_3$ accordingly is preserved. ^{13}C NMR spectra of oligo-mer I is in accordance with ^{1}H NMR spectra.

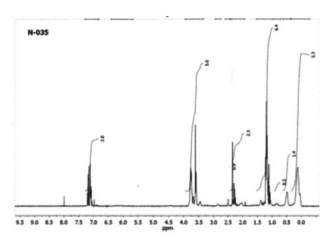

FIGURE 13.5   ¹H NMR Spectra of Polymer II.

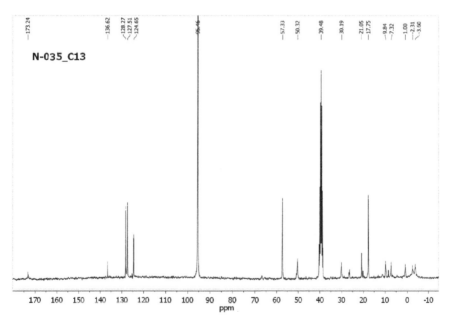

FIGURE 13.6   ¹³C NMR Spectra of Polymer II.

For obtaining polymer II, the molecular masses by ebuliometric methods have been determined by M_n = 6100. As is known from the literature data, the molecular masses of obtained polymer are sufficient for a solid polymer membranes electrolytic production.

DSC calorimetric investigation of polymer II was carried out. As it can be seen from Figure 13.7, the polymer is characterized of only one glass transition temperature, which is equal to $T_g = -81.7°C$.

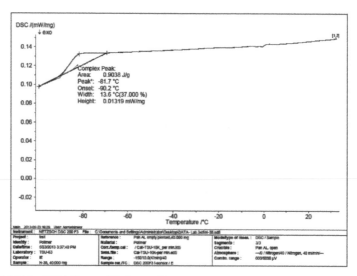

FIGURE 13.7   DSC Curve of Polymer II.

POLYMER ELECTROLYTE MEMBRANES

For obtaining the membranes, we prepared the polymer solutions of the lithium salts triflate ($LiSO_3CF_3$) and lithium-bis(trifluoromethanesul fonimide) [$(CF_3SO_2)_2NLi^+$] in the THF, where these salts were about 5–20 wt% of the polymer full mass. The solution contained 0.8 g salt in the 2 mL THF (Scheme 13.4).

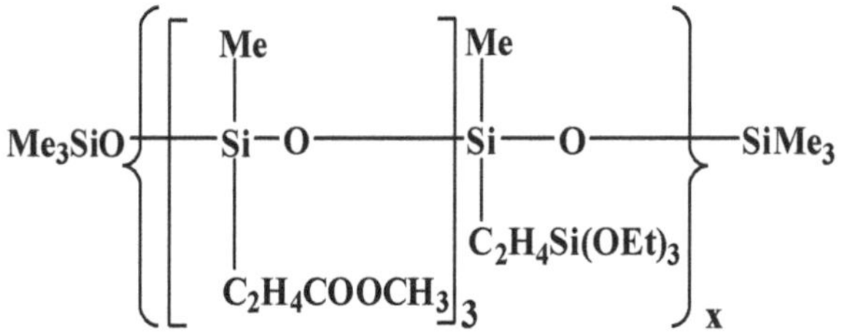

SCHEME 13.4 Experimental Solution Used.

The solution of polymer in the THF was prepared in the cylindrical form (diameter 4 cm) vessel made from Teflon. After the salt solution in the THF and 1–2 drops of 0.1 N alcohol solutions were added to the initial mixture. Mixing elapsed for 30 min. The mixture was allowed to stay in the inert atmosphere and then out-gassed; hence, the transparent yellow films were formed. The sol gel process was conducted by adding the hydrochloric acid alcohol solution, which is accompanied by cross-link reactions shown in Scheme 13.5.

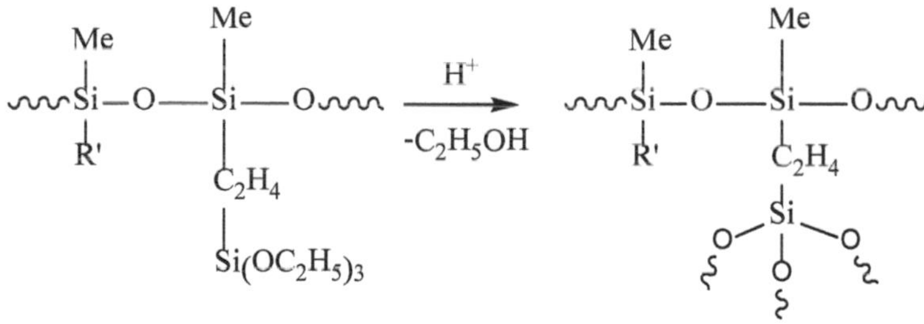

SCHEME 13.5 The Sol-Gel Cross-Linking Reaction of the Polymer.

The investigation of the electric-physical properties of synthesized polyelectrolytes was conducted. The conductivity and its dependence on the temperature were measured. We prepared the solid polymer electrolyte membranes with 5, 15, and 20 wt% of lithium salts. The dependence of the specific volumetric electric conductivity of the membranes on temperature was studied. The curves on Figure 13.8 show that their characters correspond to analogical dependences for polyelectrolytes based on silicon-organic polymers and some lithium salts and are well described by so-called "Vogel-Taman-Fulcher" formula:[8,9]

$$C(T) = a/T \exp(-E/kT),$$

where a is the pre-exponential factor, E is the activation energy of charge transfer, and k is Boltzmann constant.

It is worth mentioning that based on preliminary measures of the membranes described above, the low conductivity of membrane containing 5 wt% salt ($\sim 10^{-9}$ S/cm) is reported..

Dependence of conductivity of membranes based on polymer II with 15 (1) and 20 (2) wt% contents of lithium salt triflate on the temperature is presented in Figure 13.8.

The character of voltammograms obtained for membranes based on polymer II and 15 and 20 wt% of triflate salt is in full accordance with the values of their conductivity (Fig. 13.9).

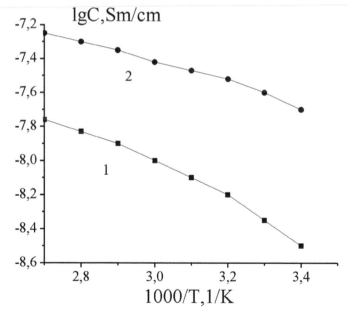

FIGURE 13.8   Dependence of Conductivity of Membranes Based on Polymer II with 15 (1) and 20 (2) wt% of Lithium Salt Triflate on the Temperature.

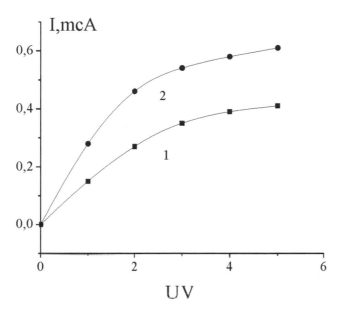

FIGURE 13.9   Voltammograms of Membranes Based on the Polymer II and 15 (1) and 20 wt% (2) of Lithium Triflate.

Thus, on the basis of the temperature dependences and corresponding voltammograms for the membranes based on polymer II and lithium triflate salt, it can be said that the ionic conducting character essentially is due to both structure of electrolytes polymer matrix and its behavior under conditions of the change of external temperature. First of all, it is expressed in the essential difference between initial and end values of the conductivity at increasing temperature. It may be proposed that microstructure of this type of membrane is highly permeable for the ions while heating, which is due to increase in the concentration of micro-empties in the polymer matrix because of high level of heterogeneity of the macromolecules with different side groups.

CONCLUSION

Hydrosilylation reaction of tetrahydrotetramethylcyclotetrasiloxane with methyl acrylate and vinyltriethoxysilane at 1:3:1 ratio of initial compounds, in the presence of Karstedt's catalyst, has been studied and corresponding additional product have been obtained. Through co-polymerization reaction of organocyclotetrasiloxane and hexamethyldisiloxane as a terminating agent, corresponding comb-type polymers have been obtained. Through sol-gel processes doped with lithium trifluoromethanesulfonate (triflate) or lithium bis(trifluoromethanesulfonyl)imide oligomer systems, solid polymer electrolyte membranes have been obtained. The dependence of ionic conductivity as a function of temperature and salt concentration was investigated. The electric conductivity of investigated polymer electrolyte membranes at room temperature changes in the range of 3×10^{-9}–2×10^{-8} S/cm.

KEYWORDS

- **Polymer electrolytes**
- **transport mechanism**
- **polysiloxane-based polymer**
- **sodium and potassium salts**
- **simulations**
- **transports of ions**

REFERENCES

1. Wright, P. V. Electrical conductivity in ionic complexes of poly(ethylene oxide). *Br. Polym. J.* 1975, 7, 319–327.
2. Armand, M. B.; Chabagno, J. M.; Duclot, M. J. Polyethers as solid electrolytes, Second International Meeting on Solid Electrolytes, St. Andrews, Scotland, 1–4, 1978.
3. Ratner, M. A.; Shriver, D. F. Ion transport in solvent-free polymers. *Chem. Rev.* 1988, 88, 109–124.
4. Nagaoka, K.; Naruse, H.; Shinohara, I.; Watanabe, M. High ionic conductivity in poly(dimethyl siloxane-co-ethylene oxide) dissolving lithium perchlorate. *J. Polym. Sci., Polym. Lett. Ed.* 1984, 22, 659–665.
5. Albinsson, I.; Mellander, B. E.; Stevens, J. R. Ionic conductivity in poly(ethylene oxide) modified poly(dimethylsiloxane) complexed with lithium salts. *Polymer.* 1991, 32, 2712–2715.

6. Fish, D.; Khan, I. M.; Smid, J. Conductivity of solid complexes of lithium perchlorate with poly{[ω-methoxyhexa(oxyethylene)ethoxy]methylsiloxane}. *Macromol. Chem. Rapid* Commun. 1986, 7, 115–120.

7. Doyle, M.; Fuller, T. F.; Newman, J. The importance of the lithium ion transference number in lithium/polymer cells. *Electrochim. Acta.* 1994, 39, 2073–2081.

8. Karatas, Yu; Kaskhedikar, N.; Wiemhofer, H. D. Synthesis of cross-linked comb polysiloxane for polymer electrolyte membranes. *Macromol. Chem. Phys.* 2006, 207, 419–425.

9. Zhang, Zh; Lyons, L. J.; Jin, J. J.; Amine, Kh; West, R. Synthesis and ionic conductivity of cyclosiloxanes with ethyleneoxy-containing substituent's. *Chem. Mater.* 2005, 17, 5646–5650.

10. Snyder, J. F.; Ratner, M. A.; Shriver, D. F. Polymer electrolytes and polyelectrolytes: Monte Carlo simulations of thermal effects on conduction. *Solid State Ionics.* 2002, 147, 249–257.

11. Mukbaniani, O.; Koynov, K.; Aneli, J.; Tatrishvili, T.; Markarashvili, E.; Chigvinadze, M. Solid polymer electrolyte membranes based on siliconorganic backbone. *Macromol. Symp.* 2013, 328(1), 38–44.

12. Mukbaniani, O.; Aneli, J.; Esartia, I.; Tatrishvili, T.; Markarashvili, E.; Jalagonia, N. Siloxane oligomers with epoxy pendant group. *Macromol. Symp.* 2013, 328(1), 25–37.

13. Iwahara, T.; Kusakabe, M.; Chiba, M.; Yonezawa, K. Synthesis of novel organic oligomers containing Si-H bonds. *J. Polym. Sci., A.* 1993, 31, 2617–2631.

14. Mukbaniani, O.; Tatrishvili, T.; Markarashvili, E.; Esartia, E. Hydrosilylation reaction of tetramethylcyclotetrasiloxane with allyl butyrate and vinyltriethoxysilane. *Georgian Chem. J.* 2011, 2(11), 153–155.

15. Uhlig, F.; Marsmann, H. Chr.*^{29}Si NMR Some Practical Aspects*; Springer: Heidelberg, 2008.

16. Khan, I. M.; Yuan, Y.; Fish, D.; Wu, E.; Smid, J. Comblike polysiloxanes with oligo(oxyethylene) side chains. *Synth. Prop. Macromol.* 1988, 21, 2684–2689.

17. Mukbaniani, O.; Aneli, J.; Tatrishvili, T.; Markarashvili, E.; Chigvinadze, M.; Abadie, M. J. M. Synthesis of cross-linked comb-type polysiloxane for polymer electrolyte membranes. *E-Polym.* 2012, 089, 1–14.

CHAPTER 14

COMPOSITES ON THE BASIS OF GLYCIDOXYGROUP CONTAINING PHENYLSILSESQUIOXANES

MARINA ISKAKOVA[1], ELIZA MARKARASHVILI[2,3], JIMSHER ANELI[3], and OMAR MUKBANIANI[2,3]

[1]Ak. Tsereteli Kutaisi state University, Department of Chemical Technology, I. Chavchavadze Ave. 1, 0179 Tbilisi, Georgia

[2]Iv. Javakhishvili Tbilisi State University, Department of Chemistry, I. Chavchavadze Ave. 1, 0179 Tbilisi, Georgia

[3]Institute of Macromolecular Chemistry and Polymeric Materials, Iv. Javakhishvili Tbilisi State University, I. Chavchavadze Ave. 13, 0179 Tbilisi, Georgia

E-mail: marinaiskakova@gmail.com

CONTENTS

INTRODUCTION

Recently, intense interest has been shown in the development of silsesquioxane-based materials because of their three-dimensional (3D) nature, ability to offer a very high degree of functionalization, ease of synthesis, and typically high thermal stability. This is evidenced by the fact that at present, there are approximately three reviews on silsesquioxanes and the related silicates.[1-3] These references describe their potential application in a broad range of areas from biomedical to organic light-emitting diodes, to nanocomposites, and so on. In the literature data, there is a lot of information about the dependence of the properties of the substance on the polyorganosiloxane organic framing at the silicon atom. For example, polyorganosiloxanes containing aromatic radicals different from polydimethylsiloxanes are characterized by high thermostability and dielectric characteristics.[4]

For improving the dielectric properties of the modified filler compounds, the oligotetraepoxysiloxane oligomers with organosilsesquioxane fragments in the chain have been synthesized. As initial components for the synthesis of tetraepoxyphenylsilsesquioxane during condensation reaction of epichlorohydrin, the cis–2.4.6.8-tetrahydroxy-2.4.6.8-tetraphenylcyclotetrasiloxane and tetrahydroxyphenyl oligomer (n = 2 10) with sodium hydroxide have been used, and phenylethoxysilsesquioxanes PhES-80 (n-1) and PhES-50 (n = 2) in the presence of catalysts iron chloride (III) 0.01% by weight have been investigated.

EXPERIMENTAL PART

MATERIALS AND TECHNIQUES

Initially, cis–2.4.6.8-tetrahydroxy-2.4.6.8-tetraphenylcyclotetrasiloxane and tetrahydroxyphenylsilsesquioxane were synthesized as given in the literature.[5,6]

FTIR spectra were recorded on a Varian 660/670/680-IR series spectrometer. Chromatographic analysis of the purity of the starting reactants and the reaction was carried out using a chromatograph grade LKhM-80, Model 2 (column 3000 chromatografic-4 mm) Media-Chromosorb W or Chromaton. For the analysis, the following phases were used: Silicone

SE-30 resin, REOPLEX-400, and carrier gas-helium. Match of the applied phase separating substances possessed the ability to OV-17 supported on Chromaton-NAW. For better separation of different monomers, the temperature was selected in the range 130–200°C.

CONDENSATION REACTION OF EPICHLOROHYDRIN WITH CIS–2.4.6.8-TETRAHYDROXY-2.4.6.8-TETRAPHENYLCYCLOTETRASILOXANE

In a four-necked flask equipped with a thermometer, mechanical stirrer, reflux condenser, and dropping funnel, 5.52 g (0.01 mol) of tetrahydroxytetraphenylcyclotetrasiloxane and 30 mL of diethyl ether were taken. While stirring at 40°C, 7.52 g (0.08 mol) of epichlorohydrin (100% excess) was added, and then distilled ether was added slowly and gradually. The reaction temperature was increased to 60–70°C and then the reaction mixture was stirred again and heated for 1 h. After carrying out these steps, 0.4 mol (16 g NaOH, 48 g H_2O) of 25% sodium hydroxide solution was added in three portions. After addition of the last portion of the alkaline solution mixture, the mixture was stirred for another 1 hour at a temperature of 70°C. Then 9.3 g of epichlorohydrin was added, and the reaction mixture was diluted with 100 mL of ether. Ether was removed by suction. Unreacted part of epichlorohydrin was evacuated at 30°C for 3 h at a residual pressure of 1 mm Hg. As a result, approximately 6.8 g of slow-moving viscous mass (compound I) with a yellowish color at 96% yield was obtained.

CONDENSATION REACTION OF EPICHLOROHYDRIN WITH OLIGOPHENYLSILSESQUIOXANE (N»3)

Compounds (II) were prepared as given in Scheme 14.2. In the flask, 15.84 g (0.01 mol) of tetrahydroxypolyphenyloligotetrole in 60 mL of dry toluene and 7.52 g (0.08 mole) epichlorohydrin (100% excess) were taken. The 16.74 g (96%) tetraepoxypolyphenylsilsesquioxane (II) was obtained. Condensation reaction of epichlorohydrin with oligotetraphenylsilsesquioxane (n≈10) was carried out in the same manner as explained in the section Condensation Reaction of Epichlorohydrin with cis–2.4.6.8-tetrahydroxy-2.4.6.8-Tetraphenylcyclotetrasiloxane.

CONDENSATION REACTION OF EPICHLOROHYDRIN WITH TETRAETHOXYTETRAPHENYLCYCLOTETRASILOXANE (PHES-80)

The reaction between an industrial product tetraethoxytetraphenylcyclo-tetrasiloxane PhES-50 (n = 1) and PhES-80 (n = 2) with epichlorohydrin was conducted in a four-necked flask equipped with a dropping funnel, a reflux condenser, a mechanical stirrer, and a thermometer. In this flask, 118 g (0.1 mol) of PhES-80 (phenethylsilsesquioxane-80), 2.4 g of iron chloride (III) 0.01% by weight of the reaction mixture, and 75.2 g (0.8 mol) epichlorohydrin (100% excess) were added. The reaction mixture was heated up to 80–85°C. The reaction was carried out for 5–6 h, and low molecular byproduct C_2H_5Cl was collected in a flask. The resulting mass was centrifuged to precipitate the catalyst. After evacuating the unreacted products, the 112.5 g of a dark brown color gummy product I' was obtained with a 95% yield.

Similarly, the condensation reaction of epichlorohydrin with PhES-50 (n = 1) in the presence of catalysts, iron (I'), and aluminum chloride (IV') have been carried out and dark brown color gummy products have been obtained.

RESULTS AND DISCUSSION

There are certain methods of synthesis of epoxyorganosiloxanes. From these methods, it is significant to denote oxidative epoxidation of unsaturated bond containing organosilanes and siloxanes,[7,8] hydrosilylation reactions of ≡Si–H bond containing silanes and siloxanes to allyl glycidyl ether in the presence of catalysts,[9,10] and the reactions of oganosilanoles or organosiloxanoles with epichlorohydrin.[11,12]

The condensation reaction of epichlorohydrin with Cis–2.4.6.8-tetra-hydroxy-2.4.6.8-tetraphenylcyclotetrasiloxane with the excess epichlorohydrin in the presence of 25% sodium hydroxide solution was performed. The reaction was carried out as given in Scheme 14.1.

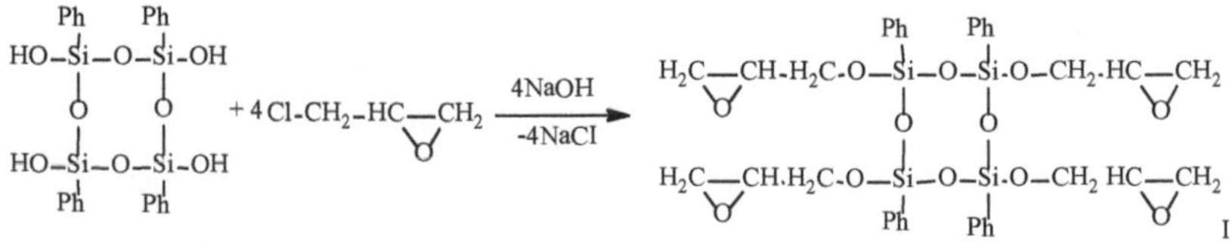

SCHEME 14.1   The Condensation Reaction of Epichlorohydrin with cis–2.4.6.8-Tetrahydroxy- 2.4.6.8-Tetraphenylcyclotetrasiloxane.

A transparent yellow viscous compound I well soluble in organic solvents was obtained. The composition and structure of the obtained compound I was studied on the basis of elemental analysis, definition of number of epoxy groups, determination of molecular masses, and FTIR spectra data. Some physicochemical data of the synthesized compounds are presented in Table 14.1.

TABLE 14.1   Some Physicochemical Data of the Synthesized Compounds

#	n	Yield, %	Amount of Epoxide[b] Groups,%	M^a_{mass}	Elemental Analysis[b],%		
					C	H	Si
I	1	96	24.16	712	60.67	5.62	15.73
			24.05	709	60.40	5.60	15.66
II	3	95	9.86	1744	57.80	4.59	19.27
			10.00	1769	58.63	4.66	19.55
III	10	96	3.23	5324	56.80	4.13	21.04
			3.18	5242	55.92	4.07	20.72

Notes: [a]Molecular masses have been determined via ebulliometric method; [b]Values above the line are calculated values and those below the line are found values.

In the FTIR spectra of obtained compounds in the range of asymmetric and symmetric valence oscillations Si–O–Si bond, the bifurcation of the strips with maximums v_{as} −1045 and 1145 cm^{-1}, and v_{as} −455 and 480 cm^{-1} were observed. In the case of phenylcyclotetrasiloxane, ring condensation in the presence of nucleophilic sodium may occur with the opening of cyclotetrasiloxane ring, which leads to the formation of the structure different from cis–configuration (see Figs. 14.1 and 14.2). This opinion is in accordance with the known data.[13,14]

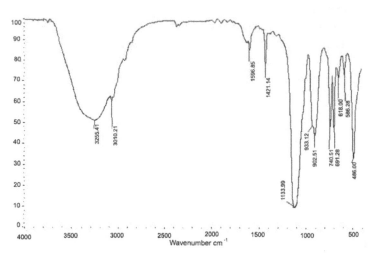

FIGURE 14.1 The FTIR spectra of initial cis–2.4.6.8-tetrahydroxy-2.4.6.8-tetraphenylcyclotetrasiloxane.

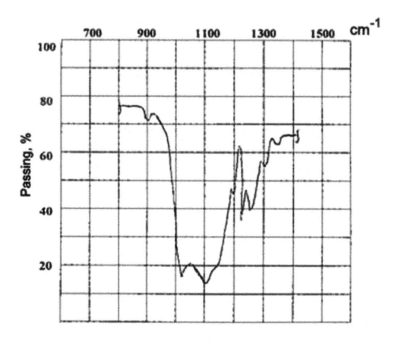

FIGURE 14.2 FTIR spectrum of 2.4.6.8-glycidoxy-2.4.6.8-tetraphenylcyclotetrasiloxane (I).

However, the realization of the cis–isotactic structure is associated with several steric hindrances and characterized by a short length of the molecules. During the tetrole polycondensation without initiator of the basis type, which allows the obtaining of the polymer without breaking the siloxane bond in the organocyclosiloxane, the conditions of synthesis ensures the perfect cyclolinear ladder structure and the macromolecules with the Kuhn segment with length about 50 Å are formed.

From the spectroscopic investigations[14] the structure of macromolecules of PPSSO, obtained by condensation of tetrole T_4, differs from the structure obtained from phenyltrichlorosilane and anion polymerization. The conducted experimental investigations suggest that the macromolecules with Kuhn segment that is approximately 50 Å have the structure somewhat similar to cis–anti–cis–tactic one:[13,14]

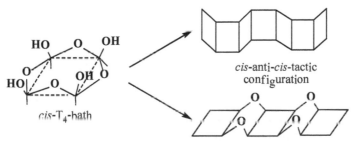

cis-anti-cis-tactic
configuration

cis-T_4-hath

cis-isotactic configuration

SCHEME 14.2   Cis–anti-cis–Tactic and cis–Isotactic Configuration of PPSSO.

Due to the combination of the T_4 fragments with transdisplacement of the functional groups, the cis–syndiotactic structure of chain may be formed with cis–syndiotactic configuration (Scheme 14.3).

trans-T_4-coach

cis-syndiotactic
configuration

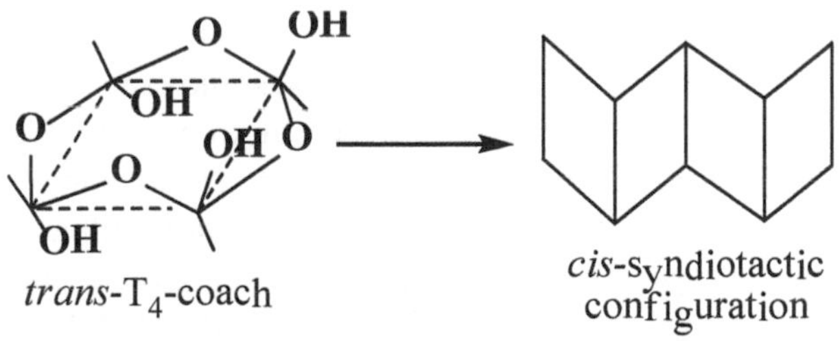

SCHEME 14.3   Cis–Syndiotactic Configuration of PPSSO.

Such structure of the chain is satirically more profitable and is characterized by high rigidity of the chain. Therefore, the dual structure of the molecules of ladder fragments lay under hydrolytic condensation organotriclorosilane and products of its partial hydrolysis and condensation.

So, it was established that in the presence of sodium hydroxide, the rearrangement of polyphenylsilsesquioxane skeleton takes place and cis–anti-cis structure of silsesquixane structure turn to cis–syndiotactic structure, which indicates the doublet character of absorption bands of ≡Si–O–Si≡ bonds.

With the aim of defining the effect of the fragment length of organocyclosiloxane in the tetraepoxy compounds based on the properties of composite materials, we investigated analogically the condensation reactions of tetrahydroxyoligoorganosilsesquioxanes with bifurcated epichlorohydrin in the presence of 25% solution of sodium hydroxide at room temperature.

The initial tetrahydroxyphenylsilsesquioxanes (n 3, 10) with *cis*–anti-*cis*–tactic structure as initial compounds obtained by thermal condensation of cis-2.4.6.8-tetrahydroxy-2.4.6.8-tetraphenylcyclotetrasiloxane were used.[5,6] Due to the interaction of the fragments of T_4 with cis–location of functional groups, polyphenylsilsesquioxanes with *cis*-isotactic configuration may be formed if the atoms of silicon in the tetrole molecules are in one and same plane and in *cis*–anti-*cis*–tactic configuration, if the T_4 cycles are in the parallel planes (Scheme 14.2).

However, the realization of cis-isotactic structure is associated with several steric hindrances and characterized by the short length of the molecules. During the tetrole polycondensation without initiator of the basis type, which allows the obtaining of the polymer without breaking of siloxane bond in the organocyclosiloxane, the conditions of synthesis ensures the perfect cyclolinear ladder structure and the macromolecules with the Kuhn segment with length about 50 Å are formed (Scheme 14.2).

Performing spectroscopic research[13] gives grounds to assume that the structure of macromolecules polyphenylsilsesquioxane obtained by the thermal condensation of tetrol T_4 toluene solution is different from the structure of the macromolecules polyphenylsilsesquioxane obtained from products phenyltrichlorsilane anionic polymerization initiator in the presence of the basic type. The experimental studies have assumed that the macromolecule with Kuhn segment with length of ~50 Å have a structure close to the cis–anti-cis–tactic configuration.[14]

So the condensation reaction of tetrahydroxyoligoorganosilsesquioxanes with epichlorohydrin was carried out according to Scheme 14.4.

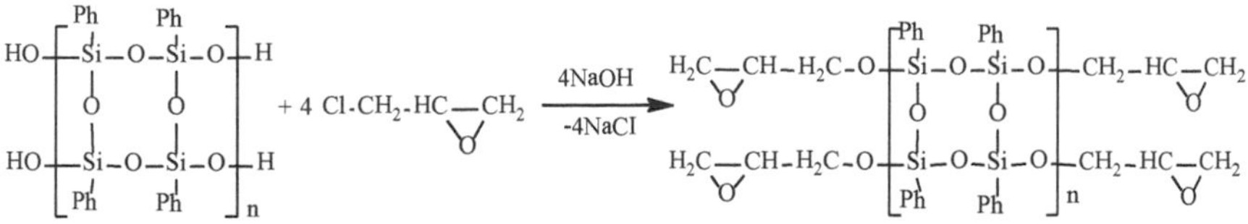

SCHEME 14.4 Condensation of Tetraethoxypolyphenylsilsesquioxane (n = 3, 10) with Epichlorohydrin in the Presence of Sodium Hydroxide.

Where n = 3 (II) and 10 (III).

The obtained compounds II and III were amber-colored viscous products well soluble in the acetone, methylethylketone, and ethyl acetate. The yield, number of epoxy groups, molecular masses, and elemental analysis of the obtained silicon-organic oligomers are presented in Table 14.1.

For obtained epoxy-phenylsilsesquioxane, the reaction between an industrial product tetraethoxyphenylsilsesquioxanes PhES-50 (n = 1) and PhES-80 (n = 2) with epichlorohydrin were investigated. The reaction was carried out at a temperature of 80–100°C in an inert gas atmosphere in the presence of a catalytic amount of iron chloride (III) or aluminum chloride, as shown in Scheme 14.5:

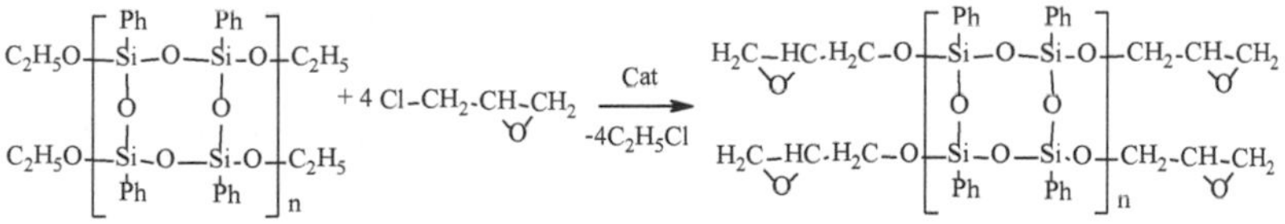

SCHEME 14.5 Condensation of Tetraethoxypolyphenylsilsesquioxane (PhES-50 and PhES- 80) with Epichlorohydrin in the Presence of Catalyst.

Where Cat-FeCl$_3$: n = 1 – I', 2 (IV); Cat – AlCl$_3$, n = 2 – IV'.

It is interesting to note that because of iron (III) chloride, the reaction mix appeared in dark brown color. Decolorizing of an obtained product was not possible either by centrifugation or by the adsorption of a solution of siliconorganic oligomers on the activated coal. Therefore, further researches were carried out in the presence of catalyst aluminum chloride.

The structure and composition of the synthesized oligomers were determined by means of elementary and functional analyzes by calculating molecular masses and FTIR and ^{1}H NMR spectra data. In the FTIR spectra of compounds I' and IV, the absorption bands for ν_{as} asymmetric and for symmetric valence oscillation of ≡Si–O–Si≡ bonds were kept in the field of absorption at 1060–1010 cm^{-1}, and also, new absorption bands at 820–840 and 917 cm^{-1} characteristic for epoxy rings were formed. The spectra did not observe the absorption bands characteristic of ethoxy groups proving that the epoxy groups have replaced the ethoxy groups. In the ^{1}H NMR spectra of synthesized oligomers, one can observe multiplet signals characteristic for methylene protons in the -CH$_2$O- group in the center of chemical shift δ = 3.3 ppm, multiplet signals characteristic for methine group of oxirane cycle in the center of chemical shift δ = 3.01 ppm, and also multiplet signals characteristic for methylene group of oxirane cycle in the center of chemical shift δ = 2.5 ppm. Yields, amount of epoxy groups, molecular masses $\overline{M}_n$, $\overline{M}_\omega$, $\overline{M}_z$ and polydispersity of synthesized epoxy group containing siliconorganic oligomers I'', IV, and IV' are presented in Table 14.2.

TABLE 14.2 Yields, Amount of Epoxy Groups, and Molecular Masses of Synthesized tetraepoxypolyphenylsilsesquioxanes (I–II)

№	Yield,%	Epoxy Group,%	M^a	$\overline{M}n$	$\overline{M}\omega$	$\overline{M}z$	$\overline{M}\omega/ \overline{M}n$
I'	94	4.74	3626	–	–	–	–
		4.80	3672				
IV	93	13.52	1272	1260	2460	6250	1.95
		13.58	1277				
IV'	95	13.52	1272	1200	2080	4100	1.83
		13.48	1268				

Notes: ^{a}The calculated values are presented in the numerator, and experimental values are presented in the denominator; Molecular masses were calculated from the values of epoxy groups.

No change in the characteristic properties of the synthesized tetraepoxypolyphenylsilsesquioxanes I' and IV for three months was observed, which implies their long viability (Table 14.3).

TABLE 14.3   Change of Specific Viscosity, Amount of Epoxy Groups for Oligomers I'
and IV Depending on Duration of their Storage

№	η_{sp} 50% Benzene Solution at 20°C					Amount of Epoxy Groups,%				
	After Reaction	After 10 Days	After 1 Month	After 2 Months	After 3 Months	After Reaction	After 10 Days	After 1 Month	After 2 Months	After 3 Months
XVI	2.6	2.8	2.8	2.8	2.9	24.30	24.29	24.28	24.28	24.28
XVII	5.2	5.2	5.3	5.5	5.5	7.19	7.18	7.17	7.17	7.17

EPOXY GROUP FUNCTIONALIZED POLYPHENYLSILSESQUIOXANE

For obtaining epoxyorganosilicon, the compounds I–III were used as modifiers for epoxy-dian resin ED-22. The compounds were prepared on the basis of ED-22 with 20–24% epoxy groups (100 mass parts) at different ratio of modifier and hardener: D-1, methylphenyldiamine (MPDA); D-2, 4,4'-diaminodiphenylsulfone (4,4'-DADPS); D-3, 4,4'-diaminodiphenylmethane (4,4'-DADPM); D-4, 4,4'-diaminotriphenyloxide (4,4'-DATPO), and D-5, metatetrahydrophtalanhydride (MTHFA). The hardening of composites was conducted both at room and high temperatures. Composition and the hardening regime of obtained composites were also conducted both at room and high temperatures. The composition and hardening regime of obtained compounds are presented in Table 14.4 and physical-mechanical and electric properties are presented in Table 14.5. The obtained compounds have the high dielectric, mechanic characteristics and thermal-oxidative stability. The results of investigation of dependence of physical-mechanical and dielectric properties on the number of the introduced modifier are presented in Table 14.6. The study of change of viscosity of the compounds in the process of hardening on temperature shows that compound differ from analogous by high viability in the range 80–90°C; however, at temperatures in the range of 150–160°C, its viability sharply decrease, which is connected with quick destruction of the functional groups and its consequent hardening.

TABLE 14.4 Composition and Hardening Regime of the Composites Based On Tetraepoxypolyphenylsilsesquioxanes and Epoxy Resin Ed-22

Composite	Composite Content	Mass Part	Composite Color	Hardening temperature, T°C	Hardening time, h
G-1	Epoxy pitch ED-22	100	Light yellow viscous mass	160	4
	Oligomer I	40–45			
	Binder D-1	13			
G-2	Epoxy pitch ED-22	100	Light yellow, transparent Mass	155	4
	Oligomer III	40–45			
	Binder D-2	13			
G-3	Epoxy pitch ED-22	100	Dark-yellow viscous mass	155	5
	Oligomer IV	40–45			
	Binder D-4	10–15			

TABLE 14.5 Mechanical and Electric Properties of Compounds Based on Tetraepoxypolyphenylsilsesquioxanes and Epoxy Resin Ed-22

№	Characteristic	Compound			Unit
		G-1	G-2	G-3	
1	Electrical strength	24.5	22	23.8	кВ/mm
2	Specific surface electrical resistivity at 20°C	1×10^6	1.2×10^6	0.6×10^{17}	Om
3	Specific volumetric electric resistivity at 20°C	1.5×10^{13}	2×10^{14}	5×10^{13}	Om/cm
4	Strength at bending	111.0	113.5	120.0	MPA
5	Heat resistance according to Martens	77.8	77.5	87.5	°C

By studying the thermal oxidation destruction of obtained composites, it was established that in comparison with initial epoxy resin, the polymers are of high thermal proof systems.

Figure 14.3 shows that increasing the ladder fragment length of a chain increases mainly the relative stability of the composites. The same case is true while studying the stability of the composites with different hardeners as in preliminary case in accordance with the Figure 14.3.

All composites G–D based on epoxy resin ED-22 modified by synthetic tetraepoxypolyphenylsilsesquioxanes and hardened by amine type hardeners are divided into two stages. In the first stage, destruction of composites is followed by high loss of the mass. This process occurs at temperatures from 180 to 420°C. Probably, on this stage, the compounds' organic groups burn out (Fig. 14.4).

In the study of polymers with different properties along with the research of new methods for the synthesis of oligomers and polymers, composites modified in optimal curing condition have gained a special attention. In this regard, elaboration of rational designs temperature of curing of composites has gained particular attention. We have studied the hardening of composite G–D with different amine hardener D-1, D-2, D-3, D-4, and D-5 at different temperatures from 80°C to 160–180°C.

TABLE 14.6 Physical-Mechanical Properties of Epoxy Resin ED-22, Modified by Epoxyorganosilicon Oligomer XII with Hardener MTGPA (D-5)

№	Containing of modifier, Mass Parts (at 100 Mass Parts of the Resin)	Destruction Stress at Stretching, MPa	Relative Elongation at Rupture, %	Hardness by Brinnel, MPa	Electric Durability, kV/mm	Tangent of Dielectric Losses Angle at 10^3Gz	Dielectric Penetration at 10^3Gz	Specific volumetric Electric Resistance, $\times 10^{13}$ Ohm $\times$ cm
1	10	7.15	10	1.0	61	0.022	5.92	4.3
2	15	8.1	12	1.0	68	0.029	6.06	4.3
3	20	6.8	10	1.2	52	0.018	5.85	4.3
4	25	5.2	9	1.4	46	0.011	5.02	4.2
5	35	5.01	8	1.5	44	0.011	5.02	4.2
6	45	5.01	7	1.7	42	0.010	5.02	4.1
7	0	3.5	0.8	1.0	20–25	0.0045	3.90	–

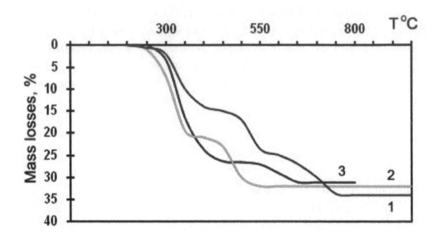

FIGURE 14.3   Thermogravimetric Curves for the Compounds G-1(1), G-2 (2), and G-3 (3) Hardened by Hardener D-1.

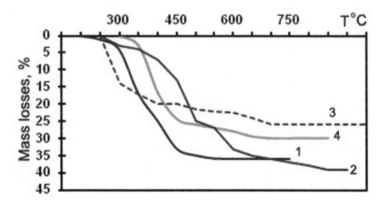

FIGURE 14.4   Thermogravimetric Curves of Thermal Oxidative Destruction for the Compound G-2 Hardened by Hardeners D-1 (1), D-2 (2), D-3 (3), and D-4 (4).

While synthesizing superposed composites of epoxyorganosilicon oligomers (I and III) and epoxy-dian resin ED-22, the chemical interaction between them and decreasing of the functional epoxy groups were not observed. However, during hardening, the result was different. Due to the presence of functional groups with high reaction ability, the presence of even the amine-type hardeners leads to chemical interaction, resulting in the step by step disappearance of the functional groups.

From Table 14.7, it can be seen that full curing of the composite G–1 with binder D-2 at a temperature of 150°C has been reached within 4–5 h.

TABLE 14.7 Yields of Gel-fraction of the Composite K-1 with Hardener D-1 in Dependence of Hardening Temperature

№	Hardening Time, min	Hardening Temperature, °C	Yield of Gel-fraction, %
1	20	18	0
2	60	45	0.1
3	120	60	0.2
4	160	80	2.8
5	180	110	10.0
6	240	120	40.0
7	270	130	60.0
8	300	140	80.0
9	360	150	100.0

The monomers containing organosilicon fragments and epoxy groups in the molecule are cured in the presence of amine curing agents that results in a three-dimensional cross-linked polymers, while the deformation heat resistance should be kept in the range 250–300°C. Combinations of different types behave differently when cured. During curing of composites by hardeners of cold cure, for example, polyethylene polyamine (PEPA), increasing of the hardening time till 3 days was observed.

The hardening kinetics for composites G–D and for pure ED-22 were studied using hardeners PEPA and MTGPA. It was shown that hardening rate for ED-22 was essentially higher compared with that of the other compounds. While heating of these composites, the fullness of hardening increases; however, the hardening rate for epoxy resin is higher than that of all organosilicon composites. This fact once more proves that the modified composites are characterized with lower reaction ability and high viability than that of epoxy resin. Therefore, these composites are more suitable for embedding materials.

Thus from all the analyses, it may be concluded that hardening of the epoxy resin modified by organosilicon oligomers (I–III) may be divided into two stages. In the first stage, the process of the slow formation of the linear polymers takes place and then increasing of the polymer grid because of the interaction between functional groups and hardener occurs, the consumption on which reaches about 30%. The temperature of reaction mass on the first stage of hardening must not exceed 60°C. This proves the application of MTGPA instead of PEPA.

In the second stage of the process, heating of the composition up to 120–150°C for 3 h takes place. Under such conditions, insoluble and infusible polymers are obtained by the interaction of the functional groups and the formation of new cross-links due to the reaction of the secondary hydroxyl groups.

In the threefold system (ED-22 + epoxyorganosilicon oligomer + hardener), interaction of hardener occurs with ED-22 and with epoxyorganosilicon modifier. And the second reaction is intense during heat treatment. The hardened material presents relatively rare grid of epoxy polymer filled with thermoplastic product epoxyorganosilicon modifier + hardener and therefore has high strengthening and thermal stability than pure epoxy polymer.

Increasing the amount of hardener in the threefold system leads to the synthesis of the polymer with high density of the grid. Such polymer is characterized by more stress structure, which leads to the decreasing of its mechanical strength because of which the polymer becomes fragile. For hardening of composites of the threefold system, the optimal amount of hardener is 10–15 mass part (20–25 wt%). For the samples hardened for 5 h at 150°C, decreasing of softening temperature and increasing of high elastic deformation were observed. This fact can be explained by the completion of the chemical processes of hardening of thermal-treated samples.

So, it appears that polymer materials based on epoxy-dian resin ED-22 modified by epoxyorganosilicon oligomers possess the complex of the valuable exploitation properties: high thermal stability and improved dielectrical, physical, and mechanical properties. The complex of these properties corresponds to modern technical demands and is perspective.

CONCLUSION

The condensation reaction of tetraethoxyphenylsilsesquioxane with the excess epichlorohydrin in the presence of catalysts has been investigated and corresponding tetraepoxy derivatives has been obtained. These composite compounds may be used as a potting material.

KEYWORDS

- **Thermal stability**
- **Biomedical**
- **Degree of functionalization**
- **Modified filler**
- **Tetraphenylcyclotetrasiloxane**
- **Phenylethoxysilsesquioxanes**
- **Catalysts iron chloride**

REFERENCES

1. Abeand, Y.; Gunji, T. *Prog. Polym.* Sci. 2004, 29, 149.
2. Bancy, R. H.; Itoh, M.; Sakakibara, A.; Suzuki, T. *Chem. Rev.* 1995, 95, 1409.
3. Cordes, D. B.; Lickissand, P. D.; Rataboul, F. *Chem. Rev.* 2010, 110, 2081.
4. Choi, J.; Yee, A. F.; Laine, R. M. *Macromolecules.* 2003, 36(15), 5666.
5. Mukbaniani, O. V.; Khananasvili, L. M.; Inaridze, I. A. *Int. J. Polym. Mater.* 1994, 24, 211.
6. Mukbaniani, O. V.; Zaikov, G. E. *New Concepts in Polymer Science.* In *Cyclolinear Organosilicon Copolymers: Synthesis, Properties, Application*; VSP: The Netherlands/Utrecht/Boston, 2003.
7. Prilezhaev, E. N. *Prilezhaev Reaction. Electrophilic Oxidation*; Nauka: Moscow, 1974; p 332. (In Russian).
8. Huirong, Y.; Richardson, D. E. *J. Am. Chem. Soc.* 2000, 122, 3220.
9. USA Patent 6124418, 2000.
10. Valetski, P. M. *Polym. Chem. Technol.* 1966, 2, 64. (In Russian).
11. Mukudan, A. L.; Balasubramian, K.; Srinisavan, K. S. V. *Polym. Commun.* 1988, 29(10), 310.
12. Iskakova, M.K.; Markarashvili, E.G.; Mindiashvili, G.S.; Shvangiradze, G.M.; Gvirgvliani, D.A.; Mukbaniani, O.V. Abstract of Communications of X All Russian Conference, Organosilicon Compounds: Synthesis, Properties and Application, Moscow, Russian Federation, 25-30 May, 3C12, 2005.
13. Volchek, B. Z.; Purkina, A. B. *Macromolecules.* 1976, 28A(6), 1203.
14. Tverdochlebova, I. I.; Mamaeva, I. I. *Macromolecules.* 1984, 26A(9), 1971.

CHAPTER 15

THE COMPARATIVE STUDY OF THERMOSTABLE PROTEIN MACROMOLECULAR COMPLEXES (CELL PROTEOMICS) FROM DIFFERENT ORGANISMS

D. DZIDZIGIRI, M. RUKHADZE, I. MODEBADZE,
N. GIORGOBIANI, L. RUSISHVILI, G. MOSIDZE, E. TAVDISHVILI,
and E. BAKURADZE

Department of Biology Faculty of Exact and Natural Sciences, Iv. Javakhishvili Tbilisi State University, Tbilisi, Georgia

Email: d_dzidziguri@yahoo.com

CONTENTS

INTRODUCTION

Identification of the functions of proteins and other polymeric complexes or cell proteomes, in which the achievements of proteomics contributes greatly, is the subject of intensive research.[1] Modern technology now allows us not only to investigate individual protein molecules in living cell, but also to understand their interaction with other macromolecules and reveal their previously unknown functions. Several facts are determined: participation of polyfunctional macromolecular protein complexes in the biosynthesis of fatty acids, involvement of erythrocyte membrane proteins macromolecular complexes in exchange of CO_2/O_2, biological effects of some growth factors (polyfunctional proteins), which sometimes is achieved by interactions of other protein complexes, and so on.[2–4]

Earlier, we have identified the protein complex with such properties in various cells of adult white rats.[5–7] The main feature of those complexes is the thermostability of the containing components. With gel electrophoreses and chromatography of hydrophobic interaction, it was determined that components with high molecular weight (45–60 kD) are hydrophobic, whereas components with low molecular weight (11–12 kD) are hydrophilic according to the column retention time. Through the inhibition of transcription, complex reduces the mitotic activity of homo- and heterotypic cells in growing animals.[5,8] Components of complex are water soluble and is not characterized by species specificity. Thus, we can assume that they are maintained in cells of phylogenetic distance organisms, and in case of confirmation of this fact, they may belong to the conservative family of proteins. In order to determine general regularities of effects of complex described by us, it is necessary to examine in detail the components phylogenesis.

AIM

The aim of this study is the extraction and comparative characterization of thermostable protein complexes (TSPCs) from phylogenetic distance organisms.

EXPERIMENTAL PART

MATERIAL AND METHODS

The thermostable protein complexes obtained by alcohol extraction from normal cells of various organisms (bacteria, snail, lizard, guinea pig, rat, and also human postsurgical material and cell culture) were used for research.

Thermostable protein fractions were obtained by the method of alcohol precipitation described by Balazs and Blazsek,[9] with modification. Animals were decapitated under diethyl ether. Organs were removed quickly, separated from capsules of connective tissues and vessels, rinsed with the physiological solution, and crushed. Aqueous homogenates were prepared in a tissue/distilled water ratio of 1:8. The homogenates were saturated stepwise with 96% ethanol to obtain 81% ethanol fraction, which was heated in a water bath (100°C) for 20 min, cooled, and centrifuged (600 g, 15 min). The supernatant was frozen in liquid nitrogen and dried in an absorptive-condensate lyophilizer. As a result, a residue of a thermostable protein complex (TSPC) was obtained, which is a light gray powder soluble in water. Samples were kept at 4°C. Protein concentration was determined by the method of Lowry et al.[10]

Hydrophobic interaction chromatography (HIC) was used for the comparative analysis of TSPC.[11] A hydrophilic polymeric sorbent, HEMA BIO Phenyl-1000 (particle size 10 mm) modified by phenyl groups, served as the stable phase. The mobile phase was phosphate buffer (pH 7.4) with ammonium sulfate. Elution was performed with the mobile phase in molar concentration range of 2.0 M–0.0 M (pure buffer) with respect to $(NH4)_2SO_4$. For coelution of hydrophilic and hydrophobic components of protein fractions, Brij-35 polyoxyethylene dodecyl ether with increasing concentration from 0 to 3% was added to the mobile phase. UV detection was usually set at 230 nm.

RESULTS AND DISCUSSION

Dynamic interaction between the protein molecules (protein–protein) determines vital activity of the cells. In the past few years, as a result of intensive research in this field, some knowledge about formation and function of protein complex has been accumulated. Dynamic interactions of proteins are studied not only in individual species but also in different types of cells and tissues. Therefore, in the first stage, the aim of the research was to obtain and compare the TSPC from organisms of different classes.

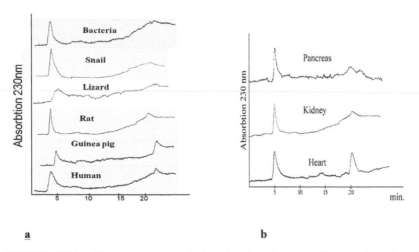

a   b

FIGURE 15.1 Chromatography of Protein Complex: (a) Protein Complex from Phylogenetically Distance Animals; (b) Protein Complex from Various Tissues of White Rat.

It was established that all the protein complex samples contain qualitatively different two groups of proteins. I group is hydrophilic, and II group is hydrophobic proteins with a column retention time of 5–6 min. and 20 min., respectively (Fig. 15.1a). The same subfractions were revealed in case of protein complexes obtained from different tissues of white rat by using this method (Fig. 15.1b).

It is known that dysfunctions of protein–protein interactions can lead to the development of various diseases, including cancer, neurodegeneration, autoimmune diseases, and so on. Therefore, the analysis of protein networks based on the protein–protein interaction may be used in developing various therapeutic approaches.[12] On the next stage of the research, we performed comparative analyses of protein complexes obtained from rat kidney and normal and transformed renal tissue of human. The differences between normal and transformed cells were revealed in this experiment. In particular, the components with low molecular weight were not observed in the protein complex obtained from human kidney cancer tissue, which indicates the changes that occur in the composition and the function of complexes during the development of the cancer (Fig. 15.2).

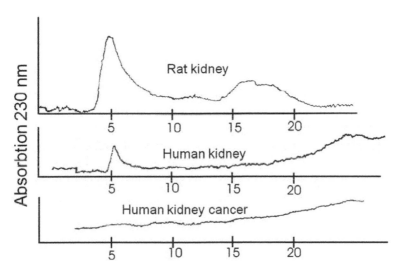

FIGURE 15.2 Chromatography of Protein Complexes Obtained from Rat Kidney and Normal and Transformed Renal Tissue of Human.

The low-molecular-weight subfraction of protein complex has mitosis-inhibitory properties, reduces number of active and moderately active nucleoli, and decreases the activation of RNA synthesis in nuclei of cardiomyocytes of newborn rat.[13] Usually, this subfraction is essential component of any TSPC obtained from various organs of adult rat. It is always seen as a major subfraction in native gel-electrophoresis (PAAG electrophoresis; Fig. 15.3).

Rat: pancreas (I), kidney (II), brain (III). Human: papillary carcinoma (IV), CLL (V), hemangioma (VI).

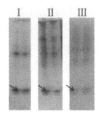
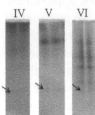

FIGURE 15.3 Electropherogram of TPC from Different Organ of White Rat and Human (Arrows Indicate Low-Molecular-Weight Subfraction of TPC).

Consequently, on the next stage of experiment, the low-molecular-weight components of protein complexes obtained from intact (rat pancreas, kidney, and brain) and transformed cells (with different degrees; human papillary carcinoma, CLL, and hemangioma) were examined.

As it is shown in Figure 15.3, component with low molecular weight are well expressed in protein complexes from various organs (kidney, pancreas, and brain) of adult rat (Fig.15.3). Different picture is shown in complexes from transforming cells. In particular, electropherograms shows that protein components with low molecular weight are manifested in the samples obtained from benign as well as malignant cells. However, intensity of silver nitrate staining is much lower compared with the norm.

CONCLUSIONS

From these results, it can be concluded that pro- and eukaryotic cells contain a TSPC that inhibits cell proliferation. Quantitative content of protein components in the complex is changing with the growth of transformation degree of cells. Currently, development of the relative proteomics allows us to determine the identity of proteins within these complexes to reliably identify the set of proteins that is responsible for and participates in the regulation of proliferation are constantly presented in the cell. With the help of comparative proteomics, it was identified that *Nilaparvata lugens* proteins that are involved in the process of proliferation and their expression change in response to insecticide treatment.[2]

KEYWORDS

- **Functions of proteins**
- **macromolecules**
- **biosynthesis**
- **interactions**
- **thermostability**
- **hydrophobic interaction**

REFERENCES

1. Amano, O.; Iseki, S. Expression and localization of cell growth factors in the salivary gland. *Kaibogaku Zasshi.* 2001, 76(2), 201–212.
2. Balazs, A.; Blazsek, I. *Control of Cell Proliferation by Endogenous Inhibitors*; Akademia Kiado: Budapest, 1979; p 302.
3. Dai, S.; Chen, T.; Chong, K.; Xue, Y.; Liu, S.; Wang, T. Proteomics identification of differentially expressed proteins associated with pollen germination and tube growth reveals characteristics of germinated, *Oryza sativa.* pollen. *Mol Cell Proteomics.* 2007, 6, 207–230.
4. Dijke, P. T.; Iwata, K. K. Growth factors for wound healing. *BioTechnology.* 1992, 7(8), 793–798.
5. Dzidziguri, D.; Iobadze, M.; Aslamazishvili, T.; Tumanishvili, G.; Bakhutashvili, V.; Chigogidze, T.; Managadze, L. The influence kidney protein factors on the proliferative activity of MDSK cells. *Tsitologiya.* 2004, 46(10), 913–914.
6. Ge, L.-Q.; Cheng, Y.; Wu, J.-C.; Jahn, G. C. Proteomic analysis of insecticide triazophos-induced mating-responsive proteins of *Nilaparvata lugens* Stal (Hemiptera: Delphacidae). *J. Proteome Res.* 2011, 10(10), 4597–4612.
7. Giorgobiani, N.; Dzidziguri, D.; Rukhadze, M.; Rusishvili, L.; Tumanishvili, G. Possible role of endogenous growth inhibitors in regeneration of organs: searching for new approaches. *Cell Biol. Int.* 2005, 29, 1047–1049. http://belki.com.ua/belki-struktura.html
8. Lowry, D. H.; Rosebrough, N. J.; Farr, A. L.; Randell, R. J. Protein measurement with the folin phenol reagent. *J. Biol. Chem.* 1951, 193, 265–275.
9. Modebadze, I.; Rukhadze, M.; Bakuradze, E.; Dzidziguri, D. Pancreatic cell protcome – qualitative characterization and function. *Georgian Med. News.* 2013, 7-8(220–221), 71–77.
10. Queiroz, J. A.; Tomaz, C. T.; Cabral, J. M. S. Hydrophobic interaction chromatography of proteins. *J. Chromatogr.* 2001, 87, 143–159.
11. Rusishvili, L.; Giorgobiani, N.; Dzidziguri, D.; Tumanishvili, G. Comparative analysis of cardiomiocyte growth-ihibitory factor in animals of different classes. *Proc. Georgian Acad. Sci., Biol. Ser. B.* 2003, 1(1–2), 42–45.

12. Terentiev, A. A.; Moldogazieva, N. T.; Shaitan, K. V. The dynamic proteomics in the modeling of living cell. Protein-protein interactions. *Success of Biological Chemistry (Adv. Biol. Chem.)*. 2009, 49, 429–480. (article in Russian).

INDEX

Milton Keynes UK
Ingram Content Group UK Ltd.
UKHW022059141024
449569UK00031B/1705

9 781774 635483